Communications and Control Engineering

For other titles published in this series, go to
www.springer.com/series/61

Series Editors

A. Isidori • J.H. van Schuppen • E.D. Sontag • M. Thoma • M. Krstic

Published titles include:

Stability and Stabilization of Infinite Dimensional Systems with Applications
Zheng-Hua Luo, Bao-Zhu Guo and Omer Morgul

Nonsmooth Mechanics (Second edition)
Bernard Brogliato

Nonlinear Control Systems II
Alberto Isidori

L_2-Gain and Passivity Techniques in Nonlinear Control
Arjan van der Schaft

Control of Linear Systems with Regulation and Input Constraints
Ali Saberi, Anton A. Stoorvogel and Peddapullaiah Sannuti

Robust and H_∞ Control
Ben M. Chen

Computer Controlled Systems
Efim N. Rosenwasser and Bernhard P. Lampe

Control of Complex and Uncertain Systems
Stanislav V. Emelyanov and Sergey K. Korovin

Robust Control Design Using H_∞ Methods
Ian R. Petersen, Valery A. Ugrinovski and Andrey V. Savkin

Model Reduction for Control System Design
Goro Obinata and Brian D.O. Anderson

Control Theory for Linear Systems
Harry L. Trentelman, Anton Stoorvogel and Malo Hautus

Functional Adaptive Control
Simon G. Fabri and Visakan Kadirkamanathan

Positive 1D and 2D Systems
Tadeusz Kaczorek

Identification and Control Using Volterra Models
Francis J. Doyle III, Ronald K. Pearson and Babatunde A. Ogunnaike

Non-linear Control for Underactuated Mechanical Systems
Isabelle Fantoni and Rogelio Lozano

Robust Control (Second edition)
Jürgen Ackermann

Flow Control by Feedback
Ole Morten Aamo and Miroslav Krstic

Learning and Generalization (Second edition)
Mathukumalli Vidyasagar

Constrained Control and Estimation
Graham C. Goodwin, Maria M. Seron and José A. De Doná

Randomized Algorithms for Analysis and Control of Uncertain Systems
Roberto Tempo, Giuseppe Calafiore and Fabrizio Dabbene

Switched Linear Systems
Zhendong Sun and Shuzhi S. Ge

Subspace Methods for System Identification
Tohru Katayama

Digital Control Systems
Ioan D. Landau and Gianluca Zito

Multivariable Computer-controlled Systems
Efim N. Rosenwasser and Bernhard P. Lampe

Dissipative Systems Analysis and Control (Second edition)
Bernard Brogliato, Rogelio Lozano, Bernhard Maschke and Olav Egeland

Algebraic Methods for Nonlinear Control Systems
Giuseppe Conte, Claude H. Moog and Anna M. Perdon

Polynomial and Rational Matrices
Tadeusz Kaczorek

Simulation-based Algorithms for Markov Decision Processes
Hyeong Soo Chang, Michael C. Fu, Jiaqiao Hu and Steven I. Marcus

Iterative Learning Control
Hyo-Sung Ahn, Kevin L. Moore and YangQuan Chen

Distributed Consensus in Multi-vehicle Cooperative Control
Wei Ren and Randal W. Beard

Control of Singular Systems with Random Abrupt Changes
El-Kébir Boukas

Nonlinear and Adaptive Control with Applications
Alessandro Astolfi, Dimitrios Karagiannis and Romeo Ortega

Stabilization, Optimal and Robust Control
Aziz Belmiloudi

Control of Nonlinear Dynamical Systems
Felix L. Chernous'ko, Igor M. Ananievski and Sergey A. Reshmin

Periodic Systems
Sergio Bittanti and Patrizio Colaneri

Discontinuous Systems
Yury V. Orlov

Constructions of Strict Lyapunov Functions
Michael Malisoff and Frédéric Mazenc

Controlling Chaos
Huaguang Zhang, Derong Liu and Zhiliang Wang

Control of Complex Systems
Aleksandar Zečević and Dragoslav D. Šiljak

Viorel Barbu

Stabilization of Navier–Stokes Flows

Springer

Prof. Viorel Barbu
Fac. Mathematics
Al. I. Cuza University
Blvd. Carol I 11
700506 Iaşi
Romania
vb41@uaic.ro

ISSN 0178-5354
ISBN 978-1-4471-2610-2 ISBN 978-0-85729-043-4 (eBook)
DOI 10.1007/978-0-85729-043-4
Springer London Dordrecht Heidelberg New York

British Library Cataloguing in Publication Data
A catalogue record for this book is available from the British Library

Softcover reprint of the hardcover 1st edition 2011

Cover design: eStudio Calamar S.L.

Printed on acid-free paper

Springer is part of Springer Science+Business Media (www.springer.com)

Preface

In the last years, notable progresses were obtained in mathematical theory of stabilization of equilibrium solution to Newtonian fluid flows as a principal tool to eliminate or attenuate the turbulence. One of the main results obtained in this direction is that the equilibrium solutions to Navier–Stokes equations are exponentially stabilizable by finite-dimensional feedback controllers with support in the interior of the domain or on the boundary. This book was completed in the idea to present these new results and techniques which are in our opinion the core of a discipline still in development and from which one might expect in the future some spectacular achievements.

Beside internal and boundary stabilization of Navier–Stokes equations, the stochastic stabilization and robustness of stabilizable feedbacks are also discussed. We had in mind a rigorous mathematical treatment of the stabilization problem, which relies on some advanced results and techniques involving the theory of Navier–Stokes equations and functional analysis as well. We tried to answer to the following questions: which is the structure of the stabilizing feedback controller and how can be designed by using a minimal set of eigenfunctions of the Stokes–Oseen operator. Though most of the feedback controllers constructed here are conceptual and their practical implementation requires a computational effort which still remains to be done, the analysis developed here provides a rigorous pattern for the design of efficient stabilizable feedback controllers in most specific problems of practical interest. To this purpose, the exposition is in mathematical style: definitions, hypotheses, theorems, proof. It should be emphasized that no rigorous stabilization theory with internal or boundary controllers is possible without unique continuation theory for the solutions to Stokes–Oseen equations.

By including a preparatory chapter on infinite-dimensional differential equations and a few appendices pertaining unique continuation properties of eigenfunctions of the Stokes–Oseen operator and stochastic processes, we have attempted to make this work essentially self-contained and so, accessible to a broad spectrum of readers. What is assumed of the reader is a knowledge of basic results in linear functional analysis, linear algebra, probability theory and general variational theory of elliptic, parabolic and Navier–Stokes equations, most of these being reviewed in Chap. 1 and in Sects. 3.8 and 4.5. An important part of the material included in this book represent the personal contribution of the author and his coworkers and, though we mentioned the basic references and a brief presentation of other significant works in this field, we did not present them, however, in details. In fact, the presentation was confined to the stabilization techniques based on the spectral decomposition of the linearized system in stable and unstable systems and so we have omitted other important results in literature.

The author is indebted to Cătălin Lefter who made pertinent observations and suggestions. I also thank Irena Lasiecka and Roberto Triggiani for useful discussions on several results presented in this book. Also, I am indebted to Mrs. Elena Mocanu from Institute of Mathematics in Iaşi who prepared this text for printing.

I wish to express my thanks to Professor Miroslav Krstic, from University of California, San Diego, for the invitation to write this book for the Springer series *Communication and Control Engineering* he is coordinating.

Special thanks are due to Mr. Oliver Jackson, Editor of Engineering at Springer, for understanding and assistance in the elaboration of this work.

Contents

Symbols and Notation

R^d	the d-dimensional Euclidean space
R	the real line $(-\infty, +\infty)$
R^+	the half line $[0, +\infty)$
$\mathbb{C}$	the complex space
$\mathbb{C}^d$	the d-dimensional complex space
$x \cdot y$	the dot product of vectors $x, y \in R^d$
$\vert \cdot \vert_X,\ \Vert \cdot \Vert_X$	the norm of a linear normed space X
$D_i y$	$\frac{\partial y}{\partial x_i},\ i = 1, \dots, d$
∇f	the gradient of the map $f : X \to Y$
$\nabla \cdot f$	the divergence of vector field $f : \mathcal{O} \to R^d \subset R^d$
$L(X, Y)$	the space of linear continuous operators from X to Y
$\Vert \cdot \Vert_{L(X,Y)}$	the norm of $L(X, Y)$
$X^*,\ X'$	the dual of the space X
$(x, y),\ (x, y)_H$	the scalar product of the vectors $x, y \in H$ (a Hilbert space). If $x \in X,\ y \in X^*$, this is the value of y at x
A^α	the fractional power of order $\alpha \in (0, 1)$ of the operator $A : D(A) \subset H \to H$
$D(A)$	the domain of the operator A
$A^*,\ A'$	the adjoint of the operator A
A^{-1}	the inverse of the operator A
$\{\Omega, \mathcal{F}, \mathbb{P}\}$	the probability space
$L^p(0, T; X)$	(X Banach space) the space of all p-sumable functions from $[0, T]$ to X
$L^p_{\text{loc}}(0, \infty; X)$	the space of measurable functions $y : (0, \infty) \to X$ which are p-integrable on each interval $(a, b) \subset (0, \infty)$
$y'(t),\ \frac{dy}{dt}(t)$	the derivative of the function $y : [0, T] \to X$
$AC([0, T]; X)$	the space of absolutely continuous functions from $[0, T]$ to X
$W^{1,p}([0, T]; X)$	the space $\{y \in AC([0, T];\ y' \in L^p(0, T; X)\}$
$C([0, T]; X)$	the space of all continuous functions from $[0, T]$ to X
$C_w([0, T]; X)$	the space of all weakly continuous functions from $[0, T]$ to X
e^{At}	the C_0-semigroup generated by A
$H^k(\mathcal{O})$	the Sobolev space of order k on $\mathcal{O} \subset R^d$

Chapter 1
Preliminaries

The purpose of this chapter is to briefly present without proofs, for later use, certain notions and fundamental results pertaining linear operators in Banach spaces, boundary value problems, nonlinear dynamics in Hilbert spaces and existence theory of Navier–Stokes equations.

1.1 Banach Spaces and Linear Operators

A Banach space is a linear normed space which is complete. The norm of the Banach space X (real or complex) is denoted by $\|\cdot\|_X$ and $L(X,X)$ is the space of all linear continuous operators from X to itself. If X is a real Banach space (that is, over the real field R), then its complexification $\widetilde{X}$ is the space $\widetilde{X}=X+iX$, that is, $\widetilde{X}=\{x+iy,\ x,y\in X\}$ with the norm $\|x+iy\|=\|x\|_X+\|y\|_X$.

If A is a linear operator from X to Y, we denote by $D(A)$ its domain, that is, $D(A)=\{x\in X;\ Ax\neq\emptyset\}$ and by $R(A)$ its range, that is, $R(A)=\{y\in Y;\ y=Ax,\ x\in D(A)\}$. The linear operator is said to be *closed* if its graph $\{(x,y)\in X\times Y;\ y=Ax\}$ is closed, that is, if $x_n\overset{X}{\longrightarrow}x$ and $y_n\in Ax_n\overset{Y}{\longrightarrow}y$ implies that $y=Ax$. Here we use the symbol $\overset{X}{\longrightarrow}$ for the convergence in the norm $\|\cdot\|_X$, that is, the strong convergence. The linear operator A is said to be densely defined if its domain $D(A)$ is dense in X. The inverse of A is denoted A^{-1}.

For each $\lambda\in\mathbb{C}$ (the complex field) denote by $(\lambda I-A)^{-1}\in L(X,X)$ the resolvent of $A: D(A)\subset X\to X$ and by $\rho(A)$ the resolvent set, $\rho(A)=\{\lambda\in\mathbb{C};\ (\lambda I-A)^{-1}\in L(X,X)\}$ and by $\sigma(A)=\mathbb{C}\setminus\rho(A)$ the spectrum of A. In each component of $\rho(A)$ the function $\lambda\to(\lambda I-A)^{-1}$ is holomorphic.

The number $\lambda\in\mathbb{C}$ is said to be *eigenvalue* of the linear operator $A: D(A)\subset X\to X$ if there is $x\in D(A)$, $x\neq 0$, such that $Ax=\lambda x$.

The corresponding vectors x are called *eigenvectors*. If λ is eigenvalue for A, then the dimension of the linear eigenvector space $\mathrm{Ker}(\lambda I-A)=\{x\in X;\ Ax=\lambda x\}$ is called the *geometric multiplicity* of λ. The vector x is called a *generalized eigenvector* corresponding to the eigenvalue λ if $(\lambda I-A)^m x=0$ for some $m\in\mathbb{N}$. The

V. Barbu, *Stabilization of Navier–Stokes Flows*,
Communications and Control Engineering,
DOI 10.1007/978-0-85729-043-4_1, © Springer-Verlag London Limited 2011

dimension of the space of generalized eigenvectors is called the *algebraic multiplicity* of the eigenvalue λ.

Theorem 1.1 is known in literature as the Riesz–Schauder–Fredholm theorem. (See, e.g., [82], p. 283.)

Theorem 1.1 *Let $T \in L(X, X)$ be a compact operator. Then its spectrum $\sigma(T)$ consists of an at most countable set of points of complex plane which has no point of accumulation except $\lambda = 0$. Moreover, every $\lambda \in \sigma(T)$ is eigenvalue of T of finite algebraic multiplicity.*

In particular, by Theorem 1.1 we have the following result.

Theorem 1.2 *Let A be a closed operator and densely defined operator in X with compact resolvent $(\lambda I - A)^{-1}$ for some $\lambda \in \rho(A)$. Then the spectrum $\sigma(T)$ consists of isolated eigenvalues $\{\lambda_j\}_{j=1}^{\infty}$ each of finite (algebraic) multiplicity m_j.*

If A is such an operator, then for each $N \in \mathbb{N}$, the spectrum $\sigma(A)$ can be written as

$$\sigma(A) = \{\lambda_j\}_{j=1}^{N} \cup \{\lambda_j\}_{j=1}^{N+1} \tag{1.1}$$

and if Γ is a closed curve in $\mathbb{C}$, which contains in interior $\{\lambda_j\}_{j=1}^{N}$, we set

$$P_N = \frac{1}{2\pi i} \int_{\Gamma} (\lambda I - A)^{-1} d\lambda \tag{1.2}$$

and $X_N^1 = P_N X$, $X_N^2 = (I - P_N)X$. Then we have a decomposition of X in the direct sum

$$X = X_N^1 \oplus X_N^2, \qquad P_N^2 = P_N, \tag{1.3}$$

and if we set

$$A_N^1 = P_N A, \qquad A_N^2 = (I - P_N)A, \tag{1.4}$$

we have the following theorem (see Theorem 6.17 in [59]).

Theorem 1.3 *Under the assumptions of Theorem 1.2,*

$$A_N^i X_N^i \subset X_N^i, \quad i = 1, 2, \tag{1.5}$$

$$\sigma(A_N^i) = \{\lambda_j\}_{j=1}^{N}, \qquad \sigma(A_N^2) = \{\lambda_j\}_{j=N+1}^{\infty}. \tag{1.6}$$

If $N = 1$, then $\dim X_N^1 = m_1$ is just the algebraic multiplicity of the eigenvalue λ_1.

Definition 1.1 An eigenvalue λ of the operator A is called *semisimple* if the algebraic multiplicity of λ coincides with the geometric multiplicity.

In general, the algebraic multiplicity is greater than the geometric multiplicity.

We note that in finite dimension the spectrum of an operator consists of semi-simple eigenvalues if its Jordan form is diagonal.

If X is a Banach space, we denote by X^* its dual space endowed with the dual norm $\|x^*\|_{X^*} = \sup({}_X(x,x^*)_{X^*};\ \|x\|_X = 1\}$. (Here, ${}_X(x,x^*)_{X^*}$ is the value of x^* at x.)

If $A : X \to Y$ is a closed and densely defined operator (X, Y are Banach spaces), then the adjoint $A^* : Y^* \to X^*$ of A is defined by

$$\begin{aligned} &{}_{X^*}(A^*y^*, x)_X = {}_{Y^*}(y^*, Ax)_Y, \quad \forall x \in D(A),\\ &D(A^*) = \{y^* \in Y^*;\ \exists C > 0, |{}_{Y^*}(y^*, Ax)_Y| \le C\|x\|_X, \ \forall x \in D(A)\}. \end{aligned}$$

The adjoint operator A^* is closed, densely defined and $(\lambda I - A^*)^{-1} = ((\bar{\lambda} I - A)^{-1})^*$, $\forall \lambda \in \rho(A)$. Moreover, if λ is eigenvalue for A, then $\bar{\lambda}$ is eigenvalue for A^* of the same multiplicity.

If A is a closed and densely defined operator from X to X, its domain $D(A)$ is a Banach space with the norm $\|x\|_{D(A)} = \|x\|_X + \|Ax\|_X$, $\forall x \in D(A)$, and we have $D(A) \subset X$ algebraically and topologically, that is, with dense and continuous embedding.

Assume now that $X = H$ is a Hilbert space with the norm $\|\cdot\|_H$ and scalar product $(\cdot,\cdot)_H$ and that there is $\lambda_0 \in \rho(A)$. Then, define the space $(D(A))'$ (the dual of $D(A)$ in the pairing $(\cdot,\cdot)$) as the completion of H in the norm

$$\|x\|_{(D(A))'} = \|(\lambda_0 I - A)^{-1}x\|_H, \quad \forall x \in H. \tag{1.7}$$

Then, we have

$$D(A) \subset H \subset (D(A))' \tag{1.8}$$

algebraically and topologically. Moreover, the operator $A : D(A) \subset H \to H$ has an extension denoted $\tilde{A} : H \to (D(A^*))'$ defined by

$${}_{(D(A^*))'}(\tilde{A}x, y)_{D(A^*)} = (x, A^*y), \quad \forall y \in D(A^*). \tag{1.9}$$

Of course, we have $\tilde{A}x = Ax$, $\forall x \in D(A)$.

Moreover, since $\tilde{A} : H \to (D(A^*))'$ is closed, by the closed graph theorem (see, e.g., [82], p. 77) we have that $\tilde{A} \in L(H, (D(A^*))')$.

In applications to partial differential equations, the extension $\tilde{A}$ of A incorporates into its domain $D(\tilde{A}) = H$ boundary value conditions. (See an example in Sect. 1.2 below.)

1.2 Sobolev Spaces and Elliptic Boundary Value Problems

Throughout this section, until further notice, we assume that $\mathcal{O}$ is an open subset of R^d. To begin with, let us briefly recall the notion of *distribution*. Let $f = f(x)$ be a continuous complex-valued function on $\mathcal{O}$. By the *support* of f, abbreviated $\operatorname{supp} f$, we mean the closure of the set $\{x \in \mathcal{O};\ f(x) \neq 0\}$ or, equivalently, the smallest closed set of $\mathcal{O}$ outside of which f vanishes identically. We will denote by $C^k(\mathscr{O})$, $0 \le k \le \infty$, the set of all complex-valued functions defined in $\mathscr{O}$ which have continuous partial derivatives of order up to and including k (of any order $< \infty$ if $k = \infty$).

Let $C_0^k(\mathscr{O})$ denote the set of all functions $\varphi \in C^k(\mathscr{O})$ with compact support in $\mathscr{O}$.

We may introduce in $C_0^\infty(\mathscr{O})$ a convergence as follows. We say that the sequence $\{\varphi_k\} \subset C_0^\infty(\mathscr{O})$ is convergent to φ, denoted $\varphi_k \Rightarrow \varphi$, if

(a) There is a compact $K \subset \mathscr{O}$ such that $\operatorname{supp} \varphi_k \subset K$ for all $k = 1, \dots$.
(b) $\lim_{k\to\infty} D^\alpha \varphi_k = D^\alpha \varphi$ uniformly on K for all $\alpha = (\alpha_1, \dots, \alpha_n)$.

Here, $D^\alpha = D_{x_1}^{\alpha} \cdots D_{x_N}^{\alpha_n}$, $D_{x_i} = \frac{\partial}{\partial x_i}$, $i = 1, \dots, n$. Equipped in this way, the space $C_0^\infty(\mathscr{O})$ is denoted by $\mathscr{D}(\mathscr{O})$. As a matter of fact, $\mathscr{D}(\mathscr{O})$ can be redefined as a locally convex, linear topological space with a suitable chosen family of seminorms.

A linear continuous functional u on $\mathscr{D}(\mathscr{O})$ is called a *distribution* on $\mathscr{O}$.

The set of all distributions on $\mathscr{O}$ is a linear space, denoted by $\mathscr{D}'(\mathscr{O})$.

The distribution is a natural extension of the notion of locally summable function on $\mathscr{O}$. Indeed, if $f \in L^1_{\text{loc}}(\mathscr{O})$, then the linear functional u_f on $C_0^\infty(\mathscr{O})$ defined by

$$u_f(\varphi) = \int_{\mathscr{O}} f(x)\varphi(x)dx, \quad \forall \varphi \in C_0^\infty(\mathscr{O})$$

is a distribution on $\mathscr{O}$, that is, $u_f \in \mathscr{D}'(\mathscr{O})$.

Given $u \in \mathscr{D}'(\mathscr{O})$, by definition, the derivative of order $\alpha = (\alpha_1, \dots, \alpha_n)$, $D^\alpha u$, of u, is the distribution

$$(D^\alpha u)(\varphi) = (-1)^{|\alpha|} u(D^\alpha \varphi), \quad \forall \varphi \in \mathscr{D}(\mathscr{O}), \text{ where } |\alpha| = \alpha_1 + \cdots + \alpha_n.$$

Let $\mathscr{O}$ be an open subset of R^d and let m be a positive integer. Denote by $H^m(\mathscr{O})$ the set of all real-valued functions $u \in L^2(\mathscr{O})$ such that distributional derivatives $D^\alpha u$ of u of order $|\alpha| \le m$ all belong to $L^2(\mathscr{O})$. In other words,

$$H^m(\mathscr{O}) = \{u \in L^2(\mathscr{O});\ D^\alpha u \in L^2(\mathscr{O}),\ |\alpha| \le m\}. \tag{1.10}$$

We present below a few basic properties of Sobolev spaces and refer to the books of Brezis [36], Adams [3], Barbu [11] for proofs.

Proposition 1.1 *$H^m(\mathscr{O})$ is a Hilbert space with the scalar product*

$$\langle u, v\rangle_m = \sum_{|\alpha| \le m} \int_{\mathscr{O}} D^\alpha u(x) \overline{D^\alpha v(x)}\, dx, \quad \forall u, v \in H^m(\mathscr{O}). \tag{1.11}$$

If $\mathscr{O} = (a, b)$, $-\infty < a < b < \infty$, $H^1(\mathscr{O})$ reduces to a subspace of absolutely continuous functions on the interval $[a, b]$.

More generally, for an integer $m \ge 1$ and $1 \le p \le \infty$, one defines the Sobolev space

$$W^{m,p}(\mathscr{O}) = \{u \in L^p(\mathscr{O});\ D^\alpha u \in L^p(\mathscr{O}),\ |\alpha| \le m\} \tag{1.12}$$

with the norm

$$\|u\|_{m,p} = \left(\sum_{|\alpha| \le m} \int_{\mathscr{O}} |D^\alpha u(x)|^p dx \right)^{1/p}. \tag{1.13}$$

For $0 < m < 1$, the space $W^{m,p}(\mathscr{O})$ is defined by

$$W^{m,p}(\mathscr{O}) = \left\{ u \in L^p(\mathscr{O});\ \frac{u(x) - u(y)}{|x - y|^{m+\frac{N}{p}}} \in L^p(\mathscr{O} \times \mathscr{O}) \right\}$$

with the natural norm. For $m > 1$, $m = s + a$, $s = [m]$, $0 < a < 1$, define

$$W^{m,p}(\mathscr{O}) = \{u \in W^{s,p}(\mathscr{O});\ D^\alpha u \in W^{a,p}(\mathscr{O});\ |\alpha| \le s\}.$$

Now, we mention without proof an important property of the space $H^1(\mathscr{O})$ known as the *Sobolev embedding theorem.*

Theorem 1.4 *Let $\mathscr{O}$ be an open subset of R^d of class C^1 with compact boundary $\partial\mathscr{O}$, or $\mathscr{O} = R^d_+$, or $\mathscr{O} = R^d$. Then, if $d > 2$,*

$$H^1(\mathscr{O}) \subset L^{p^*}(\mathscr{O}) \quad \textit{for}\ \frac{1}{p^*} = \frac{1}{2} - \frac{1}{d}. \tag{1.14}$$

If $d = 2$, then

$$H^1(\mathscr{O}) \subset L^p(\mathscr{O}) \quad \textit{for all}\ p \ge 2.$$

The inclusion relation (1.14) should be considered of course in the algebraic and topological sense, that is,

$$\|u\|_{L^{p^*}(\mathscr{O})} \le C\|u\|_{H^1(\mathscr{O})} \tag{1.15}$$

for some positive constant C independent of u.

Theorem 1.4 has a natural extension to the Sobolev space $W^{m,p}(\mathscr{O})$ for any $m > 0$ (see Adams [3], p. 217).

If $\mathscr{O}$ is an open C^1 subset of R^d with the boundary $\partial\mathscr{O}$, then each $u \in C(\overline{\mathscr{O}})$ is well-defined on $\partial\mathscr{O}$. We call the restriction of u to $\partial\mathscr{O}$ the *trace* of u to $\partial\mathscr{O}$ and it will be denoted by $\gamma_0(u)$. If $u \in L^2(\mathscr{O})$, then $\gamma_0(u)$ is no more well-defined.

We have, however, the following lemma.

Lemma 1.1 *Let $\mathscr{O}$ be an open subset of class C^1 with compact boundary $\partial\mathscr{O}$ or $\mathscr{O} = R^d_+$. Then, there is $C > 0$ such that*

$$\|\gamma_0(u)\|_{L^2(\partial\mathscr{O})} \le C\|u\|_{H^1(\mathscr{O})}, \quad \forall u \in C_0^\infty(R^d). \tag{1.16}$$

Then, a natural way to define the *trace* of a function $u \in H^1(\mathscr{O})$ is the following definition.

Definition 1.2 Let $\mathscr{O}$ be of class C^1 with compact boundary or $\mathscr{O} = R^d_+$. Let $u \in H^1(\mathscr{O})$. Then

$$\gamma_0(u) = \lim_{j\to\infty} \gamma_0(u_j) \quad \text{in } L^2(\partial\mathscr{O}),$$

where $\{u_j\} \subset C_0^\infty(R^d)$ is such that $u_j \to u$ in $H^1(\mathscr{O})$.

It turns out that the definition is consistent, that is, $\gamma_0(u)$ is independent of $\{u_j\}$. Indeed, if $\{u_j\}$ and $\{\bar{u}_j\}$ are two sequences in $C_0^\infty(R^d)$ convergent to u in $H^1(\mathcal{O})$, then, by (1.16),

$$\|\gamma_0(u_j - \bar{u}_j)\|_{L^2(\partial\mathcal{O})} \le C\|u_j - \bar{u}_j\|_{H^1(\mathcal{O})} \to 0 \quad \text{as } j \to \infty.$$

Moreover, it follows by Lemma 1.1 that the map $\gamma_0 : H^1(\mathcal{O}) \to L^2(\partial\mathcal{O})$ is continuous. As a matter of fact, it turns out that the trace operator $u \to \gamma_0(u)$ is continuous from $H^1(\mathcal{O})$ to $H^{\frac{1}{2}}(\partial\mathcal{O})$ and so it is completely continuous from $H^1(\mathcal{O})$ to $L^2(\partial\mathcal{O})$.

In general (see Adams [3], p. 114), we have $W^{m,p}(\mathcal{O}) \subset L^q(\partial\mathcal{O})$ if $mp < N$ and $p \le q \le \frac{(N-1)p}{(N-mp)}$.

Definition 1.3 Let $\mathcal{O}$ be any open subset of R^d. The space $H_0^1(\mathcal{O})$ is the closure (the completion) of $C_0^1(\mathcal{O})$ in the norm of $H^1(\mathcal{O})$.

It follows that $H_0^1(\mathcal{O})$ is a closed subspace of $H^1(\mathcal{O})$ and in general it is a proper subspace of $H^1(\mathcal{O})$. It is also clear that $H_0^1(\mathcal{O})$ is a Hilbert space with the scalar product

$$\langle u, v\rangle_1 = \sum_{i=1}^N \int_{\mathcal{O}} \frac{\partial u}{\partial x_i}\, \overline{\frac{\partial v}{\partial x_i}}\, dx + \int_{\mathcal{O}} u\bar{v}\, dx$$

with the corresponding norm

$$\|u\|_1 = \left(\int_{\mathcal{O}} (|\nabla u(x)|^2 + |u(x)|^2)dx\right)^{\frac{1}{2}}.$$

Roughly speaking, $H_0^1(\mathcal{O})$ is the subspace of functions $u \in H^1(\mathcal{O})$ which are zero on $\partial\mathcal{O}$. More precisely, we have

Proposition 1.2 *Let $\mathcal{O}$ be an open set of class C^1 and let $u \in H^1(\mathcal{O})$. Then, the following conditions are equivalent*:

(i) $u \in H_0^1(\mathcal{O})$.
(ii) $\gamma_0(u) \equiv 0$.

Proposition 1.3 below is the celebrated *Poincaré inequality*.

Proposition 1.3 *Let $\mathcal{O}$ be an open and bounded subset of R^d. Then there is $C > 0$ independent of u such that*

$$\|u\|_{L^2(\mathcal{O})} \le C\|\nabla u\|_{L^2(\mathcal{O})}, \quad \forall u \in H_0^1(\mathcal{O}).$$

In particular, Proposition 1.3 shows that if $\mathcal{O}$ is bounded, then the scalar product

$$((u, v)) = \int_{\mathcal{O}} \nabla u(x) \cdot \overline{\nabla v(x)}\, dx$$

and the corresponding norm

$$\|u\| = \left(\int_{\mathcal{O}} |\nabla u(x)|^2 dx \right)^{\frac{1}{2}}$$

define an equivalent Hilbertian structure on $H_0^1(\mathcal{O})$.

We denote by $H^{-1}(\mathcal{O})$ the dual space of $H_0^1(\mathcal{O})$, that is, the space of all linear continuous functionals on $H_0^1(\mathcal{O})$. Equivalently, $H^{-1}(\mathcal{O}) = \{u \in \mathcal{D}'(\mathcal{O});\ |u(\varphi)| \leq C_u \|\varphi\|_{H^1(\mathcal{O})},\ \forall \varphi \in C_0^\infty(\mathcal{O})\}$. The space $H^{-1}(\mathcal{O})$ is endowed with the dual norm $\|u\|_{-1} = \sup\{|u(\varphi)|;\ \|\varphi\| \leq 1\}$, $\forall u \in H^{-1}(\mathcal{O})$. By Riesz's theorem, we know that $H^{-1}(\mathcal{O})$ is isomorphic and isometric with $H_0^1(\mathcal{O})$. Note also that $H_0^1(\mathcal{O}) \subset L^2(\mathcal{O}) \subset H^{-1}(\mathcal{O})$ in algebraic and topological sense. In other words, the injections of $L^2(\mathcal{O})$ into $H^{-1}(\mathcal{O})$ and of $H_0^1(\mathcal{O})$ into $L^2(\mathcal{O})$ are continuous. Note also that the above injections are dense. Moreover, $H_0^1(R^d) = H^1(R^d)$.

The space $W_0^{1,p}(\mathcal{O})$, $p \geq 1$, is similarly defined as the closure of $C_0^1(\mathcal{O})$ into the $W^{1,p}(\mathcal{O})$ norm. The dual of $W_0^{1,p}(\mathcal{O})$ is denoted by $W^{-1,q}(\mathcal{O})$, where $\frac{1}{p} + \frac{1}{q} = 1$.

1.2.1 Variational Theory of Elliptic Boundary Value Problems

We begin by recalling an abstract existence result, the *Lax–Milgram lemma*, which is the foundation upon all the results of this section are built. Before presenting it, we had to clarify certain concepts.

Let V be a real Hilbert space and let V^* be the topological dual space of V. For each $v^* \in V^*$ and $v \in V$ we denote by (v^*, v) the value $v^*(v)$ of functional v^* at v. The functional $a : V \times V \to R$ is said to be *bilinear* if for each $u \in V$, $v \to a(u, v)$ is linear and for each $v \in V$, $u \to a(u, v)$ is linear on V. The functional a is said to be *continuous* if there exists $M > 0$ such that $|a(u, v)| \leq M \|u\|_V \|v\|_V$, $\forall u, v \in V$. The functional a is said to be *coercive* if $a(u, u) \geq \omega \|u\|_V^2$, $\forall u \in V$, for some $\omega > 0$, and *symmetric* if $a(u, v) = a(v, u)$, $\forall u, v \in V$.

Lemma 1.2 (Lax–Milgram) *Let $a : V \times V \to R$ be a bilinear, continuous and coercive functional. Then, for each $f \in V^*$, there is a unique $u^* \in V$ such that*

$$a(u^*, v) = (f, v), \quad \forall v \in V. \tag{1.17}$$

Moreover, the map $f \to u^*$ is Lipschitzian from V^* to V with Lipschitz constant ω^{-1}. If a is symmetric, then u^* minimizes the function $u \to \frac{1}{2}a(u, u) - (f, u)$ on V, that is,

$$\frac{1}{2}a(u^*, u^*) - (f, u^*) = \min\left\{\frac{1}{2}a(u, u) - (f, u);\ u \in V\right\}. \tag{1.18}$$

If a is symmetric, then the Lax–Milgram lemma is a simple consequence of Riesz's representation theorem. Indeed, in this case $(u, v) \to a(u, v)$ is an equivalent scalar product on V and so, by the Riesz theorem, the functional $v \to (f, v)$

can be represented as (1.18) for some $u^* \in V$. In the general case we proceed as follows. For each $u \in V$, the functional $v \to a(u, v)$ is linear and continuous on V and we denote it by $Au \in V^*$. Then, the equation $a(u, v) = (f, v)$, $\forall v \in V$ can be rewritten as $Au = f$. Then, the conclusion follows because $R(A)$ is simultaneously closed and dense in V^*.

Consider the Dirichlet problem

$$\begin{cases} -\Delta u + c(x)u = f & \text{in } \mathcal{O}, \\ u = 0 & \text{on } \partial\mathcal{O}, \end{cases} \tag{1.19}$$

where $\mathcal{O}$ is an open set of R^d, $c \in L^\infty(\mathcal{O})$ and $f \in H^{-1}(\mathcal{O})$ is given.

Definition 1.4 The function u is said to be *weak* or *variational solution* to the Dirichlet problem (1.19) if $u \in H_0^1(\mathcal{O})$ and

$$\int_{\mathcal{O}} \nabla u(x) \cdot \nabla \varphi(x) dx + \int_{\mathcal{O}} c(x)u(x)\varphi(x)dx = (f, \varphi) \tag{1.20}$$

for all $\varphi \in H_0^1(\mathcal{O})$ (equivalently, for all $\varphi \in C_0^\infty(\mathcal{O})$).

In (1.20), ∇u is taken in the sense of distributions and (f, φ) is the value of the functional $f \in H^{-1}(\mathcal{O})$ into $\varphi \in H_0^1(\mathcal{O})$. If $f \in L^2(\mathcal{O}) \subset H^{-1}(\mathcal{O})$, then

$$(f, \varphi) = \int_{\mathcal{O}} f(x)\varphi(x)dx.$$

By the Lax–Milgram lemma, applied to the functional

$$a(u, v) = \int_{\mathcal{O}} \nabla u(x) \cdot \nabla v(x)dx, \quad u, v \in V = H_0^1(\mathcal{O}),$$

we obtain the following result.

Theorem 1.5 *Let $\mathcal{O}$ be a bounded open set of R^d and let $c \in L^\infty(\mathcal{O})$ be such that $c(x) \geq 0$, a.e. $x \in \mathcal{O}$. Then, for each $f \in H^{-1}(\mathcal{O})$ the Dirichlet problem* (1.19) *has a unique weak solution $u^* \in H_0^1(\mathcal{O})$. Moreover, u^* minimizes on $H_0^1(\mathcal{O})$ the functional*

$$\frac{1}{2}\int_{\mathcal{O}} (|\nabla u(x)|^2 + c(x)u^2(x))dx - (f, u)$$

and the map $f \to u^*$ is Lipschitzian from $H^{-1}(\mathcal{O})$ to $H_0^1(\mathcal{O})$.

Consider the boundary value problem

$$\begin{cases} -\Delta u + cu = f & \text{in } \mathcal{O}, \\ \frac{\partial u}{\partial n} = g & \text{on } \partial\mathcal{O}, \end{cases} \tag{1.21}$$

where $c \in L^\infty(\mathcal{O})$, $c(x) \geq \rho > 0$ and $f \in L^2(\mathcal{O})$, $g \in L^2(\partial\mathcal{O})$. Here, $\frac{\partial u}{\partial n}$ is the normal derivative.

Definition 1.5 The function $u \in H^1(\mathcal{O})$ is said to be a *weak solution* to problem (1.21) if

$$\int_{\mathcal{O}} \nabla u \cdot \nabla v\, dx + \int_{\mathcal{O}} cuv\, dx = \int_{\mathcal{O}} fv\, dx + \int_{\partial\mathcal{O}} gv\, d\sigma, \quad \forall v \in H^1(\mathcal{O}). \quad (1.22)$$

Since for each $v \in H^1(\mathcal{O})$ the trace $\gamma_0(v)$ is in $L^2(\partial\mathcal{O})$, the integral $\int_{\partial\mathcal{O}} gv\, d\sigma$ is well-defined and so (1.22) makes sense.

Theorem 1.6 *Let $\mathcal{O}$ be an open subset of R^d. Then, for each $f \in L^2(\mathcal{O})$ and $g \in L^2(\partial\mathcal{O})$, problem* (1.21) *has a unique weak solution $u \in H^1(\mathcal{O})$ which minimizes the functional*

$$u \to \frac{1}{2}\int_{\mathcal{O}} (|\nabla u(x)|^2 + c(x)u^2(x))dx - \int_{\mathcal{O}} f(x)u(x)dx - \int_{\partial\mathcal{O}} gu\, d\sigma$$

on $H^1(\mathcal{O})$.

Proof One applies the Lax–Milgram lemma on the space $V = H^1(\mathcal{O})$ to the functional $a(u,v) = \int_{\mathcal{O}} (\nabla u \cdot \nabla v + cuv)dx$, $\forall u, v \in H^1(\mathcal{O})$, and $(\tilde{f}, v) = \int_{\mathcal{O}} fv\, dx + \int_{\partial\mathcal{O}} gv\, d\sigma$. □

It turns out that, if $\partial\mathcal{O}$ is smooth enough, then this solution is actually in $H^2(\mathcal{O}) \cap H_0^1(\mathcal{O})$.

Theorem 1.7 *Let $\mathcal{O}$ be a bounded and open subset of R^d of class C^2. Let $f \in L^2(\mathcal{O})$ and let $u \in H_0^1(\mathcal{O})$ be the weak solution u to problem* (1.19) *or* (1.21). *Then, $u \in H^2(\mathcal{O})$ and*

$$\|u\|_{H^2(\mathcal{O})} \le C\|f\|_{L^2(\mathcal{O})}, \quad (1.23)$$

where C is independent of f.

For proof, we refer to H. Brezis' book [36]. (See also [11].)

In particular, Theorem 1.7 implies that if $A : H_0^1(\mathcal{O}) \to H^{-1}(\mathcal{O})$ is the elliptic operator $A = -\Delta$ in $\mathcal{D}'(\mathcal{O})$, that is,

$$(Au, \varphi) = \int_{\mathcal{O}} \nabla u \cdot \nabla \varphi\, dx, \quad \forall \varphi \in H_0^1(\mathcal{O}),$$

then

$$D(A) = \{u \in H_0^1(\mathcal{O});\ Au \in L^2(\mathcal{O})\} \subset H^2(\mathcal{O})$$

and

$$\|u\|_{H^2(\mathcal{O})} \le C\|Au\|_{L^2(\mathcal{O})}, \quad \forall u \in H_0^1(\mathcal{O}) \cap H^2(\mathcal{O}),$$

and therefore

$$D(A) = H_0^1(\mathcal{O}) \cap H^2(\mathcal{O}).$$

Theorem 1.7 remains true in $L^p(\mathcal{O})$ for $p > 1$. Namely, we have the following theorem (see Agmon–Douglis–Nirenberg [4]).

Theorem 1.8 *Let $\mathcal{O}$ be a bounded open subset of R^d with smooth boundary $\partial\mathcal{O}$ and let $1 < p < \infty$. Then, for each $f \in L^p(\mathcal{O})$, the boundary value problem*

$$-\Delta u = f \quad \text{in } \mathcal{O}, \qquad u = 0 \quad \text{on } \partial\mathcal{O}$$

has a unique weak solution $u \in W_0^{1,p}(\mathcal{O}) \cap W^{2,p}(\mathcal{O})$.
Moreover, one has $\|u\|_{W^{2,p}(\mathcal{O})} \le C\|f\|_{L^p(\mathcal{O})}$, where C is independent of f.

The operator A has an extension $\widetilde{A}$ from $H = L^2(\mathcal{O})$ to

$$(D(A))' = (H_0^1(\mathcal{O}) \cap H^2(\mathcal{O}))'$$

defined by (see (1.9))

$$\langle \widetilde{A}u, v\rangle_{D(A)} = (u, Av) = -\int_{\mathcal{O}} u(x)\Delta v(x)dx,$$
$$\forall v \in H_0^1(\mathcal{O}) \cap H^2(\mathcal{O}),\ u \in L^2(\mathcal{O}). \tag{1.24}$$

As example, consider the nonhomogeneous Dirichlet problem

$$-\Delta y = f \quad \text{in } \mathcal{O}, \qquad y = u \quad \text{on } \partial\mathcal{O},$$

where $f \in L^2(\mathcal{O})$. The weak solution $y \in L^2(\mathcal{O})$ to this problem is defined by

$$-\int_{\mathcal{O}} y\Delta\varphi\, d\xi = \int_{\mathcal{O}} f\varphi\, d\xi - \int_{\partial\mathcal{O}} u\frac{\partial\varphi}{\partial n}, \quad \forall \varphi \in H^2(\mathcal{O}) \cap H_0^1(\mathcal{O}),$$

or, equivalently, $\widetilde{A}y = f + g$, where $g \in (H^2(\mathcal{O}) \cap H_0^1(\mathcal{O}))'$ is given by

$$g(\varphi) = \int_{\partial\mathcal{O}} u\frac{\partial\varphi}{\partial n}\, d\xi, \quad \forall \varphi \in H^2(\mathcal{O}) \cap H_0^1(\mathcal{O}).$$

1.2.2 Infinite-dimensional Sobolev Spaces

Let X be a real (or complex) Banach space and let $[a, b]$ be a fixed-interval on the real axis. A function $x : [a, b] \to X$ is said to be *finitely-valued* if it is constant on each of a finite number of disjoint measurable sets $A_k \subset [a, b]$ and equal to zero on $[a, b] \setminus \bigcup_k A_k$. The function x is said to be *strongly measurable* on $[a, b]$ if there is a sequence $\{x_n\}$ of finite-valued functions that converges strongly in X and almost everywhere on $[a, b]$ to x. The function x is said to be *Bochner integrable* if there exists a sequence $\{x_n\}$ of finitely-valued functions on $[a, b]$ to X that converges almost everywhere to x such that

$$\lim_{n\to\infty} \int_a^b \|x_n(t) - x(t)\|dt = 0.$$

The space of all Bochner integrable functions $x : [a, b] \to X$ is a Banach space with the norm

$$\|x\|_1 = \int_a^b \|x(t)\|dt,$$

and is denoted $L^1(a, b; X)$.

More generally, the space of all (classes) of strongly measurable functions x on $[a, b]$ to X such that

$$\|x\|_p = \left(\int_a^b \|x(t)\|^p dt \right)^{1/p} < \infty$$

for $1 \leq p < \infty$ and $\|x\|_\infty = \operatorname{ess\,sup}_{t \in [a,b]} \|x(t)\| < \infty$, will be denoted $L^p(a, b; X)$. This is a Banach space in the norm $\|\cdot\|_p$.

If X is reflexive, then the dual of $L^p(a, b; X)$ is the space $L^q(a, b; X^*)$, where $p < \infty$, $\frac{1}{p} + \frac{1}{q} = 1$. (See [49].)

An X-valued function x defined on $[a, b]$ is said to be *absolutely continuous* on $[a, b]$ if for each $\varepsilon > 0$ there exists $\delta(\varepsilon)$ such that $\sum_{n=1}^N \|x(t_n) - x(s_n)\| \leq \varepsilon$, whenever $\sum_{n=1}^N |t_n - s_n| \leq \delta(\varepsilon)$ and $(t_n, s_n) \cap (t_m, s_m) = \emptyset$ for $m \neq n$. Here, (t_n, s_n) is an arbitrary subinterval of (a, b).

Let us denote, as above, by $\mathscr{D}(a, b)$ the space of all infinitely differentiable real-valued functions on $[a, b]$ with compact support in (a, b), and by $\mathscr{D}'(a, b; X)$ the space of all continuous operators from $\mathscr{D}(a, b)$ to X. An element u of $\mathscr{D}'(a, b; X)$ is called an X-valued distribution on (a, b). If $u \in \mathscr{D}'(a, b; X)$ and j is a natural number, then

$$u^{(j)}(\varphi) = (-1)^j u(\varphi^{(j)}), \quad \forall \varphi \in \mathscr{D}(a, b),$$

defines another distribution $u^{(j)}$, which is called the derivative of order j of u.

We note that every element $u \in L^1(a, b; X)$ defines uniquely the distribution (again denoted u)

$$u(\varphi) = \int_a^b u(t)\varphi(t)dt, \quad \forall \varphi \in \mathscr{D}(a, b),$$

and so $L^1(a, b; X)$ can be regarded as a subspace of $\mathscr{D}'(a, b; X)$. In all what follows, we identify a function $u \in L^1(a, b; X)$ with the distribution u defined above.

Let k be a natural number and $1 \leq p \leq \infty$. We denote by $W^{k,p}([a, b]; X)$ the space of all X-valued distributions $u \in \mathscr{D}'(a, b; X)$ such that

$$u^{(j)} \in L^p(a, b; X) \quad \text{for } j = 0, 1, \ldots, k.$$

Here, $u^{(j)}$ is the derivative of order j of u in the sense of distributions.

We denote by $A^{1,p}([a, b]; X)$, $1 \leq p \leq \infty$, the space of all absolutely continuous functions u from $[a, b]$ to X having the property that they are a.e. differentiable on (a, b) and $\frac{du}{dt} \in L^p(a, b; X)$. If the space X is reflexive, it follows by Theorem 1.9 that $u \in A^{1,p}([a, b]; X)$ if and only if u is absolutely continuous on $[a, b]$ and $\frac{du}{dt} \in L^p(a, b; X)$.

It turns out that the space $W^{1,p}$ can be identified with $A^{1,p}$. More precisely, we have the following theorem.

Theorem 1.9 *Let X be a Banach space and let $u \in L^p(a, b; X)$, $1 \leq p \leq \infty$. Then the following conditions are equivalent:*

(i) $u \in W^{1,p}([a,b];X)$;
(ii) *There is* $u^0 \in A^{1,p}([a,b];X)$ *such that* $u(t) = u^0(t)$, *a.e.*, $t \in (a,b)$.
Moreover, $u' = \frac{du^0}{dt}$, *a.e. in* (a,b).

We note also the following result. (See [35].)

Theorem 1.10 *Let* X *be reflexive Banach and let* $u \in L^p(a,b;X)$, $1 < p \leq \infty$. *Then the following two conditions are equivalent:*

(i) $u \in W^{1,p}([a,b];X)$;
(ii) *There is* $C > 0$ *such that*

$$\int_a^{b-h} \|u(t+h) - u(t)\|^p dt \leq C|h|^p, \quad \forall h \in [0, b-a]$$

with usual modification in the case $p = \infty$.

Let V be a reflexive Banach space and H be a real Hilbert space such that $V \subset H \subset V'$ in the algebraic and topological senses. Here, V' is the dual space of V and H is identified with its own dual. Denote by $|\cdot|$ and $\|\cdot\|$ the norms of H and V, respectively, and by $(\cdot,\cdot)$ the duality between V and V'. If $v_1, v_2 \in H$, then (v_1, v_2) is the scalar product in H of v_1 and v_2. Denote by $W_p([a,b];V)$, $1 < p < \infty$, the space

$$W_p([a,b];V) = \{u \in L^p(a,b;V);\ u' \in L^q(a,b;V')\}, \qquad \frac{1}{p} + \frac{1}{q} = 1,$$

where u' is the derivative of u in the sense of $\mathscr{D}'(a,b;V)$. By Theorem 1.9, we know that every $u \in W_p([a,b];V)$ can be identified with an absolutely continuous function $u^0 : [a,b] \to V'$. However, we have a more precise result. (See [62].)

Theorem 1.11 *Let* $u \in W_p([a,b];V)$. *Then there is a continuous function* $u^0 : [a,b] \to H$ *such that* $u(t) = u^0(t)$, *a.e.*, $t \in (a,b)$. *Moreover, if* $u, v \in W_p([a,b];V)$, *then the function* $t \to (u(t), v(t))$ *is absolutely continuous on* $[a,b]$ *and*

$$\frac{d}{dt}(u(t), v(t)) = (u'(t), v(t)) + (u(t), v'(t)), \quad \text{a.e. } t \in (a,b).$$

1.3 The Semigroups of Class C_0

Definition 1.6 Let X be a real or complex Banach space and $S(t)$, $t \geq 0$, a one parameter family of linear continuous operators in $L(X,X)$. $\{S(t),\ t \geq 0\}$ is called a *semigroup of class* C_0 (or C_0*-semigroup*) if

$$S(t)S(s) = S(t+s), \quad \forall t, s \geq 0, \tag{1.25}$$

$$S(0) = I, \tag{1.26}$$

$$\lim_{t \to 0} S(t)x = x, \quad \forall x \in X. \tag{1.27}$$

It turns out that a semigroup of class C_0 satisfies the condition

$$\|S(t)\|_{L(X,X)} \le Ce^{\omega t}, \quad \forall t \ge 0,$$

for some $\omega \in R$.

The infinitesimal generator of the C_0-semigroup $S(t)$ is defined by

$$A_0 x = \lim_{t\to 0} \frac{S(t)x - x}{t}, \quad \forall x \in D(A_0),$$

and it turns out that A_0 is a linear closed and densely defined operator in X. The infinitesimal generator of a C_0-semigroup is characterized by the famous Hille–Yosida theorem. (See, e.g., [67, 82].)

Theorem 1.12 *The linear, closed and densely defined operator A_0 is the infinitesimal generator of a C_0-semigroup $S(t)$ in $L(X, X)$ if and only if for there is $\omega \in R$ such that*

$$\|(\lambda I - A_0)^{-n}\|_{L(X,X)} \le \frac{M}{(\operatorname{Re}\lambda - \omega)^n} \tag{1.28}$$

for all $\lambda \in \mathbb{C}$ with $\operatorname{Re}\lambda > \omega$.

If A_0 is the infinitesimal generator of a C_0-semigroup $S(t)$, then the Cauchy problem

$$\begin{aligned} \frac{du(t)}{dt} &= A_0 u(t), \quad \forall t \ge 0, \\ u(0) &= x \end{aligned} \tag{1.29}$$

has for each $x \in D(A)$ a unique solution $u \in C^1([0,T]; X)$ with $A_0 u \in C([0,T]; X)$. Moreover,

$$S(t)x = u(t), \quad \forall t \ge 0, \ \forall x \in D(A_0) \tag{1.30}$$

and

$$\|S(t)x\|_X \le Me^{\omega t}\|x\|_X, \quad \forall x \in X.$$

A standard example is $X = L^p(\mathcal{O})$, $p > 1$, $A_0 = \Delta$, $D(A_0) = W_0^{1,p}(\mathcal{O}) \cap W^{2,p}(\mathcal{O})$.

The semigroup $S(t)$ generated by A_0 is also denoted by $e^{A_0 t}$.

Definition 1.7 The C_0-semigroup $S(t)$ is said to be analytic if it admits an analytic extension $S(\lambda)$ in the complex plane given by

$$S(\lambda)x = \sum_{j=0}^{\infty} \frac{(\lambda - t)^j}{j!} \frac{d^j}{dt^j} S(t)x \quad \text{for } |\arg(\lambda - \lambda_0)| < C, \text{ where } \lambda_0 \in R.$$

We have the following theorem (see, e.g., [67, 82]).

Theorem 1.13 *The linear, closed and densely defined operator A_0 is the infinitesimal generator of a C_0-analytic semigroup if and only if*

$$\|(\lambda I - A_0)^{-1}\|_{L(X,X)} \le \frac{C}{|\lambda - \lambda_0|}, \quad \forall \lambda \in \mathbb{C},\ \operatorname{Re}\lambda > \lambda_0. \tag{1.31}$$

If A_0 generates a C_0-analytic semigroup, then $S(t)x = e^{A_0 t}x$ is differentiable on $(0, \infty)$ for all $x \in X$.

If A_0 generates a C_0-semigroup and $f \in C^1([0, T]; X)$, then the Cauchy problem

$$\begin{aligned} &\frac{du}{dt}(t) = A_0 u(t) + f(t), \quad t \in [0, T], \\ &u(0) = u_0, \end{aligned} \tag{1.32}$$

has for each $u_0 \in D(A_0)$ a unique solution $u \in C^1([0, T]; X) \cap C([0, T]; D(A_0))$ given by the variation of the constant formula

$$u(t) = S(t)u_0 + \int_0^t S(t - s)f(s)ds, \quad \forall t \in [0, T]. \tag{1.33}$$

If $u_0 \in X$ or $f \in L^1(0, T; X)$, then the function u defined by (1.33) is continuous only and is called a "mild" solution to Cauchy problem (1.32).

However, if A_0 generates an analytic C_0-semigroup and $u_0 \in X$, $f \in L^2(0, T; X)$, then the mild solution u is almost everywhere differentiable on $(0, T)$ and satisfies (1.32), a.e., on $(0, T)$. Moreover,

$$u \in W^{1,2}([\delta, T]; X), \quad \forall \delta \in (0, T).$$

Definition 1.8 The infinitesimal generator A_0 of a C_0-semigroup $S(t) = e^{A_0 t}$ is said to have the *growth logarithmic property* if $\sigma(A_0) \subset \{\lambda;\ \operatorname{Re}\lambda < \omega\}$ implies that

$$\|S(t)\|_{L(X,X)} \le Me^{\omega t}, \quad \forall t \ge 0.$$

It should be said that contrary to the finite-dimensional situation, this property is not satisfied by all the infinitesimal generators A_0. We have, however, the theorem below (see, e.g., [32], p. 120).

Theorem 1.14 *If A_0 is the infinitesimal generator of a C_0-analytic semigroup, then A_0 has the growth logarithmic property.*

This means that in this case the exponential long time behaviour of $S(t) = e^{A_0 t}$ is precisely described by the spectral properties of A_0.

1.4 The Nonlinear Cauchy Problem

Throughout this section, X will be a real Banach space with the norm $\|\cdot\|$, X' will be its dual space and $(\cdot, \cdot)$ the pairing between X and X'. We will denote as usual $J : X \to X'$ the duality mapping of the space X, that is,

$$J(x) = \{x^* \in X';\ (x, x^*) = \|x\|^2 = \|x^*\|^2_{X'}\}.$$

Definition 1.9 A subset A of $X \times X$ (equivalently, a multivalued operator from X to X) is called *accretive* if for every pair $[x_1, y_1], [x_2, y_2] \in A$ there is $z \in J(x_1 - x_2)$ such that

$$(y_1 - y_2, z) \geq 0. \tag{1.34}$$

If $X = H$ is a Hilbert space, then Condition (1.34) reduces to $(y_1 - y_2, x_1 - x_2) \geq 0$.

An accretive set A is said to be *m-accretive* if

$$R(I + A) = X. \tag{1.35}$$

Here, we have denoted by I the unity operator in X.

Finally, A is said to be *ω-accretive* (*ω-m-accretive*), where $\omega \in R$, if $A + \omega I$ is accretive (*m*-accretive, respectively).

The accretiveness of A can be equivalently expressed as

$$\|x_1 - x_2\| \leq \|x_1 - x_2 + \lambda(y_1 - y_2)\|, \quad \forall \lambda > 0,\ [x_1, y_i] \in A,\ i = 1, 2. \tag{1.36}$$

Proposition 1.4 *A subset A of $X \times X$ is accretive if and only if inequality* (1.36) *holds for all $\lambda > 0$ and all $[x_i, y_i] \in A$, $i = 1, 2$.*

In particular, it follows that A is ω-accretive if and only if

$$\|x_1 - x_2 + \lambda(y_1 - y_2)\| \geq (1 - \lambda\omega)\|x_1 - x_2\|$$
$$\text{for } 0 < \lambda < 1/\omega \text{ and } [x_i, y_i] \in A,\ i = 1, 2. \tag{1.37}$$

Hence, if A is accretive, then the operator $(I + \lambda A)^{-1}$ is nonexpansive on $R(I + \lambda A)$, that is, $\|(I + \lambda A)^{-1}x - (I + \lambda A)^{-1}y\| \leq \|x - y\|, \forall \lambda > 0,\ x, y \in R(I + \lambda A)$. If A is ω-accretive, then it follows by (1.37) that $(I + \lambda A)^{-1}$ is Lipschitz with Lipschitz constant $1/(1 - \lambda\omega)$ on $R(I + \lambda A), 0 < \lambda < 1/\omega$. Let us define the operators J_λ and A_λ

$$J_\lambda x = (I + \lambda A)^{-1}x, \quad x \in R(I + \lambda A); \tag{1.38}$$
$$A_\lambda x = \lambda^{-1}(x - J_\lambda x), \quad x \in R(I + \lambda A). \tag{1.39}$$

The operator A_λ is called the *Yosida approximation* of A.

Proposition 1.5 *An accretive set $A \subset X \times X$ is m-accretive if and only if $R(I + \lambda A) = X$ for all (equivalently, for some) $\lambda > 0$.*

Note that an m-accretive set of $H \times H$ is maximal accretive, that is, it has no proper extension to accretive sets. Indeed, if $[x, y] \in X \times X$ is such that

$$\|x - u\| \leq \|x + \lambda y - (u + \lambda v)\|, \quad \forall [u, v] \in A,\ \lambda > 0,$$

the choosing $[u, v] \in A$ such that $u + \lambda v = x + \lambda y$, we see that $x = u$ and so $v = y \in Ax$.

In particular, for every $x \in D(A)$ we have $Ax = \{y \in X';\ (y - v, J(x - u)) \geq 0, \forall [u, v] \in A\}$. Hence, Ax is a closed convex subset of X.

We notice also the following proposition.

Proposition 1.6 *Let X be a Banach space. Then any m-accretive linear operator $A : X \to X$ is densely, that is, $\overline{D(A)} = X$.*

A standard example of a nonlinear m-accretive operator is that given by

$$\begin{aligned} &X = L^2(\mathcal{O}), \qquad Ay = -\Delta y + \beta(y), \quad \forall y \in D(A), \\ &D(A) = \{y \in H_0^1(\mathcal{O}) \cap H^2(\mathcal{O});\ \beta(y) \in L^2(\mathcal{O})\}, \end{aligned} \tag{1.40}$$

where $\beta : R \to R$ is a monotonically nondecreasing continuous function and $\mathcal{O}$ is a bounded and open subset of R^d, with smooth boundary.

Let X be a real Banach space with the norm $\|\cdot\|$ and dual X^* and let $A \subset X \times X$ be a quasi-m-accretive set of $X \times X$. Consider the Cauchy problem

$$\begin{cases} \frac{dy}{dt}(t) + Ay(t) \ni f(t), & t \in [0, T], \\ y(0) = y_0, \end{cases} \tag{1.41}$$

where $y_0 \in X$ and $f \in L^1(0, T; X)$.

Definition 1.10 A strong solution to (1.41) is a function $y \in W^{1,1}((0,T]; X) \cap C([0, T]; X)$ such that $f(t) - \frac{dy}{dt}(t) \in Ay(t)$, a.e., $t \in (0, T)$, $y(0) = y_0$.

Here,

$$W^{1,1}((0, T]; X) = \{y \in L^1(0, T; X);\ y' \in L^1(\delta, T; X),\ \forall \delta \in (0, T)\}.$$

The main existence result is Theorem 1.15 below. (See [15, 43].)

Theorem 1.15 *Let X be a reflexive Banach space and let A be a quasi-m-accretive operator. Then, for each $y_0 \in D(A)$ and $f \in W^{1,1}([0, T]; X)$, Problem (1.41) has a unique strong solution $y \in W^{1,\infty}([0, T]; X)$.*

More can be said about the regularity of a strong solution to Problem (1.41) if the space X is uniformly convex.

Theorem 1.16 *Let A be quasi-m-accretive, that is, $A + \omega I$ m-accretive, $f \in W^{1,1}([0, T]; X)$, $y_0 \in D(A)$ and let X be uniformly convex along with the dual X^*. Then, the strong solution to Problem (1.41) is everywhere differentiable from the right, $(d^+/dt)y$ is right continuous, and*

$$\frac{d^+}{dt} y(t) + (Ay(t) - f(t))^0 = 0, \quad \forall t \in [0, T), \tag{1.42}$$

$$\left\| \frac{d^+}{dt} y(t) \right\| \le e^{\omega t} \|(Ay_0 - f(0))^0\| + \int_0^t e^{\omega(t-s)} \left\| \frac{df}{ds}(s) \right\| ds, \quad \forall t \in [0, T). \tag{1.43}$$

Here, $(Ay - f)^0$ is the element of minimum norm in the set $Ay - f$.

In particular, it follows that, if A is linear and m-accretive, then $-A$ generates a C_0-semigroup of contractions e^{-At} on X.

In Theorem 1.16, the strong solution $y \in W^{1,\infty}([0,T];X)$ to Problem (1.41) can be obtained as

$$y(t) = \lim_{\lambda\to 0} y_\lambda(t) \quad \text{in } X \text{, uniformly on } [0,T],$$

where $y_\lambda \in C^1([0,T];X)$ are the solutions to the Yosida approximating equation

$$\begin{cases} \frac{dy_\lambda}{dt}(t) + A_\lambda y_\lambda(t) = f(t), & t \in [0,T], \\ y_\lambda(0) = y_0, \end{cases}$$

where $A_\lambda = \lambda^{-1}(I - (I + \lambda A)^{-1})$ for $0 < \lambda < \lambda_0$. Theorems 1.15 and 1.16 apply neatly to the parabolic boundary value problems of the form

$$\begin{aligned} &\frac{\partial y}{\partial t} - \Delta y + \beta(y) = f(t,x), \quad \forall (t,x) \in (0,T) \times \mathcal{O}, \\ &y(0,x) = y_0(x), \quad x \in \mathcal{O}, \\ &y = 0 \quad \text{on } (0,T) \times \partial\mathcal{O}, \end{aligned} \tag{1.44}$$

where β is a continuous and monotonically nondecreasing function on R (more generally, a maximal monotone graph in $R \times R$). In this case, X and A are as in (1.40).

1.5 Strong Solutions to Navier–Stokes Equations

The classical Navier–Stokes equations

$$\begin{cases} y_t(x,t) - \nu\Delta y(x,t) + (y\cdot\nabla)y(x,t) = f(x,t) + \nabla p(x,t), \\ \quad x \in \mathcal{O},\ t \in (0,T), \\ (\nabla\cdot y)(x,t) = 0, \quad \forall (x,t) \in \mathcal{O} \times (0,T), \\ y = 0 \quad \text{on } \partial\mathcal{O} \times (0,T), \\ y(x,0) = y_0(x), \quad x \in \mathcal{O} \end{cases} \tag{1.45}$$

describe the non-slip motion of a viscous, incompressible, Newtonian fluid in an open domain $\mathcal{O} \subset R^d$, $d = 2,3$. Here $y = (y_1, y_2, \dots, y_d)$ is the velocity field, p is the pressure, f is the density of an external force, and $\nu > 0$ is the viscosity of the fluid.

We have used the following standard notation

$$\begin{cases} \nabla\cdot y = \operatorname{div} y = \sum_{i=1}^d D_i y_i, \quad D_i = \frac{\partial}{\partial x_i},\ i = 1,\dots,d, \\ (y\cdot\nabla)y = \sum_{i=1}^d y_i D_i y_j, \quad j = 1,\dots,d. \end{cases}$$

By a classical device due to J. Leray, the boundary value problem (1.45) can be written as an infinite-dimensional Cauchy problem in an appropriate function space on $\mathcal{O}$. To this end we introduce the following spaces

$$H = \{y \in (L^2(\mathcal{O}))^d;\ \nabla\cdot y = 0,\ y\cdot n = 0 \text{ on } \partial\mathcal{O}\}, \tag{1.46}$$

$$V = \{y \in (H_0^1(\mathcal{O}))^d;\ \nabla\cdot y = 0\}. \tag{1.47}$$

Here n is the outward normal to $\partial\mathcal{O}$.

The space H is a closed subspace of $(L^2(\mathcal{O}))^d$ and it is a Hilbert space with the scalar product

$$(y,z)=\int_{\mathcal{O}} y\cdot z\,dx \tag{1.48}$$

and the corresponding norm $|y|=(\int_{\mathcal{O}}|y|^2dx)^{\frac{1}{2}}$. (We denote by the same symbol $|\cdot|$ the norm in R^d, $(L^2(\mathcal{O}))^d$, and H, respectively.) The norm of the space V is denoted by $\|\cdot\|$:

$$\|y\|=\left(\int_{\mathcal{O}}|\nabla y(x)|^2dx\right)^{\frac{1}{2}}. \tag{1.49}$$

We denote by $P:(L^2(\mathcal{O}))^d\to H$ the orthogonal projection of $(L^2(\mathcal{O}))^d$ onto H (the Leray projector) and set

$$a(y,z)=\int_{\mathcal{O}}\nabla y\cdot\nabla z\,dx,\quad \forall y,z\in V, \tag{1.50}$$

$$A=-P\Delta,\qquad D(A)=(H^2(\mathcal{O}))^d\cap V. \tag{1.51}$$

Equivalently,

$$(Ay,z)=a(y,z),\quad \forall y,z\in V.$$

The *Stokes operator* A is self-adjoint in H, $A\in L(V,V')$ (V' is the dual of V with the norm denoted by $\|\cdot\|_{V'}$) and

$$(Ay,y)=\|y\|^2,\quad \forall y\in V. \tag{1.52}$$

Finally, consider the trilinear functional

$$b(y,z,w)=\int_{\mathcal{O}}\sum_{i,j=1}^{d} y_iD_iz_jw_j\,dx,\quad \forall y,z,w\in V, \tag{1.53}$$

and we denote by $S:V\to V'$ the nonlinear operator defined by

$$Sy=P(y\cdot\nabla)y \tag{1.54}$$

or, equivalently,

$$(Sy,w)=b(y,y,w),\quad \forall w\in V.$$

Let $f\in L^2(0,T;V')$ and $y_0\in H$. The function $y:[0,T]\to H$ is said to be a *weak solution* to equation (1.45) if

$$y\in L^2(0,T;V)\cap C_w([0,T];H)\cap W^{1,1}([0,T];V'), \tag{1.55}$$

$$\begin{cases}\frac{d}{dt}(y(t),\psi)+\nu a(y(t),\psi)+b(y(t),y(t),\psi)=(f(t),\psi), & \text{a.e., } t\in(0,T),\\ y(0)=y_0,\quad \forall\psi\in V.\end{cases} \tag{1.56}$$

Here $(\cdot,\cdot)$ is, as usual, the pairing between V, V' and the scalar product of H. By $C_w([0,T];H)$ we denoted the space of weakly continuous functions $y : [0,T] \to H$.

Equation (1.56) can be equivalently written as

$$\begin{cases} \frac{dy}{dt}(t) + \nu Ay(t) + Sy(t) = f(t), & \text{a.e., } t \in (0,T), \\ y(0) = y_0, \end{cases} \tag{1.57}$$

where $\frac{dy}{dt}$ is the strong derivative of function $y : [0,T] \to V'$.

The function y is said to be the *strong solution* to (1.45) if $y \in W^{1,1}([0,T];H) \cap L^2(0,T;D(A))$ and (1.57) holds with $\frac{dy}{dt} \in L^1(0,T;H)$ the strong derivative of function $y : [0,T] \to H$.

The existence result given here for (1.57) follows the approach developed in [15].

Before proceeding with the existence for problem (1.1), we pause briefly to present some fundamental properties of the trilinear functional b defining the inertial operator S (see [41, 73]).

Proposition 1.7 *Let* $1 \le d \le 3$. *Then*

$$b(y,z,w) = -b(y,w,z), \quad \forall y,z,w \in V, \tag{1.58}$$

$$|b(y,z,w)| \le C\|y\|_{m_1}\|z\|_{m_2+1}\|w\|_{m_3}, \quad \forall u \in V_{m_1},\ v \in V_{m_2},\ w \in V_{m_3}, \tag{1.59}$$

where $m_i \ge 0$, $i = 1,2,3$ *and*

$$\begin{aligned} m_1 + m_2 + m_3 &\ge \frac{d}{2} \quad \text{if } m_i \ne \frac{d}{2}, \ \forall i = 1,2,3, \\ m_1 + m_2 + m_3 &> \frac{d}{2} \quad \text{if } m_i = \frac{d}{2}, \text{ for some } i = 1,2,3. \end{aligned} \tag{1.60}$$

Here $V_{m_i} = V \cap (H_0^{m_i}(\mathcal{O}))^d$ and $\|\cdot\|_m$ is the norm in V_m.

Proof It suffices to prove (1.59) for $y,z,w \in \{y \in (C_0^\infty(\mathcal{O}))^d;\ \nabla \cdot y = 0\}$. We have

$$\begin{aligned} b(y,z,w) &= \int_{\mathcal{O}} y_i D_i z_j w_j dx = \int_{\mathcal{O}} (y_i D_i(z_j w_j) - y_i D_i w_j z_j) dx \\ &= -\int_{\mathcal{O}} y_i D_i w_j z_j dx = -b(y,z,w) \end{aligned}$$

because $\nabla \cdot y = 0$. By Hölder's inequality we have

$$|b(y,z,w)| \le |y_i|_{q_1} |D_i z_j|_{q_2} |w_j|_{q_3}, \quad \frac{1}{q_1} + \frac{1}{q_2} + \frac{1}{q_3} \le 1. \tag{1.61}$$

(Here $|\cdot|_q$ is the norm of $L^q(\mathcal{O})$.) On the other hand, by the Sobolev embedding theorem we have (see Theorem 1.5) $H^{m_i}(\mathcal{O}) \subset L^{q_i}(\mathcal{O})$ for $\frac{1}{q_i} = \frac{1}{2} - \frac{m_i}{d}$ if $m_i < \frac{d}{2}$. Then, (1.61) yields $|b(y,z,w)| \le C\|y\|_{m_1}\|z\|_{m_2+1}\|w\|_{m_3}$ if $m_i < \frac{d}{2}$, $i = 1,2,3$.

If one m_i is larger than $\frac{d}{2}$ the previous inequality still remains true because, in this case, $H^{m_i}(\mathcal{O}) \subset L^\infty(\mathcal{O})$.

If $m_i = \frac{d}{2}$ then $H^{m_i}(\mathcal{O}) \subset \bigcap_{q>2} L^q(\mathcal{O})$ and so (1.61) holds for $\frac{1}{q_2} + \frac{1}{q_3} < 1$ and $q_1 = \varepsilon$ where $\frac{1}{\varepsilon} = 1 - \frac{1}{q_2} - \frac{1}{q_3}$. Then (1.59) follows for $m_1 + m_2 + m_3 > \frac{d}{2}$ as claimed.

We have also the interpolation inequality

$$\|u\|_m \le c\|u\|_\ell^{1-\alpha}\|u\|_{\ell+1}^\alpha, \quad \text{for } \alpha = \ell + 1 - m \in [0, 1]. \tag{1.62}$$

It should be noticed that if A^m, $0 < m < 1$, is the fractional power of the operator A, then we have

$$\|y\|_m = |A^{\frac{m}{2}} y|$$

and so, estimate (1.59) can be equivalently expressed in terms of $|A^{\frac{m_i}{2}} y|$.

In particular, it follows by Proposition 1.7 that S is continuous from V to V'. Indeed, we have

$$(Sy - Sz, w) = b(y, y - z, w) + b(y - z, z, w), \quad \forall w \in V,$$

and this yields (notice that $\|\cdot\| = \|\cdot\|_1$ and $|Ay| = |y|_2$)

$$|(Sy - Sz, w)| \le C(\|y\|\|y - z\|\|w\| + \|y - z\|\|z\|\|w\|).$$

Hence

$$\|Sy - Sz\|_{V'} \le C\|y - z\|(\|y\| + \|z\|), \quad \forall y, z \in V. \tag{1.63}$$

We would like to treat (1.57) as a nonlinear Cauchy problem in the space H. However, because the operator $\nu A + S$ is not quasi-m-accretive in H, we first consider a quasi-m-accretive approximation of the operator S.

For each $M > 0$, define the operator $S_M : V \to V'$

$$S_M y = \begin{cases} Sy & \text{if } \|y\| \le M, \\ \frac{M^2}{\|y\|^2} Sy & \text{if } \|y\| > M, \end{cases}$$

and consider the operator $\Gamma_M : D(\Gamma_M) \subset H \to H$

$$\Gamma_M = \nu A + S_M, \qquad D(\Gamma_M) = D(A). \tag{1.64}$$

Let us show that Γ_M is well-defined. Indeed, we have

$$|\Gamma_M y| \le \nu|Ay| + |S_M y|, \quad \forall y \in D(A).$$

On the other hand, by (1.59) for $m_1 = 1$, $m_2 = \frac{1}{2}$, $m_3 = 0$, we have for $\|y\| \le M$

$$|(S_M y, w)| = |b(y, y, w)| \le C\|y\|^{\frac{3}{2}}|Ay|^{\frac{1}{2}}|w|$$

because $\|y\|_{3/2} \le \|y\|^{\frac{1}{2}}|Ay|^{\frac{1}{2}}$. Hence $|S_M y| \le C|Ay|^{\frac{1}{2}}\|y\|^{\frac{3}{2}}$, $\forall y \in D(A)$. Similarly, we get, for $\|y\| > M$,

$$|S_M y| \le \frac{CM^2}{\|y\|^2}|Ay|^{\frac{1}{2}}\|y\|^{\frac{3}{2}} \le C|Ay|^{\frac{1}{2}}\|y\|^{\frac{3}{2}}.$$

This yields

$$|\Gamma_M y| \le \nu |Ay| + C|Ay|^{\frac{1}{2}} \|y\|^{\frac{3}{2}}, \quad \forall y \in D(A), \tag{1.65}$$

as claimed. □

Moreover, we have (see Lemma 5.2 in [15]).

Lemma 1.3 *There is α_M such that $\Gamma_M + \alpha_M I$ is m-accretive in $H \times H$.*

For each $M > 0$, consider the equation

$$\begin{cases} \frac{dy}{dt}(t) + \nu Ay(t) + S_M y(t) = f(t), & t \in (0,T), \\ y(0) = y_0. \end{cases} \tag{1.66}$$

Proposition 1.8 *Let $y_0 \in D(A)$ and $f \in W^{1,1}([0,T];H)$ be given. Then there is a unique solution $y_M \in W^{1,\infty}([0,T];H) \cap L^\infty(0,T;D(A)) \cap C([0,T];V)$ to* (1.66). *Moreover, $(d^+/dt)y_M(t)$ exists for all $t \in [0,T)$ and*

$$\frac{d^+}{dt} y_M(t) + \nu A y_M(t) + S_M y_M(t) = f(t), \quad \forall t \in [0,T). \tag{1.67}$$

Proof It follows by Theorem 1.16. Because $\Gamma_M y_M = \nu A y_M + S_M y_M \in L^\infty(0,T;H)$, by (1.65) we infer that $Ay_M \in L^\infty(0,T;H)$. As $\frac{dy_M}{dt} \in L^\infty(0,T;H)$, we conclude also that $y_M \in C([0,T];V) \cap L^\infty(0,T;D(A))$, as claimed. □

Now we are ready to formulate the main existence result for the strong solutions to the Navier–Stokes equation (1.45).

Theorem 1.17 *Let $d = 2,3$ and $f \in W^{1,1}([0,T];H)$, $y_0 \in D(A)$ where $0 < T < \infty$. Then there is a unique function $y \in W^{1,\infty}([0,T^*);H) \cap L^\infty(0,T^*;D(A)) \cap C([0,T^*];V)$ such that*

$$\begin{cases} \frac{dy(t)}{dt} + \nu Ay(t) + Sy(t) = f(t), & \text{a.e., } t \in (0,T^*), \\ y(0) = y_0, \end{cases} \tag{1.68}$$

for some $T^* = T^*(\|y_0\|) \le T$. If $d = 2$, then $T^* = T$. Moreover, $y(t)$ is right differentiable and

$$\frac{d^+}{dt} y(t) + \nu Ay(t) + Sy(t) = f(t), \quad \forall t \in [0,T^*). \tag{1.69}$$

Proof The idea of the proof is to show that for M sufficiently large the flow $y_M(t)$, defined by Proposition 1.8, is independent of M on each interval $[0,T]$ if $d = 2$ or on $[0, T(y_0)]$ if $d = 3$.

If we multiply (1.67) by y_M and integrate on $(0,t)$, we get

$$|y_M(t)|^2 + \nu \int_0^t \|y_M(s)\|^2 ds \le C\left(|y_0|^2 + \frac{1}{\nu}\int_0^T \|f(t)\|_{V'}^2 dt\right), \quad \forall M. \tag{1.70}$$

Next, we multiply (1.67) (scalarly in H) by $Ay_M(t)$. We get

$$\frac{1}{2}\frac{d}{dt}\|y_M(t)\|^2 + \nu|Ay_M(t)|^2 \le |(S_M y_M(t), Ay_M(t))| + |f(t)||Ay_M|,$$
$$\text{a.e., } t \in (0,T).$$

This yields

$$\|y_M(t)\|^2 + \nu \int_0^t |Ay_M(s)|^2 ds \le C\left(\|y_0\|^2 + \frac{1}{\nu}\int_0^T |f(t)|^2 dt + \int_0^t |(S_M y_M, Ay_M)| ds\right). \quad (1.71)$$

On the other hand, for $d = 3$, by (1.59) we have (the case $d = 2$ is treated separately below)

$$\begin{aligned}|(S_M y_M, Ay_M)| &< |b(y_M, y_M, Ay_M)| \\ &\le C\|y_M\|\|y_M\|_{3/2}|Ay_M| \\ &\le C\|y_M\|^{\frac{3}{2}}|Ay_M|^{\frac{3}{2}}, \quad \text{a.e., } t \in (0,T).\end{aligned}$$

(Everywhere in the following C is independent of M, ν.)

Then, by (1.71) we have

$$\begin{aligned}&\|y_M(t)\|^2 + \nu \int_0^t |Ay_M(s)|^2 ds \\ &\le C\left(\|y_0\|^2 + \frac{1}{\nu}\int_0^T |f(t)|^2 dt + \int_0^t |Ay_M(s)|^{\frac{3}{2}}\|y_M(s)\|^{\frac{3}{2}} ds\right) \\ &\le C\left(\|y_0\|^2 + \frac{1}{\nu}\int_0^T |f(t)|^2 dt + \frac{1}{\nu}\int_0^t \|y_M(s)\|^6 ds\right) + \frac{\nu}{2}\int_0^t |Ay_M(s)|^2 ds, \\ &\forall t \in [0,T].\end{aligned}$$

Finally,

$$\|y_M(t)\|^2 + \frac{\nu}{2}\int_0^t |Ay_M(s)|^2 ds \le C_0\left(\|y_0\|^2 + \frac{1}{\nu}\int_0^T |f(s)|^2 ds + \frac{1}{\nu}\int_0^t \|y_M(s)\|^6 ds\right). \quad (1.72)$$

Next, we consider the integral inequality

$$\|y_M(t)\|^2 \le C_0\left(\|y_0\|^2 + \frac{1}{\nu}\int_0^T |f(s)|^2 ds + \frac{1}{\nu}\int_0^t \|y_M(s)\|^6 ds\right). \quad (1.73)$$

We have

$$\|y_M(t)\|^2 \le \varphi(t), \quad \forall t \in (0,T),$$

where

$$\varphi' \le \frac{C_0}{\nu}\varphi^3, \quad \forall t \in (0,T),$$

and

$$\varphi(0) = C_0\left(\|y_0\|^2 + \frac{1}{\nu}\int_0^T |f(s)|^2 ds\right).$$

Hence

$$\|y_M(t)\|^2 \le \left(\frac{\nu}{\varphi}^3(0)\nu - 3t\varphi^3(0)\right)^{1/3}, \quad \forall t \in (0, T^*), \tag{1.74}$$

where $T^* = \frac{\nu}{3C_0^3(\|y_0\|^2 + \frac{1}{\nu}\int_0^T |f(s)|^2 ds)^3}$. Then, we get

$$\|y_M(t)\|^2 + \frac{\nu}{2}\int_0^t |Ay_M(s)|^2 ds \le C_1(\delta)\left(\|y_0\|^2 + \frac{1}{\nu}\int_0^T |f(t)|^2 dt\right),$$
$$0 < t < T^* - \delta. \tag{1.75}$$

For $d = 2$, we have $|(S_M y_M, Ay_M)| \le C|y_M|^{\frac{1}{2}}\|y_M\||Ay_M|^{\frac{3}{2}} \le \frac{\nu}{2}|Ay_M|^2 + \frac{C}{\nu}\|y_M\|^4$. This yields

$$\begin{aligned}\|y_M(t)\|^2 &+ \frac{\nu}{2}\int_0^t |Ay_M(s)|^2 ds \\ &\le C\left(\|y_0\|^2 + \frac{1}{\nu}\int_0^T |f(t)|^2 dt + \frac{1}{\nu}\int_0^t \|y_M(s)\|^4 ds\right).\end{aligned}$$

Then, by the Gronwall lemma, we obtain

$$\begin{aligned}\|y_M(t)\|^2 &+ \frac{\nu}{2}\int_0^t |Ay_M(s)|^2 ds \\ &\le C\left(\|y_0\|^2 + \frac{1}{\nu}\int_0^T |f(t)|^2 dt\right), \quad \forall t \in (0, T).\end{aligned} \tag{1.76}$$

By (1.74), (1.76) we infer that for M large enough, $\|y_M(t)\| \le M$ on $(0, T^*)$ if $d = 3$ or on the whole of $(0, T)$ if $d = 2$. Hence $S_M y_M = Sy_M$ on $(0, T^*)$ (respectively on $(0, T)$) and so $y_M = y$ is a solution to (1.68). This completes the proof of existence. The uniqueness is immediate. □

Theorem 1.18 *Let $y_0 \in H$ and $f \in L^2(0, T; V')$, $d = 2, 3$. Then there is at least one weak solution y^* to* (1.45) *given by*

$$y^* = w - \lim_{n\to\infty} y_{M_n} \quad \text{in } L^2(0, T; V), \text{ weak-star in } L^\infty(0, T; H),$$
$$\frac{dy^*}{dt} = w - \lim_{n\to\infty} \frac{d}{dt} y_{M_n} \quad \text{in } L^{\frac{4}{3}}(0, T; V'),$$

for some $n \to \infty$.

Proof One uses Estimate (1.70) and

$$|(S_M y_M, w)| \le C|y_M|^{\frac{1}{2}}\|y_M\|^{\frac{3}{2}}\|w\|, \quad \forall w \in V,$$

which implies that $\|S_M y_M\|_{V'} \leq C\|y_M\|^{\frac{3}{2}} |y_M|^{\frac{1}{2}}$ and, therefore,

$$\int_0^T \left\| \frac{dy_M}{dt} \right\|_{V'}^{\frac{4}{3}} dt \leq C \left(|y_0|^2 + \int_0^T |f(t)|_{V'}^2 dt \right)^{\frac{2}{3}}.$$

Then one passes to limit $M \to \infty$ into (1.67). For details, see [15], pp. 264–265. □

Chapter 2
Stabilization of Abstract Parabolic Systems

We discuss here a few stabilization techniques for nonlinear parabolic-like equations in Hilbert spaces. The abstract theory of stabilization presented below captures most of the techniques to be developed for the specific problems which are treated in the next chapters. As a matter of fact, most of the stabilization results for Navier–Stokes equations can be formulated and proven for control systems in Hilbert spaces governed by so-called *abstract parabolic* systems to be defined below.

2.1 Nonlinear Parabolic-like Systems

Consider a real Hilbert space H with the scalar product $(\cdot,\cdot)$ and norm $|\cdot|_H$ and $F : D(F) \subset H \to H$ a nonlinear operator on H with domain $D(F)$. In almost all the situations considered in the following, F is of the form

$$Fy = Ay + F_0(y), \quad \forall y \in D(A), \tag{2.1}$$

where A is a closed and densely defined linear operator on H with domain $D(A)$ and $F_0 : D(F_0) \subset H \to H$ is a nonlinear operator.

We assume that

(i) $-A$ generates a C_0-analytic semigroup on H.
(ii) F_0 is Gâteaux differentiable on $D(A)$, that is,

$$F_0'(y^*)(z) = \lim_{\lambda\to 0} F_0(y^* + \lambda z) - F_0(y^*) \tag{2.2}$$

exists in H for all $y^*, z \in D(A)$, $F_0(0) = 0$, and for some $\alpha \in (0, 1)$

$$|F_0'(y^*)z|_H \le \alpha|Az|_H + C|z|, \quad \forall z \in D(A). \tag{2.3}$$

It is easily seen that, for each $y \in D(A)$, the operator

$$\mathscr{A} = A + F_0'(y^*), \qquad D(\mathscr{A}) = D(A) \tag{2.4}$$

V. Barbu, *Stabilization of Navier–Stokes Flows*,
Communications and Control Engineering,
DOI 10.1007/978-0-85729-043-4_2, © Springer-Verlag London Limited 2011

is closed, densely defined and $-\mathscr{A}$ generates a C_0-semigroup on H. (See Theorem 1.12.) The linear operator $\mathscr{A}$ can be viewed as the linearization of F in y^*.

A nonlinear operator of the form $\frac{d}{dt} + F$ with F satisfying Conditions (i), (ii) is called *abstract parabolic* operator.

The standard example is $H = L^2(\mathscr{O})$ and $F : L^2(\mathscr{O}) \to L^2(\mathscr{O})$ defined by

$$F(y)(x) = -\Delta y(x) + \beta(y(x)) + g(\nabla y(x)), \quad \text{a.e. } x \in \mathscr{O}, \tag{2.5}$$

where $\mathscr{O}$ is an open, bounded domain of R^d with smooth boundary $\partial\mathscr{O}$, $1 \le d \le 3$, and $\beta \in C^1(R)$, $g \in C^1(R^d)$, $\beta' \in L^\infty(R)$, $g' \in L^\infty(R^d)$, $\beta(0) = 0$, $g(0) = 0$. In this case, we have for all $y \in D(A) = H_0^1(\mathscr{O}) \cap H^2(\mathscr{O})$,

$$F_0'(y)z = \beta'(y)\, z + g'(\nabla y)z, \quad \text{a.e. } x \in \mathscr{O}, \forall z \in L^2(\mathscr{O}),$$

and Conditions (i), (ii) are obviously satisfied. More general, C^1-functions β, g with polynomial growth satisfy (ii) if the growth of β and the dimension d of the space are correlated via Sobolev imbedding theorem but the details are omitted.

Consider the Cauchy problem

$$\begin{aligned} &\frac{dy}{dt} + F(y) = 0, \quad t \ge 0, \\ &y(0) = y_0. \end{aligned} \tag{2.6}$$

Under appropriate conditions on F, this problem is well-posed, but we do not discuss this here since Cauchy problems of this type were presented in Sect. 1.4. We simply assume that (2.6) generates a semigroup (semiflow)

$$y(t) = y(t, y_0), \quad \forall t \ge 0. \tag{2.7}$$

An equilibrium (steady-state) solution y_e to system (2.6) is a solution to the stationary equation

$$F(y_e) = 0. \tag{2.8}$$

The equilibrium solution y_e is said to be *stable* or, more precisely, *asymptotically stable* if

$$\lim_{t \to \infty} y(t, y_0) = y_e,$$

for all y_0 in a neighborhood $\mathscr{V}$ of y_e. As is well-known, the stability of y_e can be reduced to the stability of the solution $y = 0$ to the system

$$\begin{aligned} &\frac{dy}{dt} + \mathscr{A}y + G(y) = 0, \quad t \ge 0, \\ &y(0) = y_0 - y_e, \end{aligned} \tag{2.9}$$

where $\mathscr{A}$ is defined by (2.4) with $y^* = y_e$

$$G(y) = F_0(y + y_e) - F_0(y_e) - F_0'(y_e)(y). \tag{2.10}$$

If the equilibrium solution y_e to system (2.6) is not stable (that is, asymptotically stable), the standard way to stabilize it is to associate with (2.6) a *control system*

$$\begin{aligned} &\frac{dy}{dt} + F(y) = Bu, \quad t \geq 0, \\ &y(0) = y_0, \end{aligned} \tag{2.11}$$

where the controller function $u : [0, \infty) \to U$ takes values in another space U (which is assumed Hilbert everywhere in the following) and B is a linear, closed operator from U to H.

The stabilization problem is to find a controller $u \in L^2(0, \infty; U)$ such that the corresponding solution $y = y(t, y_0, u)$ to system (2.11) has the property that

$$\lim_{t \to \infty} y(t, y_0, u) = y_e,$$

for y_0 in a neighborhood of y_e. In such a case, system (2.11) is said to be *stabilizable*. If the *stabilizable controller* u is in feedback form, that is,

$$u(t) = K(y(t)), \quad \forall t \geq 0, \tag{2.12}$$

where K is a given operator form H to U, then system (2.11) is said to be *feedback stabilizable*.

The *stabilization problem* for such a system is to find a feedback controller of the form (2.12) which stabilizes the equilibrium solution y_e, that is, in a neighborhood $\mathscr{V}(y_e)$ of y_e we have that

$$\lim_{t \to \infty} y_K(t, y_0) = y_e, \quad \forall y_0 \in \mathscr{V}(y_e), \tag{2.13}$$

where $y = y_K$ is the solution to the *closed-loop system*

$$\begin{aligned} &\frac{dy}{dt} + F(y) - BKy = 0, \quad \forall t \geq 0, \\ &y(0) = y_0. \end{aligned} \tag{2.14}$$

We do not discuss here the existence of solutions y to (2.14) which follows under specific assumptions on F and K by the general results presented in Sect. 1.4. It should be emphasized, however, that the true controller in (2.11) is the "acting" controller $v(t) = Bu(t)$ which is the realization of the input controller u under the operator B. Larger is the space $R(B)$, more probably is the stabilization effect but a large space $R(B)$ means also a large space of controllers u, which implies of course an expensive stabilization procedure. The true objective of the stabilization theory is to obtain stabilization via a "minimal" class of input controllers u. Roughly speaking, this means that the space $\{v = Bu\}$ of "acting" controllers should be a proper subspace of H or, as the case will be in the examples presented below, it has zero element intersection with the space H. (This happens, for instance, if B is unbounded.)

Two classes of controllers (or, more precisely, of control systems of the form (2.11)), are largely used in stabilization theory of *parameter distributed systems*, that is, of infinite-dimensional systems represented by partial differential equations: *internal controllers* and *boundary controllers*.

1° **Internal control systems** are control systems of the form (2.11), where B is a linear continuous operator form U to H (that is, $B \in L(U, H)$). A typical example is

$$(Bu)(x) = \mathbf{1}_{\mathcal{O}_0}(x)u(x), \quad \forall x \in \mathcal{O}, \tag{2.15}$$

where $\mathbf{1}_{\mathcal{O}_0}$ is the characteristic function of a subdomain $\mathcal{O}_0 \subset \mathcal{O}$, that is,

$$\mathbf{1}_{\mathcal{O}_0}(x) = 1, \quad \forall x \in \mathcal{O}_0; \qquad \mathbf{1}_{\mathcal{O}_0}(x) = 0, \quad \forall x \in \mathcal{O}_0^c = \mathcal{O} \setminus \mathcal{O}_0. \tag{2.16}$$

This means that the corresponding control system with F given by (2.5) is

$$\begin{aligned} &\frac{\partial y}{\partial t} - \Delta y + \beta(y) + g(\nabla y) = \mathbf{1}_{\mathcal{O}_0} u && \text{in } (0, \infty) \times \mathcal{O}, \\ &y(0, x) = y_0(x), && x \in \mathcal{O}, \\ &y(t, x) = 0, && \forall t \geq 0,\ x \in \partial\mathcal{O}. \end{aligned} \tag{2.17}$$

In this case, the acting controller $v = \mathbf{1}_{\mathcal{O}} u$ is active on the subset $\mathcal{O}_0$ of $\mathcal{O}$ only. In terms of automatic control theory, this means that the control actuation is on the subset $\mathcal{O}_0$. So, the objective of the stabilization problem in this case is to construct a controller u (in feedback form) such that $y(t, y_0) \to y_e$ in $L^2(\mathcal{O})$ as $t \to \infty$. Of course, on this line other types of internal stabilizable controllers are relevant, but that presented above is most important.

2° **Boundary control systems.** An abstract boundary control problem is that in which $B \in L(U, X')$ where X is a Hilbert space such that $X \subset H$ algebraically and topologically and X' is the dual of X in the duality induced by H, that is, with H as pivot space. In other words, $X \subset H \subset X'$ algebraically and topologically. The precise description of this functional setting will be given in Sect. 2.2.

A typical example of such a control system, if one invokes once again the parabolic operator (2.5) is,

$$\begin{aligned} &\frac{\partial y}{\partial t} - \Delta y + \beta(y) + g(\nabla y) = 0 && \text{in } (0, \infty) \times \mathcal{O}, \\ &y(0, x) = y_0(x), && x \in \mathcal{O}, \\ &\frac{\partial y}{\partial n} = u && \text{on } (0, \infty) \times \partial\mathcal{O}, \end{aligned} \tag{2.18}$$

where the flux $u \in L^2_{\text{loc}}(0, \infty; L^2(\partial\mathcal{O}))$, is a boundary controller. This is a control system with flux actuation on the boundary $\partial\mathcal{O}$. Such a system can be written as (2.11), where $H = L^2(\mathcal{O})$, $U = L^2(\partial\mathcal{O})$ and $B \in L(U, (H^1(\mathcal{O}))')$ given by

$$(Bu, \psi) = \int_{\partial\mathcal{O}} u(\xi)\psi(\xi)d\xi, \quad \forall \psi \in H^1(\mathcal{O}). \tag{2.19}$$

(Here, $(H^1(\mathcal{O}))'$ is the dual of $H^1(\mathcal{O}) \subset L^2(\mathcal{O})$ in the pairing $(\cdot,\cdot)$ induced by scalar product $(\cdot,\cdot)$ of $L^2(\mathcal{O})$.)

A more delicate problem arises in the case of Dirichlet boundary control system

$$\begin{aligned} &\frac{\partial y}{\partial t} - \Delta y + \beta(y) + g(\nabla y) = 0 && \text{in } (0,\infty)\times\mathcal{O},\\ &y(0,x) = y_0(x), && x\in\mathcal{O},\\ &y = u, && \text{on } (0,\infty)\times\partial\mathcal{O}, \end{aligned} \tag{2.20}$$

where $u \in L^2_{\text{loc}}(0,\infty; L^2(\partial\mathcal{O}))$.

In order to represent (2.20) into form (2.11), we consider first the so-called *Dirichlet map* $\widetilde{y} = Du$ which is defined as the solution to the Dirichlet problem

$$\begin{aligned} -\Delta\widetilde{y} &= 0 \quad \text{in } \mathcal{O},\\ \widetilde{y} &= u \quad \text{on } \partial\mathcal{O}. \end{aligned} \tag{2.21}$$

It turns out (see, e.g., [60]) that $D \in L(L^2(\partial\mathcal{O}), H^{\frac{1}{2}}(\mathcal{O}))$. Then, subtracting (2.20) and (2.21), we obtain that

$$\begin{aligned} &\frac{\partial}{\partial t}(y - Du) - \Delta(y - Du) + \beta(y) + g(\nabla y) = -\frac{\partial}{\partial t}Du, \quad t > 0, \text{ in } (0,\infty)\times\mathcal{O},\\ &y - Du = 0, \quad \text{on } \partial\mathcal{O},\\ &(y - Du)(0) = y_0 - Du(0), \quad \text{in } \mathcal{O}. \end{aligned}$$

Substituting $y - Du = z$, we reduce the latter to the differential equation in $H = L^2(\mathcal{O})$,

$$\begin{aligned} &\frac{d}{dt}z + A_0z + \beta(z + Du) + g(\nabla z + \nabla Du) = -\frac{d}{dt}Du, \quad t \geq 0,\\ &z(0) = y_0 - Du(0), \end{aligned}$$

where $A_0 = -\Delta$, $D(A_0) = H^1_0(\mathcal{O}) \cap H^2(\mathcal{O})$.

Equivalently,

$$\begin{aligned} z(t) = e^{-A_0t}(y_0 - Du(0)) &- \int_0^t e^{-A_0(t-s)}(\beta(z(s) + Du(s)) + g(\nabla z + \nabla Du))ds\\ &- \int_0^t e^{-A_0(t-s)}\frac{d}{ds}Du(s)ds, \quad t \geq 0. \end{aligned}$$

Integrating by parts, we obtain that

$$\begin{aligned} &y(t) = e^{-A_0t}y_0 + \int_0^t A_0e^{-A_0(t-s)}Du(s)ds - \int_0^t e^{-A_0(t-s)}(\beta(y(s)) + g(\nabla y(s)))ds,\\ &\forall t \geq 0, \end{aligned}$$

that is,

$$\frac{dy}{dt} + A_0 y + \beta(y + g(\nabla y)) = A_0 Du, \quad t \geq 0,$$
$$y(0) = y_0, \tag{2.22}$$

where the operator $Bu = A_0 Du$, $\forall u \in L^2(\partial\mathcal{O})$ is defined from $U = L^2(\partial\mathcal{O})$ to $(D(A_0))'$ by (see (1.9))

$$Bu(\psi) = (Du, A_0\psi), \quad \forall \psi \in D(A_0) = H_0^1(\mathcal{O}) \cap H^2(\mathcal{O}). \tag{2.23}$$

Clearly, $B \in L(U, (D(A_0))')$. Hence, we are in the general situation presented above where $U = L^2(\partial\mathcal{O})$, $H = L^2(\mathcal{O})$ and $X = D(A_0) = H_0^1(\mathcal{O}) \cap H^2(\mathcal{O})$, $X' = (D(A_0))'$.

The general feature of boundary control systems of the form (2.18) and (2.22) is that the control operator B is unbounded from U to H and so the "acting" controller $v = Bu$ takes values in a larger space $X' \supset H$. In the case of Dirichlet boundary control system (2.20) (equivalently (2.22)) the space X' is a distribution space, while in the case of Neumann boundary control system (2.18) it is an abstract space $(H^1(\mathcal{O}))'$. It is useful to notice that in both cases (and this is a general property of abstract boundary control systems) the space $\{v = Bu;\ u \in U\}$ of "acting" controllers has zero element intersection with H and so it is a "meager" control set.

The first step to stabilization of steady-state solution y_e to system (2.6) is of course the stabilization of the linearized system

$$\frac{dy}{dt} + \mathcal{A} y = 0, \quad t \geq 0,$$
$$y(0) = y_0, \tag{2.24}$$

where $\mathcal{A}$ is given by (2.4) with $y^* = y_e$.

We discuss this problem separately for the internal and the boundary stabilization case.

2.2 Internal Stabilization of Linearized System

Consider the controlled system

$$\frac{dy}{dt} + \mathcal{A} y = Bu, \quad y \geq 0,$$
$$y(0) = y_0, \tag{2.25}$$

where, in agreement with Hypotheses (i) and (ii), we assume that

(H1) $-\mathcal{A}$ *generates a C_0-analytic semigroup and the resolvent $(\lambda I - \mathcal{A})^{-1}$ of $\mathcal{A}$ is compact in H.*

As regards the operator $B : U \to H$, we assume that

(H2) $B \in L(U, H)$.

Hypothesis (H1) implies, via Fredholm–Riesz theory (see Theorem 1.1), that the operator $\mathscr{A}$ has a countable set of eigenvalues λ_j and corresponding eigenvectors φ_j, that is,

$$\mathscr{A}\varphi_j = \lambda_j \varphi_j, \quad j = 1, \ldots. \tag{2.26}$$

We recall (see Sect. 1.1) that, for each λ_j, there is a finite number m_j of linear independent eigenvectors $\{\varphi_j^i\}_{i=1}^{m_j}$ and m_j is called the multiplicity of λ_j. It should be emphasized that some of the eigenvectors φ_j^i, $i = 1, \ldots, m_j$, might be generalized eigenvectors (that is, $(\mathscr{A} - \lambda_j)^k \varphi_j^k = 0$, $1 < k \le m_j$). The algebraic multiplicity m_j of λ_j is the number of generalized eigenvectors while the geometric multiplicity $\widetilde{m}_j$ is the number of proper vectors φ_j^k (that is, $\mathscr{A}\varphi_j^k = \lambda_j \varphi_j^k$, $1 \le k \le \widetilde{m}_j$). In general, we have $1 < \widetilde{m}_j \le m_j$, for all j. An eigenvalue λ_j is called *semisimple* if all the eigenvectors are proper (that is, $\widetilde{m}_j = m_j$).

The spectrum $\sigma(\mathscr{A}) = \{\lambda_j\}_{j=1}^{\infty}$ is said to be *semisimple* if all the eigenvalues λ_j are semisimple. An eigenvalue λ_j is said to be simple if $m_j = 1$.

From now on, each eigenvalue λ_j will be repeated according to its multiplicity m_j in order to have a correspondence $\lambda_j \to \varphi_j$, $j = 1, \ldots$.

Taking into account that some of the eigenvalues λ_j might be complex, it will be convenient in the sequel to view $\mathscr{A}$ as a linear operator (again denoted $\mathscr{A}$) in the complexified space $\widetilde{H} = H + iH$. We denote by $\langle \cdot, \cdot \rangle$ the scalar product of $\widetilde{H}$ and by $|\cdot|_{\widetilde{H}}$ its norm. We denote again by $\sigma(\mathscr{A})$ the spectrum of $\mathscr{A}$ and notice that each finite part of the spectrum, let say $\{\lambda_j\}_{j=1}^N$, can be separated from the rest of spectrum by a rectifiable contour Γ_N in the complex space $\mathbb{C}$. If we denote by $\mathscr{X}_N$ the linear space generated by eigenvectors $\{\varphi_j\}_{j=1}^N$, that is,

$$\mathscr{X}_N = \text{lin span}\{\varphi_j\}_{j=1}^N,$$

then the operator $P_N = \widetilde{H} \to \mathscr{X}_N$ defined by

$$P_N = \frac{1}{2\pi i} \int_{\Gamma_N} (\lambda I - \mathscr{A})^{-1} d\lambda \tag{2.27}$$

is the algebraic projection of $\widetilde{H}$ onto $\mathscr{X}_N$ (that is, $\mathscr{X} = P_N \widetilde{H}$). (See, Theorem 1.3.) Moreover, the operator

$$\mathscr{A}_N = P_N \mathscr{A} \tag{2.28}$$

maps the space $\mathscr{X}_N$ into itself and $\sigma(\mathscr{A}_N) = \{\lambda_j\}_{j=1}^N$. In fact, $\mathscr{A}_N : \mathscr{X}_N \to \mathscr{X}_N$ is finite-dimensional and can be represented by an $N \times N$ matrix.

If $\mathscr{A}^*$ is the dual operator of $\mathscr{A}$, then its eigenvalues are precisely $\overline{\lambda}_j$, $j = 1, \ldots,$ and the corresponding eigenvectors

$$\mathscr{A}^* \varphi_j^* = \overline{\lambda}_j \varphi_j^*, \quad j = 1, \ldots$$

have the same properties as φ_j. In particular, the multiplicity of φ_j^* coincides with the multiplicity m_j of φ_j and the dual operator P_N^* of P_N is given by

$$P_N^* = \frac{1}{2\pi} \int_{\Gamma_N} (\lambda I - \mathscr{A}^*)^{-1} d\lambda,$$

while $\mathscr{X}_N^* = \text{lin span}\{\varphi_j^*\}_{j=1}^N = P_N^* \widetilde{H}$.

We have also the following proposition.

Proposition 2.1 *Assume that the spectrum $\sigma(\mathscr{A})$ is semisimple. Then there is a biorthogonal system $\{\varphi\}_{j=1}^\infty$, $\{\varphi_j^*\}_{j=1}^\infty$ of eigenfunctions, that is,*

$$\langle \varphi_j, \varphi_i^* \rangle = \delta_{ij}, \quad i, j = 1, \dots, \tag{2.29}$$

$$\mathscr{A}\varphi_j = \lambda_j \varphi_j, \qquad \mathscr{A}^* \varphi_j^* = \overline{\lambda}_j \varphi_j^*. \tag{2.30}$$

Here and everywhere in the following, $\langle \cdot, \cdot \rangle$ is the scalar product in the complexified space $\widetilde{H} = H + iH$.

The proof of Proposition 2.1 follows immediately if taking into account that for $\lambda_j \neq \lambda_i$ we have by (2.30) that

$$\lambda_j \langle \varphi_j, \varphi_i^* \rangle = \langle \varphi_j, \mathscr{A}^* \varphi_i^* \rangle = \lambda_i \langle \varphi_j, \varphi_i^* \rangle.$$

If $\lambda_j = \lambda_i$, then (2.29) follows via the Schmidt orthogonalization procedure.

Let $\gamma > 0$ and let $N = \inf\{j;\ \text{Re}\,\lambda_j \geq \gamma\}$. By Assumption (H1) it follows that $N < \infty$. Let

$$M = \max\{m_j;\ j = 1, \dots, N\}. \tag{2.31}$$

First, we study the stabilization of System (2.25) under Hypotheses (H1), (H2) and

(H3) *The eigenvalues $\{\lambda_j,\ j = 1, \dots, N\}$ are semisimple.*

Let $\mathscr{B}$ be the $N \times M$ matrix

$$\mathscr{B} = \| \langle B\varphi_j^*, \varphi_i^* \rangle \|_{i=1}^{N}{}_{j=1}^{M} \tag{2.32}$$

and let D_k, $k = 1, \dots, \ell$, be the matrices

$$\begin{aligned} D_1 &= \left\| \langle \varphi_i^*, B^* \varphi_j^* \rangle \right\|_{i=1\, j=1}^{m_1, M}, \\ D_2 &= \left\| \langle \varphi_i^*, B^* \varphi_j^* \rangle \right\|_{i=m_1+1, j=1}^{m_1+m_2, M}, \dots, \\ D_\ell &= \left\| \langle \varphi_i^*, B^* \varphi_j^* \rangle \right\|_{i=m_{\ell-1}+1, j=1}^{m_{\ell-1}+m_\ell, M}. \end{aligned} \tag{2.33}$$

Theorem 2.1 *Assume that Hypotheses* (H1)∼(H3) *hold. Assume also that*

$$\text{rank}\, D_k = m_k, \quad \forall k = 1, 2, \dots, \ell. \tag{2.34}$$

Then there is a controller $u = u(t)$ *of the form*

$$u(t) = \sum_{j=1}^{M} v_j(t)\varphi_j^*, \quad v_j \in L^2(0, \infty), \tag{2.35}$$

which stabilizes exponentially the complexified system (2.25) *with exponent decay* $-\gamma$. *Moreover, the controller* $v = \{v_j\}_{j=1}^M$ *can be chosen in the feedback form*

$$v_j(t) = -\langle B\varphi_j^*, R_0 y^*(t)\rangle, \quad j = 1, \ldots, M, \ t \geq 0, \tag{2.36}$$

where $R_0 \in L(\widetilde{H}, \widetilde{H})$, $R_0 = R_0^*$, $R_0 \geq 0$ *is the solution to the algebraic Riccati equation* (2.48).

It should be said that M is the minimal dimension of the stabilizable controller u. As a matter of fact, we can replace M in Theorem 2.1 by any number $M \leq \widetilde{M} \leq N$ for which (2.34) holds. In particular, if

$$\det \| \langle B\varphi_j^*, \varphi_i^* \rangle \|_{i=1}^{N} {}_{j=1}^{N} \neq 0,$$

then one might take $\widetilde{M} = N$. However, depending on the multiplicity of eigenvalues λ_j, this number $\widetilde{M}$ might be $< N$. For instance, we have

Corollary 2.1 *Assume that the eigenvalues* $\{\lambda_j\}_{j=1}^N$ *are simple and* $\langle B\varphi_j^*, \varphi_1^* \rangle \neq 0$, $\forall j = 1, \ldots, N$. *Then the stabilizable controller* u *can be chosen of the form*

$$u(t) = v(t)\varphi_1^*, \quad \forall t \geq 0,$$

where $v(t) = -\langle B\varphi_1^*, R_0 y^*(t)\rangle$.

In Theorem 2.1, $\{\varphi_j^*\}$ is the dual system of eigenvectors satisfying (2.29) and (2.30).

Proof of Theorem 2.1 We represent the solution y to System (2.25) as $y = y_u + y_s$, where $y_u = P_N y$, $y_s = (I - P_N)y$. Recalling Notation (2.28), we may rewrite System (2.25) with controller (2.35) as

$$\begin{aligned} &\frac{dy_u}{dt} + \mathscr{A}_u y_u = \sum_{j=1}^{M} v_j(t) P_N B\varphi_j^*, \quad t \geq 0, \\ &y_u(0) = P_N y_0, \end{aligned} \tag{2.37}$$

$$\begin{aligned} &\frac{dy_s}{dt} + \mathscr{A}_s y_s = \sum_{j=1}^{M} v_j(t)(I - P_N) B\varphi_j^*, \quad t \geq 0, \\ &y_s(0) = (I - P_N) y_0, \end{aligned} \tag{2.38}$$

where $\mathscr{A}_u = P_N \mathscr{A}$, $\mathscr{A}_s = (I - P_N)\mathscr{A}$.

Recalling that spaces $X_u = P_N \widetilde{H}$ and $X_s = (I - P_N)\widetilde{H}$ are invariant to $\mathscr{A}$, we have that $\sigma(\mathscr{A}_u) = \{\lambda_j\}_{j=1}^N$, $\sigma(\mathscr{A}_s) = \{\lambda_j\}_{j=N+1}^\infty$. Moreover, (2.37) is finite-dimensional while (2.38) is an infinite-dimensional system. We note however that $-\mathscr{A}_s$ generates a C_0-analytic semigroup in X_s and together with $\sigma(\mathscr{A}_s) \subset \{\lambda;\ \operatorname{Re}\lambda > \gamma\}$, this implies that (see Theorem 1.14)

$$\|e^{-\mathscr{A}_s t}\|_{L(\widetilde{H},\widetilde{H})} \le Ce^{-\gamma t}, \quad \forall t \ge 0. \tag{2.39}$$

Now, coming back to System (2.37) and representing the solution y_u as

$$y_u(t) = \sum_{j=1}^N y_j(t)\varphi_j,$$

by (2.29) we may rewrite it as

$$\begin{aligned} & y_i'(t) + (\Lambda y(t))_i = \sum_{j=1}^M v_j(t)\langle B\varphi_j^*, \varphi_i^*\rangle, \quad i = 1, \dots, N, \\ & y_i(0) = \langle y_0, \varphi_i^*\rangle, \end{aligned} \tag{2.40}$$

where Λ is the matrix $\|\langle \mathscr{A}\varphi_j, \varphi_i^*\rangle\|_{i,j=1}^N$. Equivalently,

$$\begin{aligned} & y'(t) + \Lambda y(t) = \mathscr{B}v(t), \quad t \ge 0, \\ & y(0) = \{y_i(0)\}_{i=1}^N \end{aligned} \tag{2.41}$$

where $v(t) = \{v_j(t)\}_{j=1}^M$, $y(t) = \{y_i(t)\}_{i=1}^N$ and $\mathscr{B}$ is the matrix (2.32).

We note that, by virtue of Assumption (H3), Λ is a diagonal matrix of the form

$$\Lambda = \left\| \begin{matrix} J_1 & & & \\ & J_2 & & 0 \\ 0 & & \ddots & \\ & & & J_\ell \end{matrix} \right\|$$

where

$$J_j = \left\| \begin{matrix} \lambda_j & & \\ & \ddots & \\ & & \lambda_j \end{matrix} \right\|$$

has the dimension $m_j \times m_j$, $m_1 + m_2 + \cdots + m_\ell = N$.

Lemma 2.1 is the main step of the proof.

Lemma 2.1 *System* (2.40) *is exactly null controllable, that is, there is*

$$v(t) = \{v_j(t)\}_{j=1}^M \subset L^2(0, T; \mathbb{C}^M)$$

such that $y_i(T) = 0$ *for* $i = 1, \dots, N$.

Proof It is well-known that the finite-dimensional system (2.40) is exactly controllable if and only if

$$\mathscr{B}^* e^{-\Lambda t} x = 0, \quad \forall t \geq 0, \tag{2.42}$$

implies $x = 0$. (This is a variant of the Kalman controllability criterion.)

Taking into account that $\mathscr{B}^* = \|\langle \varphi_j^*, B^* \varphi_i^* \rangle\|_{i,j=1}^{N,M}$ and

$$e^{-\Lambda t} x = \begin{Vmatrix} e^{-\lambda_1 t} x_1 \\ \vdots \\ e^{-\lambda_1 t} x_{m_1} \\ e^{-\lambda_2 t} x_{m_1+1} \\ \vdots \\ e^{-\lambda_\ell t} x_N \end{Vmatrix}, \quad x = \begin{Vmatrix} x_1 \\ \vdots \\ x_N \end{Vmatrix},$$

(2.42) reduces to

$$e^{-\lambda_1 t} \sum_{i=1}^{m_1} b_{ij} x_i + e^{-\lambda_2 t} \sum_{i=m_1+1}^{m_2} b_{ij} x_i + \cdots + e^{-\lambda_\ell t} \sum_{i=m_{\ell-1}+1}^{m_\ell} b_{ij} x_i = 0,$$
$$\forall t \geq 0, \ j = 1, 2, \ldots, M,$$

where $b_{ij} = \langle \varphi_i^*, B \varphi_j^* \rangle$.

This yields

$$\sum_{i=1}^{m_1} b_{ij} x_i = 0, \quad \sum_{i=m_1+1}^{m_2} b_{ij} x_i = 0, \quad \ldots, \quad \sum_{i=m_{\ell-1}+1}^{b} b_{ij} x_i = 0, \quad j = 1, \ldots, M$$

and by Assumption (2.34) of the theorem the latter implies $x \equiv 0$, as claimed. □

Hence, there is a system $\{v_i\}_{i=1}^M \subset L^2(0, T; \mathbb{C}^M)$ such that the corresponding solution $y_u \in C([0, T]; \mathbb{C}^N)$ to (2.37) satisfies

$$y_u(0) = P_N y_0, \qquad y_u(T) = 0, \tag{2.43}$$

where T is arbitrary but fixed. Without loss of generality, we may assume that $v_j(t) = 0$, $\forall t \geq T$. If we plug this controller in (2.38), it follows by (2.39) that

$$\begin{aligned} |y_s(t)|_{\widetilde{H}} &\leq C e^{-\gamma t} |(I - P_N) y_0|_{\widetilde{H}} + C \int_0^T e^{-\gamma(t-s)} \left(\sum_{j=1}^M |v_j(s)| \right) ds \\ &\leq C_1 e^{-\gamma t} |y_0|_{\widetilde{H}}, \quad \forall t \geq 0. \end{aligned} \tag{2.44}$$

(The latter is the consequence of the fact that the controller $v = \{v_j\}_{j=1}^N$ can be chosen in such a way that $\int_0^T |v(t)|^2 dt \le c|P_N y_0|_{\widetilde{H}}$.)

It is useful to notice for later use that starting from the controller $\{v_i\}_{i=1}^M = v$, which steers $P_N y$ into origin, we may construct via the algebraic Riccati equation associated with the stabilizable finite-dimensional system (2.37) a feedback controller $v^*(t) = Ry_u^*(t)$, which exponentially stabilizes (2.37) and $v^* \in C^1([0,\infty), \mathbb{C}^M)$

$$|y_u^*(t)| + |v^*(t)| + |(v^*)'(t)| \le Ce^{-\gamma t}|P_N y_0|_{\widetilde{H}}.$$

(For internal stabilization, this choice of v^* is not relevant, but it is however so in boundary stabilization.) Then, by (2.43) and (2.44) we see that there is a controller u of the form (2.35) which stabilizes the asymptotically system (2.25) with exponential rate $-\gamma$.

In order to find a stabilizable feedback controller $u = K(y)$ for (2.25), we proceed in a standard way (see, e.g., [32, 60]). Namely, we associate with (2.25) the infinite horizon optimal control problem

Minimize

$$\int_0^\infty (|y(t)|_{\widetilde{H}}^2 + |v(t)|_{\mathbb{C}^M}^2)dt \tag{2.45}$$

subject to

$$\frac{dy}{dt} + \mathscr{A}y - \gamma y = \sum_{i=1}^M v_j B\varphi_j^* \stackrel{\text{def}}{=} Dv, \quad t \ge 0. \tag{2.46}$$

By the first part of the proof, System (2.46) is stabilizable and so (2.45) has a unique solution $\{y^* = v^*\}$. Moreover, this optimal controller $v^* = \{v_j^*\}$ is given in the feedback form

$$v^*(t) = -D^* R_0 y^*(t), \quad \forall t \ge 0, \tag{2.47}$$

where D^* is the dual operator of D, that is, $D^* p = \{\langle B\varphi_j^*, p\rangle\}_{j=1}^M$, $\forall p \in \widetilde{H}$, and $R_0 \in L(\widetilde{H}, \widetilde{H})$ is the self-adjoint positive solution to the algebraic Riccati equation

$$\langle \mathscr{A}y - \gamma y, R_0 y\rangle + \frac{1}{2}|D^* R_0 y|_{\widetilde{H}}^2 = \frac{1}{2}|y|_H^2, \quad \forall y \in D(\mathscr{A}). \tag{2.48}$$

In fact, R_0 is given by

$$\langle R_0 y_0, y_0\rangle = \int_0^\infty (|y^*(t)|_{\widetilde{H}}^2 + |v^*(t)|_{\mathbb{C}^M}^2)dt, \quad \forall y_0 \in \widetilde{H}. \tag{2.49}$$

Substituting (2.47) into (2.35), we obtain the desired result. □

2.2.1 The Case of Not Semisimple Eigenvalues

It turns out that in the case where some of the eigenvalues λ_j, $j=1,\dots,N$, are not semisimple, that is, the corresponding eigenvectors $\{\varphi_j^i\}_{i=1}^{m_j}$ are generalized,

$$(\mathscr{A}-\lambda_j)^i\varphi_j^i=0, \quad i=1,\dots,m_j,\ j=1,\dots,N,$$

Theorem 2.1 still remains true but the argument becomes very technical in absence of a biorthogonal system of the eigenfunctions $\{\varphi_j\}_{j=1}^N$, $\{\varphi_j^*\}_{j=1}^N$ (see Sect. 3.3 for the treatment of this case for Navier–Stokes systems).

In order to avoid a tedious argument, we establish here a weaker form (as regards the dimension of the controller) of Theorem 2.1 in this general case.

Theorem 2.2 *Assume that Hypotheses* (H1) *and* (H2) *hold and that*

$$\det \|\langle B\varphi_j^*, P_N^*\varphi_i\rangle\|_{i,j=1}^N \neq 0. \tag{2.50}$$

Then there is a controller u of the form

$$u(t)=\sum_{j=1}^N v_j(t)\varphi_j^*, \quad t\geq 0,\ v_j\in L^2(0,\infty), \tag{2.51}$$

which stabilizes the exponentially system (2.25) *with exponent* $-\gamma$*. Moreover, the stabilizing controller* $v=\{v_j\}_{j=1}^N$ *can be chosen in the feedback form* (2.36).

Proof As in the previous case, it suffices to show that the finite-dimensional control system (2.37) where $M=N$ is exactly controllable on some interval $[0,T]$. If we represent y_u as

$$y_u=\sum_{i=1}^N y_i\varphi_i,$$

we obtain as above that

$$\sum_{i=1}^N y_i'(t)\langle\varphi_i,\varphi_j^*\rangle+\sum_{i=1}^N\langle\mathscr{A}_u\varphi_i,\varphi_j^*\rangle y_i(t)=\sum_{i=1}^N\langle B\varphi_i^*, P_N^*\varphi_j^*\rangle v_i(t), \quad j=1,\dots,N.$$

If we set

$$\Lambda=\|\langle\mathscr{A}_u\varphi_i,\varphi_j^*\rangle\|_{i,j=1}^N, \qquad L=\|\langle\varphi_i,\varphi_j^*\rangle\|_{i,j=1}^N$$

and

$$\widetilde{\mathscr{B}}=\|\langle B\varphi_i^*, P_N^*\varphi_j\rangle\|_{i,j=1}^N,$$

we obtain that $y(t) = \{y_i(t)\}_{i=1}^N$ and $v(t) = \{v_i(t)\}_{i=1}^N$ satisfy the system

$$\begin{aligned} &\frac{dy}{dt} + L^{-1}\Lambda y = L^{-1}\widetilde{\mathscr{B}}v, \quad t \geq 0, \\ &y(0) = P_N y_0. \end{aligned} \tag{2.52}$$

(We note that since the systems $\{\varphi_j\}_{j=1}^N$ and $\{\varphi_j^*\}_{j=1}^N$ are linearly independent, the Gram matrix L is not singular.)

Since by Assumption (2.50) the matrix $\widetilde{\mathscr{B}}$ and, consequently, $L^{-1}\widetilde{\mathscr{B}}$ are nonsingular, we conclude that System (2.52) is exactly null controllable on each interval $(0, T]$ and from now on the proof is exactly the same as that of Theorem 2.1. □

Remark 2.1 The difference between Theorems 2.1 and 2.2 is that the latter provides stabilization but with a larger dimension of the controller u.

It should be said that, in specific situations, Condition (2.50) as well as Assumption (2.34) of Theorem 2.1 regarding the non zero minors D_k, $m \leq k \leq M$, of the matrix (2.33) are checked via unique continuation results for eigenfunctions or solutions to homogeneous partial differential equations of elliptic type. To be more specific, let us come back to the parabolic system (2.17). Then the corresponding linearized system is

$$\begin{aligned} &\frac{\partial y}{\partial t} - \Delta y + \beta'(y_e)y + g'(\nabla y_e)\cdot\nabla y = u\mathbf{1}_{\mathscr{O}_0} && \text{in } (0,\infty)\times\mathscr{O}, \\ &y = 0 && \text{on } (0,\infty)\times\partial\mathscr{O}, \\ &y(0,x) = y_0(x) && \text{in } \mathscr{O}. \end{aligned} \tag{2.53}$$

In this case, as seen earlier, $Bu = u\mathbf{1}_{\mathscr{O}_0}$, $\mathscr{A}y = -\Delta y + \beta'(y_e)y + g'(\nabla y_e)\cdot\nabla y$, $D(\mathscr{A}) = H_0^1(\mathscr{O}) \cap H^2(\mathscr{O})$, and the dual eigenfunctions φ_j^* are solutions to the elliptic equation

$$\begin{aligned} &-\Delta\varphi_j^* + \beta'(y_e)\varphi_j^* - \operatorname{div}(g'(\nabla y_e)\varphi_j^*) = \overline{\lambda}_j\varphi_j^* && \text{in } \mathscr{O}, \\ &\varphi_j^* = 0 && \text{on } \partial\mathscr{O}. \end{aligned} \tag{2.54}$$

(Or, eventually, $(\mathscr{A}^* - \overline{\lambda}_j)^k\varphi_j^* = 0$, $k = 1, 2, \dots, m_j$.) We have

$$\mathscr{B} = \left\| \int_{\mathscr{O}_0} \varphi_j^*\overline{\varphi}_i^*\,dx \right\|_{i,j=1}^{N,M}$$

and D_k are of the form

$$D_k = \left\| \int_{\mathscr{O}_0} \varphi_j^*\overline{\varphi}_i^*\,dx \right\|_{i=m_{k-1}+1,\,j=1}^{m_{k-1}+m_k,\,M}.$$

Then Condition (2.34) reduces to: for each p and $p+k \le M$, the system $\{\varphi_j^*\}_{j=p}^{p+k}$ is linearly independent on $\mathcal{O}_0 \subset \mathcal{O}$. But the latter condition is automatically satisfied by the solutions φ_j^* to (2.54) because, by the unique continuation property of solutions to elliptic equations, each φ_j^* which is zero on $\mathcal{O}_0$ is zero everywhere. By a simple induction argument, this implies that if the system $\{\varphi_j^*\}_{j=p}^{k+p}$ is linearly dependent on $\mathcal{O}_0$, then it is linearly dependent on $\mathcal{O}$, which is of course absurd. (See also Sect. 3.8.)

Hence Theorem 2.1 applies in the present case to stabilize (2.53). More about this subject will be said in the next chapter. If the eigenvalues λ_j are not semisimple, one must invoke Theorem 2.2 with Condition (2.50) which is also automatically satisfied.

Theorems 2.1 and 2.2 are proven in the complexified space $\widetilde{H} = H + iH$. However, if we set

$$\psi_j^1 = \operatorname{Re}\varphi_j^*, \quad \psi_j^2 = \operatorname{Im}\varphi_j^*, \quad j = 1, \dots, N,$$

then it follows that there is a controller of the form

$$\widetilde{u}(t) = \sum_{j=1}^{M} (v_j^1(t)\psi_j^1 + v_j^2(t)\psi_j^2),$$

which stabilizes the real system (2.25) and which can be written as in the proof of Theorem 2.1 in the feedback form

$$v_j^i(t) = -\langle B\psi_j^1, Ry^*(t)\rangle, \quad i = 1, 2.$$

It should be noticed that if $M = N$, then the dimension of the controller remains the same because a complex eigenvalue λ_j arises always in the system together with its conjugate $\overline{\lambda}_j$. Thus the dimension M^* of the real controller $\widetilde{u}$ is dependent of the maximum multiplicity m_j of complex eigenvalues $\lambda_j, \lambda = 1, \dots, \ell$. More precisely, we have

$$M^* = 2M \quad \text{if } M \text{ is equal to } \max_j\{m_j;\ \lambda_j \text{ complex}\}. \tag{2.55}$$

Therefore, we have

1° $M^* = N$ if $M = N$.
2° $M^* = M$ if one of the eigenvalues λ_j of maximum multiplicity is real. (In particular, if all the eigenvalues are real.)
3° $M^* = 2$ if all the eigenvalues are simple but complex-valued.
4° $M^* = 1$ if all the eigenvalues are simple and real.

We have, therefore, the following stabilization result for the real system (2.25).

Corollary 2.2 *Assume that $U = H$ and that Hypotheses* (H1)∼(H3) *and* (2.34) *hold. Then there is a real-valued controller u of the form*

$$u = \sum_{j=1}^{M^*} v_j(t)\psi_j, \quad v_j \in L^2(0,\infty),$$

where M^* is defined by (2.55) *and ψ_j is either* $\operatorname{Re}\varphi_j^*$ *or* $\operatorname{Im}\varphi_j^*$*, which stabilizes exponentially System* (2.25) *with decaying rate* $-\gamma$.

Similarly, under the hypotheses of Theorem 2.2 we have the following corollary.

Corollary 2.3 *Assume that $U = H$ and that Hypotheses* (H1)∼(H2) *and* (2.50) *hold. Then there is a real-valued controller u of the form*

$$u(t) = \sum_{j=1}^{N} v_j(t)\psi_j, \quad v_j \in L^2(0,\infty),$$

which stabilizes System (2.25). *Here ψ_j is either* $\operatorname{Re}\varphi_j^*$ *or* $\operatorname{Im}\varphi_j^*$.

2.2.2 Direct Proportional Stabilization of Unstable Modes

The previous method of stabilization of the linear system (2.25) might be called *spectral controllability-based* approach. Its advantage is that it provides a linear stabilizing feedback controller with a minimal dimension M which depends on spectral properties of unstable eigenvalues. On the other hand, the construction of this feedback controller in the form (2.47) involves an infinite-dimensional Riccati equation (see (2.48)). Below, we describe a simpler design of stabilizing feedback controller which is conceptually different from the previous one.

We assume that the operator $\mathcal{A}$ satisfies Assumptions (H1)–(H3) and let N be such that $\operatorname{Re}\lambda_j \le \gamma$, $j = 1,\dots,N$, where $\gamma > 0$ is arbitrary but fixed. If $\{\varphi_j\}_{j=1}^N$ and $\{\varphi_j^*\}_{j=1}^N$ are the corresponding eigenvectors of $\mathcal{A}$ and $\mathcal{A}^*$, respectively, we consider the feedback controller

$$u(t) = -\eta \sum_{i=1}^{N} \langle y(t), \varphi_i^* \rangle \phi_i \tag{2.56}$$

where $\{\phi_j\}$ is a system of functions such that

$$\langle \phi_i, B^*\varphi_j^* \rangle = \delta_{ij}, \quad i,j = 1,\dots,N. \tag{2.57}$$

Such a system can be found of the form

$$\phi_i = \sum_{k=1}^{N} \alpha_{ki}\varphi_k^*, \quad i = 1,\dots,N,$$

where $\alpha_{ki} \in \mathbb{C}$ are chosen from the system

$$\sum_{k=1}^{N} \alpha_{ki} \langle \varphi_k^*, B^*\varphi_j^* \rangle = \delta_{ij}, \quad i,j = 1,\dots,N.$$

Assuming that

$$\det \|\langle \varphi_k^*, B^*\varphi_j^*\rangle\|_{k,j=1}^N \neq 0, \tag{2.58}$$

then clearly there is a system $\{\alpha_{ki}\}$ such that $\{\phi_i\}$ satisfy Condition (2.57).

Assume also that

$$\eta \geq \gamma - \operatorname{Re}\lambda_j, \quad j = 1, \dots, N. \tag{2.59}$$

Theorem 2.3 *Under Assumptions* (H1), (H2), (H3) *and* (2.58), (2.59), *the solution* y *to the closed-loop system*

$$\begin{aligned} &y' + \mathscr{A}y + \eta \sum_{i=1}^N \langle y, \varphi_j^*\rangle B\phi_j = 0, \quad t \geq 0, \\ &y(0) = y_0, \end{aligned} \tag{2.60}$$

satisfies $|y(t)| \leq Ce^{-\gamma t}|y_0|$, $t \geq 0$.

Proof As in the proof of Theorem 2.1, we rewrite System (2.60) as (see (2.37), (2.38))

$$\begin{aligned} &\frac{dy_u}{dt} + \mathscr{A}_u y_u = -\eta P_N \sum_{i=1}^N \langle y, \varphi_i^*\rangle B\phi_i, \\ &y_u(0) = P_N y_0, \\ &\frac{dy_s}{dt} + \mathscr{A}_s y_s = -\eta(I - P_N) \sum_{i=1}^N \langle y, \varphi_i^*\rangle B\phi_i, \\ &y_s(0) = P_N y_0. \end{aligned}$$

Setting

$$y_u = \sum_{j=1}^N y_j \varphi_j$$

and taking into account (2.29), (2.57), we obtain that

$$y_i' + \lambda_i y_i = -\eta y_i, \quad i = 1, \dots, N,$$

and, therefore,

$$|y_u(t)| \leq e^{-(\operatorname{Re}\lambda_i + \eta)t}|y_0| \leq e^{-\gamma t}|y_0|.$$

Then, substituting y into the right-hand side of the system in y_s, and taking into account that

$$|e^{-\mathscr{A}_s t} y_0| \leq Ce^{-\gamma t}|y_0|, \quad \forall t \geq 0,$$

we conclude the proof. □

It should be said that in the above construction Assumption (H3) can be dispensed with. Indeed, we may replace the system $\{\varphi_j\}_{j=1}^N$ by an orthonormal system $\{\widetilde{\varphi}_j\}_{j=1}^N$ obtained by Schmidt's algorithm and choose the controller u of the form

$$u(t) = -\eta \sum_{i=1}^{N} \langle y(t), \widetilde{\varphi}_i \rangle \widetilde{\phi}_i, \tag{2.61}$$

where $\{\widetilde{\phi}_i\}_{i=1}^N$ are chosen such that

$$\langle \widetilde{\phi}_i, B^* P_N^* \widetilde{\varphi}_j \rangle = \delta_{ij} \quad \text{for } i, j = 1, \dots, N.$$

The latter is possible if one assumes

$$\det \|\langle P_N^* \widetilde{\varphi}_i, B^* \widetilde{\phi}_j \rangle\| \neq 0. \tag{2.62}$$

We obtain, therefore, the following theorem.

Theorem 2.4 *Under Assumptions* (H1), (H2) *and* (2.58), (2.59), *the closed-loop*

$$y' + \mathscr{A} y + \eta \sum_{i=1}^{N} \langle y, \widetilde{\varphi}_i \rangle B \widetilde{\phi}_i = 0,$$

$$y(0) = y_0,$$

is exponentially stable.

Coming back to Example (2.53), we note that Condition (2.62) is obviously satisfied in this case by the unique continuation property of eigenfunctions φ_j.

Remark 2.2 It should be noticed that though the structure of the controller (2.56) is very simple, it is not however robust. More precisely, it might be very sensitive to structural perturbations of the system which modify the spectrum and, implicitly, the basic system (2.57) from which the controller (2.61) is derived.

2.3 Boundary Stabilization of Linearized System

We consider System (2.25) under the following assumptions on the operator B.

(H4) $B \in L(U, (D(\mathscr{A}^*))')$.

Here, $D(\mathscr{A}^*)$ is the domain of adjoint operator $\mathscr{A}^*$ endowed with the graph norm and $(D(\mathscr{A}^*))'$ is its dual in the pairing $\langle \cdot, \cdot \rangle$ with the pivot space $\widetilde{H}$.

More precisely, $(D(\mathscr{A}^*))'$ is the completion of the space $D(\mathscr{A}^*)$ in the norm $\|y\|_{(D(\mathscr{A}^*))'} = |(\lambda I - \mathscr{A}^*)^{-1}y|$, $y \in \widetilde{H}$, where $\lambda \in \rho(\mathscr{A}^*)$ is arbitrary but fixed. (See (1.7)–(1.9).) Then, for each $y_0 \in \widetilde{H}$, $u \in L^2(0,T;U)$ and $T > 0$, the function

$$y(t) = e^{-\mathscr{A}t}y_0 + \int_0^t e^{-\mathscr{A}(t-s)}Bu(s)ds, \quad t \geq 0, \tag{2.63}$$

belongs to $C([0,T];(D(\mathscr{A}^*))')$ and it is a generalized (mild) solution to System (2.25) under Assumptions (H1) and (H4). In general, y does not belong to $C([0,T];\widetilde{H})$, but this happens, however, under additional assumptions on B (see [32, 60]).

It should be emphasized that in this formulation the space $(D(\mathscr{A}^*))'$ becomes the basic space of the system. The operator $\mathscr{A} : H \to (D(\mathscr{A}^*))'$ (or, more exactly, its extension to $(D(A^*))'$, $\widetilde{\mathscr{A}}$), defined by $\langle \widetilde{\mathscr{A}}y, \psi\rangle = \langle y, \mathscr{A}^*\psi\rangle$, $\forall \psi \in D(\mathscr{A}^*)$ generates a C_0-analytic semigroup on $(D(\mathscr{A}^*))'$, again denoted by $e^{-\mathscr{A}t}$. Moreover, the spectrum of this extension coincides with that of the original operator.

We denote by $\mathscr{D}$ the matrix

$$\mathscr{D} = \|\langle B^*\varphi_j^*, B^*\varphi_i\rangle_U\|_{i,j=1}^{N\ M} \tag{2.64}$$

and

$$\mathscr{D}_k = \|\langle B^*\varphi_i^*, B^*\varphi_j\rangle_U\|_{i=m_{k-1}+1,j=1}^{m_{k-1}+m_k,M}, \quad k = 1,\dots,\ell.$$

Here, $\{\varphi_j^*\}_{j=1}^N$ are, as in the previous case, the eigenvectors of $\mathscr{A}^*$ corresponding to the eigenvalues $\{\overline{\lambda}_j,\ 1 \leq j \leq N\}$, $\langle\cdot,\cdot\rangle_U$ is the scalar product of U and $B^*: D(\mathscr{A}^*) \to U$ the dual operator.

Theorem 2.5 *Assume that Hypothesis* (H1), (H3), *and* (H4) *hold and also that*

$$\operatorname{rank} \mathscr{D}_k = m_k, \quad k = 1,\dots,\ell. \tag{2.65}$$

Then there is a controller

$$u(t) = \sum_{j=1}^M v_j(t)B^*\varphi_j^* \tag{2.66}$$

which stabilizes exponentially System (2.25) *in* $(D(\mathscr{A}^*))'$. *Moreover,* $v_j(t)$ *can be chosen in feedback form*

$$v_j(t) = R_j(y(t)), \quad j = 1,\dots,N. \tag{2.67}$$

Proof We proceed as in Theorem 2.1. Namely, we write System (2.25) as

$$\frac{dy_u}{dt} + \mathscr{A}_u y_u = \sum_{j=1}^M v_j(t)P_N BB^*\varphi_j^*, \tag{2.68}$$

$$\frac{dy_s}{dt} + \mathscr{A}_s y_s = \sum_{j=1}^{M} v_j(t)(I - P_N)BB^*\varphi_j^*, \tag{2.69}$$

where $\mathscr{A}_u = P_N\mathscr{A}$ and $\mathscr{A}_s$ is the extension of $(I - P_N)\mathscr{A}$ to all of $\widetilde{H}$ and with values in $(D(\mathscr{A}^*))'$.

As in the previous case, System (2.68) can be put in the form (see (2.40))

$$\begin{aligned} &y' + \Lambda y = \mathscr{D}v, \quad t \in (0, T),\\ &y(0) = P_N y_0 \end{aligned} \tag{2.70}$$

and, by Assumption (2.65), Lemma 2.1 remains true in the present case and we may conclude, as in the proof of Theorem 2.2, that there is a stabilizing controller u of the form (2.66). (We note that Estimate (2.39) remains valid here in $(D(\mathscr{A}^*))'$ for the extended semigroup $e^{-\mathscr{A}_s t}$.) Hence

$$\|y(t)\|_{(D(\mathscr{A}^*))'} \le Ce^{-\gamma t}|y_0|_{\widetilde{H}}, \quad \forall t \ge 0. \qquad \square$$

Remark 2.3 It should be said that, under additional assumptions, one has strong stabilization of (2.25) in the space $\widetilde{H}$. Indeed, we have by (2.63) and (2.69) that

$$y_s(t) = e^{-\mathscr{A}_s t} P_N y_0 + \int_0^t e^{-\mathscr{A}(t-s)} \sum_{j=1}^{M} v_j(s)(I - P_N)BB^*\varphi_j ds.$$

Since v_j can be taken in such a way that $|v_j'(t)| \le Ce^{-\delta t}$, $\forall t > 0$, $j = 1, \ldots, M$, then, if $\mathscr{A}^{-1}B \in L(\widetilde{H}, \widetilde{H})$, we see that

$$|y_s(t)|_{\widetilde{H}} \le Ce^{-\gamma t}|y_0|_{\widetilde{H}}.$$

The construction of the feedback controller is similar to that from the proof of Theorem 2.1, so it will be omitted.

We must remark that the stabilization effect of Controller (2.66) is in a weaker topology than that of Controller (2.35) designed in Theorem 2.1. This is due to the singularity of the operator B and, in particular, of the weaker regularity property of the function $t \to e^{-\mathscr{A}t}Bu$. However, as we see later in some specific situations and, in particular, to that of boundary control systems governed by the Stokes–Oseen operator which will be treated in Sect. 3.4, this stabilization result can be strengthen to the strong topology of H. We come back to Theorem 2.5 and to the boundary control problem (2.20) or, more precisely, to its linearization

$$\begin{cases} \dfrac{\partial y}{\partial t} - \Delta y + \beta'(y_e)y + g'(\nabla y_e)\cdot\nabla y = 0 & \text{in } (0,\infty)\times\mathscr{O},\\ y = u & \text{on } (0,\infty)\times\partial\mathscr{O},\\ y(0,x) = y_0(x), & x \in \mathscr{O}. \end{cases} \tag{2.71}$$

As seen earlier, (2.71) can be written as (2.25) in the space $H = L^2(\mathcal{O})$, where $U = L^2(\partial\mathcal{O})$, $Bu = A_0 Du$, $A_0 = -\Delta$ with $D(A_0) = H_0^1(\mathcal{O}) \cap H^2(\mathcal{O})$ and $D : L^2(\partial\mathcal{O}) \to H^{\frac{1}{2}}(\mathcal{O})$ is the Dirichlet map defined by (2.21). Then, we have

$$B^* p = -\frac{\partial p}{\partial n}, \quad \forall p \in D(\mathcal{A}^*) = D(A_0)$$

and so the stabilizing controller u in (2.71) with actuation on the boundary is of the form

$$u(t,x) = -\sum_{j=1}^{M} v_j(t) \frac{\partial \varphi_j^*}{\partial n}(x), \quad t \geq 0,\ x \in \partial\mathcal{O}. \tag{2.72}$$

As regards Condition (2.65), in this case it reduces to

$$\det \left\| \int_{\partial\mathcal{O}} \frac{\partial \varphi_j^*}{\partial n} \frac{\partial \overline{\varphi}_i^*}{\partial n}\, dx \right\|_{i=m_{k-1},\, j=1}^{i=m_{k-1}+m_k,M} \neq 0. \tag{2.73}$$

Condition (2.73) is equivalent to the linear independence of the system $\{\frac{\partial \varphi_j^*}{\partial n}\}_p^{p+k}$ in $L^2(\partial\mathcal{O})$, a condition which automatically holds for solutions φ_j^* of (2.54). In fact, it is a consequence of the fact that if a solution to (2.54) has zero normal derivative on all of $\partial\mathcal{O}$ or on some part of it with nonempty interior it is everywhere zero. (The uniqueness of the Cauchy problem.)

It should be mentioned that, in this case, Remark 2.3 applies and so, Controller (2.72) stabilizes (2.71) in $L^2(\mathcal{O})$ topology.

As in the previous section, one might construct in this case a stabilizing feedback of the form (2.56) or (2.61) for System (2.25). Namely,

$$u(t) = -\eta \sum_{j=1}^{N} \langle y, \widetilde{\varphi}_j \rangle B^* \widetilde{\phi}_j \tag{2.74}$$

where $\{\widetilde{\varphi}_j\}$ and $\{\widetilde{\phi}_j\}$ are chosen as in Theorem 2.4.

We have, therefore, as in the previous case, the following theorem.

Theorem 2.6 *Under Assumptions* (H1), (H3), (H4) *and*

$$\det \| \langle B^* \phi_j, P_N^* B^* \widetilde{\varphi}_i \rangle \|_{i,j=1}^{N} \neq 0,$$

if $\eta > 0$ *is sufficiently large, the feedback controller* (2.74) *stabilizes exponentially System* (2.25) *in* $(D(\mathcal{A}^*))'$.

2.4 Stabilization by Noise of the Linearized Systems

Here, we study the stabilization by noise of the linear system (2.25), where the operator $\mathcal{A}$ satisfies Assumptions (H1), (H3).

Roughly speaking, the noise stabilization of (2.25) means to design a stochastic controller of the form $\sum_{i=1}^{N} \psi_i \dot{\beta}_i$, where $\dot{\beta}_i$ are white noises, which stabilizes the system in probability.

In other words, the solution X to the system

$$\begin{aligned} &\dot{X}(t) + \mathscr{A}X(t) = \sum_{j=1}^{N} B\psi_j(t)\dot{\beta}_j(t), \quad t > 0, \\ &X(0) = x \end{aligned} \tag{2.75}$$

is asymptotically convergent to zero in probability as $t \to \infty$.

Here, $\{\beta_j\}_{j=1}$ is an independent system of complex Brownian motions in a probability space $\{\Omega, \mathbb{P}, \mathscr{F}, \mathscr{F}_t\}_{t>0}$ and $\{\psi_j\} \subset L^\infty(0, \infty; U)$.

Equation (2.75) should be taken of course in Ito's sense, that is (see Sect. 4.5),

$$\begin{aligned} &dX(t) + \mathscr{A}X(t)dt = \sum_{j=1}^{N} B\psi_j(t)d\beta_j(t), \quad t > 0, \\ &X(0) = x, \end{aligned} \tag{2.76}$$

or, equivalently,

$$X(t) = e^{-\mathscr{A}t}x + \int_0^t \sum_{j=1}^{N} e^{-\mathscr{A}(t-s)} B\psi_j(s)d\beta_j(s). \tag{2.77}$$

Here, $B \in L(U, H)$.

We see below that, under quite general assumptions, such a stabilizable feedback controller exists and has a simple form.

Let $\{\varphi_j\}_{j=1}^N$, $\{\varphi_j^*\}_{j=1}^N$ be the eigenvectors system satisfying (2.29), (2.30) and we also assume that

$$\det \|\langle B\varphi_i^*, \varphi_j^*\rangle\|_{i,j=1}^N \neq 0. \tag{2.78}$$

Consider the system $\{\phi_j\}_{j=1}^N \subset H$ defined by

$$\phi_j = \sum_{i=1}^{N} \alpha_{ij}\varphi_i^*, \quad j = 1, \ldots, N, \tag{2.79}$$

where α_{ij} are chosen in such a way that

$$\sum_{i=1}^{N} \alpha_{ik}\langle B\varphi_i^*, \varphi_j^*\rangle = \delta_{jk}, \quad j, k = 1, \ldots, N. \tag{2.80}$$

By Condition (2.78) it is clear that (2.80) has solution and, by (2.79) and (2.80), we see that

$$\langle B\phi_j, \varphi_i^*\rangle = \delta_{ij}, \quad i, j = 1, \ldots, N. \tag{2.81}$$

We consider the stochastic feedback controller

$$u(t) = \eta \sum_{i=1}^{N} \langle X(t), \varphi_i^* \rangle \phi_i \dot{\beta}_i$$

and we show that it stabilizes in probability the control system (2.25). Namely, one has the following theorem.

Theorem 2.7 *For each $x \in \widetilde{H}$ and $\eta^2 > 2(\gamma - \operatorname{Re} \lambda_j)$, $\forall j = 1, \dots, N$, the equation*

$$\begin{aligned} &dX(t) + \mathscr{A} X(t)dt = \eta \sum_{i=1}^{N} \langle X(t), \varphi_i^* \rangle B\phi_i d\beta_i \quad \text{in } (0, \infty), \ \mathbb{P}\text{-a.s.}, \\ &X(0) = x, \end{aligned} \tag{2.82}$$

has a unique solution $X \in C_W([0, T]; L^2(\Omega, \widetilde{H}))$, $\forall T > 0$, such that

$$\mathbb{P}\Big[\lim_{t \to \infty} e^{\gamma t} |X(t)|_{\widetilde{H}} = 0\Big] = 1. \tag{2.83}$$

Here, $C_W([0, T]; L^2(\Omega, \widetilde{H}))$ is the space of all adapted square-mean $\widetilde{H}$-valued continuous processes on $[0, T]$ and, as mention earlier, (2.82) is understood in the following "mild" sense

$$X(t) = e^{-\mathscr{A} t} x + \eta \sum_{i=1}^{N} \int_0^t \langle X(s), \varphi_i^* \rangle e^{-\mathscr{A}(t-s)} (B\phi_i)(s) d\beta_i(s), \quad t \geq 0. \tag{2.84}$$

(See Sect. 4.5.)

Proof of Theorem 2.7 The idea, already used before, is to decompose (2.82) in a finite-dimensional system and an infinite-dimensional exponentially stable system. To this end, we set $X_u = P_N X$, $X_s = (I - P_N)X$ and we rewrite (2.82) as

$$\begin{aligned} &dX_u(t) + \mathscr{A}_u X_u(t)dt = \eta P_N \sum_{i=1}^{N} \langle X_u(t), \varphi_i^* \rangle B\phi_i d\beta_i(t), \quad \mathbb{P}\text{-a.s., } t \geq 0, \\ &X_u(0) = P_N x, \end{aligned} \tag{2.85}$$

$$\begin{aligned} &dX_s(t) + \mathscr{A}_s X_s(t)dt = \eta (I - P_N) \sum_{i=1}^{N} \langle X_u(t), \varphi_i^* \rangle B\phi_i d\beta_i(t), \quad \mathbb{P}\text{-a.s., } t \geq 0, \\ &X_s(0) = (I - P_N)x. \end{aligned} \tag{2.86}$$

Then, we may represent X_u as $\sum_{i=1}^N y_i(t)\varphi_i$ and so reduce (2.85) via the biorthogonal relations (2.29) and (2.81) to the finite-dimensional stochastic system

$$\begin{aligned} &dy_j + \lambda_j y_j dt = \eta y_j d\beta_j, \quad j = 1, \dots, N, \ t \geq 0, \ \mathbb{P}\text{-a.s.}, \\ &y_j(0) = y_j^0, \end{aligned} \tag{2.87}$$

where $y_j^0 = \langle P_N x, \varphi_j^* \rangle$.

It is well-known that the solution y_j to the stochastic differential equation (2.87) is given by

$$y_j(t) = e^{-\lambda_j t - \frac{\eta^2}{2} t + \eta\beta_j(t)} y_j^0 = 1, \dots, N, \tag{2.88}$$

and, therefore, there is $\varepsilon > 0$ such that

$$|y_j(t)|e^{(\varepsilon+\gamma)t} \le e^{\eta\beta(t)}|y_j^0|, \quad \mathbb{P}\text{-a.s.}$$

Taking into account that, for each $\lambda > 0$ and $r > 0$, we have (see Lemma 4.6 in Sect. 4.5)

$$\mathbb{P}\left(\sup_{t\ge 0} e^{\beta_j(t)-\lambda t} \ge r\right) = r^{-2\lambda}. \tag{2.89}$$

We infer, therefore, by (2.88) that for each $r > 0$ there is $\Omega_r \subset \Omega$ such that

$$|y_j(t)|e^{\gamma t} \le C r^{\eta} |y_j^0| \quad \text{in } \Omega_r,$$

where C is independent of r and $\mathbb{P}(\Omega_r) \ge 1 - r^{-2\varepsilon}$. This implies that

$$\lim_{t\to\infty} |y_j(t)|e^{\gamma t} = 0, \quad \mathbb{P}\text{-a.s.}$$

and also

$$\int_0^\infty |y_j(t)|^2 e^{2\gamma t} dt < \infty, \quad \mathbb{P}\text{-a.s.}$$

We have therefore that

$$\lim_{t\to\infty} e^{2\gamma t} |X_u(t)|^2_{\widetilde{H}} = 0, \quad \mathbb{P}\text{-a.s.}, \tag{2.90}$$

$$\int_0^\infty e^{2\gamma t} |X_u(t)|^2_{\widetilde{H}} dt < \infty, \quad \mathbb{P}\text{-a.s.} \tag{2.91}$$

Next, we come back to the infinite-dimensional system (2.86). Since, as seen earlier, the operator $-\mathscr{A}_s$ generates a γ-exponentially stable C_0-semigroup on $\widetilde{H}$, by the Lyapunov theorem there is $Q \in L(\widetilde{H}, \widetilde{H})$, $Q = Q^* \ge 0$ such that

$$\text{Re}\langle Qx, \mathscr{A}_s x - \gamma x\rangle = \frac{1}{2}|x|^2_{\widetilde{H}}, \quad \forall x \in D(\mathscr{A}_s).$$

(We note that, though Q is not positively definite in the sense that

$$\inf\{\langle Qx, x\rangle; |x| = 1\} > 0,$$

we have, nevertheless, that $\langle Qx, x\rangle > 0$ for all $x \ne 0$.)

Applying Ito's formula in (2.86) (see Theorem 4.8) to the function

$$\varphi(x) = \frac{1}{2}\langle Qx, x\rangle,$$

we obtain that

$$\begin{aligned}
&\frac{1}{2}\, d\langle QX_s(t), X_s(t)\rangle + \frac{1}{2}\,|X_s(t)|^2_{\tilde{H}}dt + \gamma\langle QX_s(t), X_s(t)\rangle dt\\
&\quad = \frac{1}{2}\,\eta^2\sum_{i=1}^{N}\langle QY_i(t), Y_i(t)\rangle_H dt + \eta\sum_{i=1}^{N}((\mathrm{Re}(QX_s(t)), \mathrm{Re}\, Y_i(t))_H\\
&\qquad + (\mathrm{Im}(QX_s(t)), \mathrm{Im}\, Y_i(t))_H)d\beta_i(t),
\end{aligned}$$

where Y_i are stochastic processes defined by

$$Y_i(t) = \langle X_u(t), \varphi_i^*\rangle(I - P_N)B\phi_i, \quad i = 1, \dots, N.$$

This yields

$$\begin{aligned}
&e^{2\gamma t}\langle QX_s(t), X_s(t)\rangle + \int_0^t e^{2\gamma s}|X_s(s)|^2_{\tilde{H}}ds\\
&\quad = \langle Q(I - P_N)x, (I - P_N)x\rangle\\
&\qquad + \eta^2\sum_{i=1}^{N}\int_0^t e^{2\gamma s}\langle QY_i(s), Y_i(s)\rangle ds\\
&\qquad + 2\eta\sum_{i=1}^{N}\int_0^t e^{2\gamma s}((\mathrm{Re}(QX_s(s)), \mathrm{Re}\, Y_i(s))_H\\
&\qquad + (\mathrm{Im}(QX_s(s)), \mathrm{Im}\, Y_i(s)))_H)d\beta_i(s), \quad t \ge 0,\ \mathbb{P}\text{-a.s.}
\end{aligned} \tag{2.92}$$

Now, we apply Lemma 4.5 in Sect. 4.5 to the stochastic processes Z, I, I_1, M defined below

$$\begin{aligned}
&Z(t) = e^{2\gamma t}\langle QX_s(t), X_s(t)\rangle,\\
&I(t) = \int_0^t e^{2\gamma s}|X_s(s)|^2_{\tilde{H}}ds, \qquad I_1(t) = \eta^2\sum_{i=1}^{N}\int_0^t e^{2\gamma s}\langle QY_i, Y_i\rangle ds,\\
&M(t) = 2\eta\sum_{i=1}^{N}\int_0^t e^{2\gamma s}((\mathrm{Re}(QX_s(s)), \mathrm{Re}, Y_i(s))_H\\
&\qquad \times (\mathrm{Im}(QX_s(s)), \mathrm{Im}\, Y_i(s)))_H)d\beta_i(s),\\
&\mathbb{P}\text{-a.s.},\ t \ge 0.
\end{aligned}$$

Because, by the first step of the proof (see (2.91)), $I_1(\infty) < \infty$, we conclude that

$$\lim_{t\to\infty} e^{2\gamma t}\langle QX_s(t), X_s(t)\rangle = 0, \quad \mathbb{P}\text{-a.s.},$$

and, since Q is positive definite in the sense that $\langle Qx, x\rangle = (Qx, x)_H > 0$ for all $x \in \widetilde{H}$, we have that

$$\lim_{t\to\infty} e^{\gamma t}|X_s(t)|_{\widetilde{H}} = 0, \quad \mathbb{P}\text{-a.s.}$$

Recalling that $X = X_u + X_s$ and again invoking (2.91), the latter implies (2.83), thereby completing the proof of Theorem 2.7. □

Now, we illustrate Theorem 2.7 on Example (2.53). (Other more sophisticated examples will be discussed in Sect. 4.1.) In this case, $Bu = \mathbf{1}_{\mathscr{O}_0}u$ where $\mathbf{1}_{\mathscr{O}_0}$ is the characteristic func tion on some open subdomain $\mathscr{O}_0 \subset \mathscr{O}$. Then, Condition (2.78) reduces to

$$\det\left\| \int_{\mathscr{O}_0} \varphi_i^* \overline{\varphi_j^*} \right\|_{i,j=1}^N \neq 0,$$

which clearly holds because as noticed earlier the eigenvalue system $\{\varphi_i^*\}_{i=1}^N$ is linearly independent on $\mathscr{O}_0$. Then the stochastic feedback controller is, in this case, of the form

$$\eta \sum_{i=1}^N \left(\int_{\mathscr{O}} X(t,\zeta)\varphi_i^*(\zeta)d\zeta \mathbf{1}_{\mathscr{O}_0}(\xi)\phi_i(\xi) \right) \dot{\beta}_i(t) \tag{2.93}$$

and by Theorem 2.7 it stabilizes exponentially in probability equation (2.53), that is,

$$\begin{aligned}
&dX - \Delta X\,dt + \beta'(y_e)X\,dt + g'(\nabla y_e)\nabla X\,dt \\
&\quad = \eta \sum_{i=1}^N \int_{\mathscr{O}} X(t,\zeta)\varphi_i^*(\zeta)d\zeta \mathbf{1}_{\mathscr{O}_0}(\xi)\phi_i(\xi)d\beta_i(t), \quad t \ge 0,\ \xi \in \mathscr{O}, \\
&X = 0 \quad \text{on } (0,\infty)\times\partial\mathscr{O}, \\
&X(0,\xi) = x(\xi), \quad \xi \in \mathscr{O}.
\end{aligned}$$

Moreover, the feedback controller (2.93) has the support in $\mathscr{O}_0$.

2.4.1 The Boundary Stabilization by Noise

We consider here System (2.25), where $\mathscr{A}$ satisfies Assumptions (H1), (H3), and B satisfies (H4). We consider the stochastic differential equation

$$\begin{aligned}
&dX + \mathscr{A}X\,dt = \eta \sum_{i=1}^N BB^*\phi_i\langle X, \varphi_i^*\rangle d\beta_i(t), \quad t \ge 0, \\
&X(0) = x.
\end{aligned} \tag{2.94}$$

Here, ϕ_i are defined by (2.79) where α_{ij} are chosen as in (2.80), that is,

$$\sum_{i=1}^{N} \alpha_{ij} \langle B^* \varphi_i^*, B^* \varphi_k^* \rangle = \delta_{jk}, \quad j,k = 1,\dots,N. \tag{2.95}$$

We assume that

$$\det \| \langle B^* \varphi_i^*, B^* \varphi_k^* \rangle \|_{i,k=1}^{N} \neq 0 \tag{2.96}$$

and so α_{ij} are well-defined. By (2.95) we have, therefore, that

$$\langle BB^* \phi_i, \varphi_j^* \rangle = \delta_{ij}, \quad i,j = 1,\dots,N. \tag{2.97}$$

(As in the previous case, we refer to Sect. 4.5 for the existence of a solution $X \in C_W([0,T]; L^2(\widetilde{\mathscr{O}}, H))$ to (2.94).)

We have the following theorem.

Theorem 2.8 *For $|\eta|$ large enough, we have*

$$\mathbb{P}\Big[\lim_{t\to\infty} e^{\gamma t} \|X(t)\|_{(D(\mathscr{A}^*))'} = 0\Big] = 1.$$

Proof We argue as into the proof of Theorem 2.7. Namely, we decompose System (2.94) in two parts

$$dX_u + \mathscr{A}_u X_u dt = \eta P_N \sum_{i=1}^{N} BB^* \phi_i \langle X_u, \varphi_i^* \rangle d\beta_i \tag{2.98}$$

and

$$dX_s + \mathscr{A}_s X_s dt = \eta (I - P_N) \sum_{i=1}^{N} BB^* \phi_i \langle X_u, \varphi_i^* \rangle d\beta_i$$

and treat the finite-dimensional stochastic system (2.98) exact as in the previous case. After that, the proof continues exactly as in the proof of Theorem 2.7. The details are omitted. □

In the case of the boundary control system (2.71), (2.94) has the form

$$\begin{aligned} dX + \widetilde{\mathscr{A}} X\, dt &= -\eta \sum_{i=1}^{N} B\left(\frac{\partial \phi_i}{\partial n}\right) \int_{\mathscr{O}} X \overline{\varphi}_i^* d\xi d\beta_i, \\ X(0) &= x, \end{aligned} \tag{2.99}$$

where $\widetilde{\mathscr{A}} : L^2(\mathscr{O}) \to (D(\mathscr{A}^*))'$ is the extension of $\mathscr{A}$ on all of $\widetilde{H} = L^2(\mathscr{O})$ and $B = A_0 D$.

In terms of boundary control system, this equation can be, equivalently, written as

$$\begin{aligned} &dX - \Delta X\,dt + \beta'(y_e)X\,dt + g'(\nabla y_e)\cdot\nabla X\,dt = 0 && \text{in } (0,\infty)\times\mathscr{O},\\ &X(0,\xi) = x(\xi), && \xi\in\mathscr{O},\\ &X(t,\xi) = -\eta\sum_{i=1}^{N}\frac{\partial\phi_i}{\partial n}\left(\int_{\mathscr{O}} X\overline{\varphi}_i^*\,d\xi\right)\dot{\beta}_i(t) && \text{on } (0,\infty)\times\partial\mathscr{O}. \end{aligned}$$

Then, by Theorem 2.8, for $|\eta|$ large enough, the stochastic boundary controller

$$u(t) = -\eta\sum_{i=1}^{N}\frac{\partial\phi_i}{\partial n}\left(\int_{\mathscr{O}} X(t,\xi) - y_e(\xi)\varphi_i^*\,d\xi\right)\dot{\beta}_i(t) \quad \text{on } (0,\infty)\times\partial\mathscr{O}$$

stabilizes exponentially in probability the equilibrium state $X = y_e(t)$ of System (2.20). (As a matter of fact, the above stochastic feedback controller stabilizes the linearization of (2.20).)

Remark 2.4 We notice that the noise controller arising in Theorem 2.7 has a similar structure as the deterministic feedback controller (2.56) and, apparently, the latter is simpler. However, as remarked earlier, the noise controller is robust, which is not the case with (2.56). We come back later on to this discussion.

2.5 Internal Stabilization of Nonlinear Parabolic-like Systems

We come back to the nonlinear system (2.11) or, more precisely, to

$$\begin{aligned} &\frac{dy}{dt} + \mathscr{A}y + G(y) = 0, \quad t\ge 0,\\ &y(0) = y_0, \end{aligned} \tag{2.100}$$

where $\mathscr{A} : D(\mathscr{A})\subset H\to H$ is of the form (2.4), that is

$$\mathscr{A} = A + F_0'(y_e), \tag{2.101}$$

and G is given by (2.10). We assume everywhere in this section that

(j) *A is a self-adjoint positive definite linear operator with domain $D(A)$, that is,*

$$(Ay,y)\ge\delta|y|^2,\quad \forall y\in H \text{ for some } \delta>0.$$

Moreover, assume that the space $V = D(A^{\frac{1}{2}})$ with the norm $\|y\|_V = |A^{\frac{1}{2}}y|^2$ is compactly imbedded in H.

(jj) *$F_0'(y_e) : H\to H$ is linear, closed, densely defined and*

$$|F_0'(y_e)y|\le C|A^{\frac{1}{2}}y|,\quad \forall y\in D(A^{\frac{1}{2}}).$$

As regards the operator $G : D(G) \subset H \to H$, it is made precise later on. Here, H is a real Hilbert space with the scalar product denoted by $(\cdot, \cdot)$ and the norm $|\cdot|$.

Under Assumptions (j) and (jj), it is easily seen that $\mathscr{A}$ defined by (2.101) satisfies Assumption (H1) and also Assumptions (i), (ii) from Sect. 2.1.

We keep the notation of Sect. 2.2 for the spectrum and eigenvectors $\{\varphi_j\}$ of $\mathscr{A}$. Also, N is the number of eigenvalues λ_j with $\operatorname{Re}\lambda_j \le \gamma$.

Our goal here is to construct a stabilizable feedback controller u for System (2.11), that is, a map

$$u = -K(y - y_e), \tag{2.102}$$

such that the solution $y = y(t)$ to the system

$$\begin{aligned} &\frac{dy}{dt} + F(y) = -BK(y - y_e), \quad t \ge 0, \\ &y(0) = \widetilde{y}_0, \end{aligned} \tag{2.103}$$

has the property that

$$|y(t) - y_e| \le C|\widetilde{y}_0 - y_e|e^{-\gamma t}, \quad \forall t \ge 0, \tag{2.104}$$

for all $\widetilde{y}_0$ in a neighborhood of y_e. (Here, $B \in L(H, H)$.) This means that the feedback controller (2.102) stabilizes exponentially the equilibrium solution y_e and the corresponding system (2.103) is the *closed-loop system* associated with feedback law (2.102).

If we translate y into $y - y_e$, this reduces to the stabilization of null solution to (2.100), where G is given by (2.10) and the corresponding closed-loop system is

$$\begin{aligned} &\frac{dy}{dt} + \mathscr{A}y + Gy = -BKy, \quad t \ge 0, \\ &y(0) = y_0 = \widetilde{y}_0 - y_e. \end{aligned} \tag{2.105}$$

Here, we prove that, under the above assumptions, there is a feedback controller $u = -Ky$ which stabilizes System (2.105) or, more precisely, its zero solution.

In fact, the feedback controller $u = -Ky$ will be a stabilizable feedback controller for the linearized equation (2.25) associated with (2.105). We have shown in Sect. 2.2 that such a feedback controller can be obtained from an infinite horizon linear quadratic problem associated with the control system

$$\begin{aligned} &\frac{dy}{dt} + \mathscr{A}y - \gamma y = \sum_{j=1}^{M^*} v_j B\psi_j, \quad t \ge 0, \\ &y(0) = y_0. \end{aligned} \tag{2.106}$$

Here and everywhere in the following, M^* is determined by (2.55) under the assumptions of Theorem 2.1 and $M^* = N$ under that of Theorem 2.2 (see Corollaries 2.2 and 2.3). The system $\{\psi_j\}_{j=1}^{M^*}$ is that made precise in Corollary 3.1, respectively Corollary 3.2. We need, however, a sharper feedback controller and this can be obtained in a similar way analyzing more closely under present assumptions on $\mathscr{A}$ the solution R to the corresponding Riccati equation (2.48). It is clear that the properties of R will depend also of the structure of the linear quadratic cost functional we associate to the control system (2.106). Here, it is of the form

$$J_\alpha(y,v)=\frac{1}{2}\int_0^\infty(|A^\alpha y(t)|^2+|v(t)|^2_{M^*})dt, \tag{2.107}$$

where $|v|^2_{M^*}=\sum_{j=1}^{M^*}|v_j|^2$ and $A^\alpha,\ 0\le\alpha\le 1$, is the fractional power of order α of the operator A. In examples to partial differential equations A is an operator of elliptic type and $|y|_\alpha=|A^\alpha y|$ is the Sobolev norm of order 2α of y. In particular, $|A^0y|=|y|$ is the L^2-norm. We may view J_α as a cost functional with $D(A^\alpha)$-gain.

We consider here two situations.

1° $\alpha=\frac{3}{4}$ (high-gain Riccati-based feedback)
2° $\alpha=0$ (low-gain Riccati-based feedback)

In both cases, we construct a Riccati-based linear feedback operator $u=-Ky$ which inserted into the nonlinear system

$$\begin{aligned}&\frac{dy}{dt}+\mathscr{A}y+Gy=Bu,\quad t\ge 0,\\&y(0)=y_0,\end{aligned} \tag{2.108}$$

stabilizes exponentially the zero solution in a neighborhood of origin.

2.5.1 High-gain Riccati-based Stabilizable Feedback

Let

$$\Phi_\alpha(y_0)=\inf\{J_\alpha(y,v);\ (y,v)\text{ subject to (2.106)}\}. \tag{2.109}$$

We have the following proposition.

Proposition 2.2 *Let $\alpha=\frac{3}{4}$. Then there is a linear self-adjoint operator $R:D(R)\subset H\to H$ such that*

$$\frac{1}{2}(Ry_0,y_0)=\Phi_\alpha(y_0),\quad \forall y_0\in D(A^{\frac{1}{4}}), \tag{2.110}$$

$$a_1|A^{\frac{1}{4}}y_0|^2\le(Ry_0,y_0)\le a_2|A^{\frac{1}{4}}y_0|^2,\quad \forall y_0\in D(A^{\frac{1}{4}}), \tag{2.111}$$

$$R\in L(D(A^{\frac{3}{4}}),D(A^{\frac{1}{4}}))\cap L(D(A^{\frac{1}{2}}),H)\cap L(D(A^{\frac{1}{4}}),(D(A^{\frac{1}{4}}))'), \tag{2.112}$$

$$(\mathscr{A}y_0-\gamma y_0,Ry_0)+\frac{1}{2}\sum_{i=1}^{M^*}(B\psi_j,Ry_0)^2=\frac{1}{2}\,|A^{\frac{3}{4}}y_0|^2,\quad \forall y_0\in D(A), \tag{2.113}$$

where $a_i > 0,\ i = 1, 2$. *Moreover, the corresponding feedback controller*

$$u = -\sum_{i=1}^{M^*} (B\psi_j, Ry)\psi_j$$

stabilizes exponentially the linearized system (2.25), *that is*,

$$\begin{aligned} &\int_0^\infty e^{2\gamma t} |A^{\frac{3}{4}} y(t)|^2 dt \le C \|y_0\|_W^2, \quad \forall y_0 \in W, \\ &\|y(t)\|_W \le Ce^{-\gamma t} \|y_0\|_W, \qquad \forall y_0 \in W, \end{aligned} \tag{2.114}$$

where $W = D(A^{\frac{1}{4}})$.

Proof By Corollaries 2.2 and 2.3, we know that $\Phi_\alpha(y_0) < \infty$ for each $y_0 \in H$ and, therefore, there is a pair $(y^*, v^*) \in L^2(0, \infty; D(A^\alpha)) \cap L^2(0, \infty; R^{M^*})$ satisfying System (2.106) and such that $J_\alpha(y^*, v^*) = \Phi_\alpha(y_0)$.

By (2.106), that is,

$$\begin{aligned} &\frac{d}{dt} y^* + Ay^* + F_0'(y_e)y^* - \gamma y^* = Dv^*(t), \quad t \ge 0, \\ &y^*(0) = y_0, \end{aligned} \tag{2.115}$$

where $Dv = \sum_{j=1}^{M^*} v_j B\psi_j,\ v = \{v_j\}_{j=1}^{M^*}$, we obtain by multiplication with $A^{\frac{1}{2}} y^*$ that

$$\begin{aligned} &\frac{1}{2}\frac{d}{dt} |A^{\frac{1}{4}} y^*(t)|^2 + |A^{\frac{3}{4}} y^*(t)|^2 + (F_0'(y_e)y^*(t), A^{\frac{1}{2}} y^*(t)) - \gamma(y^*(t), A^{\frac{1}{2}} y^*(t)) \\ &\quad = (Dv^*(t), A^{\frac{1}{2}} y^*(t)), \quad \text{a.e., } t > 0. \end{aligned}$$

Taking into account (j) and (jj), we obtain that

$$a_1 |A^{\frac{1}{4}} y_0|^2 \le \Phi_{\frac{3}{4}}(y_0) \le a_2 |A^{\frac{1}{4}} y_0|^2 \quad \text{for } y_0 \in D(A^{\frac{1}{4}}), \text{ where } a_1, a_2 > 0.$$

If we denote by R the Gâteaux derivative of the function $\Phi_{\frac{3}{4}}$ on $D(A^{\frac{1}{4}})$, we have

$$R \in L(D(A^{\frac{1}{4}}), (D(A^{\frac{1}{4}}))') \quad \text{and} \quad \frac{1}{2}(Ry_0, y_0) = \Phi_{\frac{3}{4}}(y_0), \quad \forall y_0 \in D(A^{\frac{1}{4}}).$$

Then, (2.110) and (2.111) follows.

By the dynamic programming principle, we have that, for all $0 < T < \infty$, (y^*, v^*) is also optimal in the problem

$$\text{Min}\left\{ \frac{1}{2} \int_t^T (|A^{\frac{3}{4}} y(s)|^2 + |v(s)|_{M^*}^2) ds + \Phi_{\frac{3}{4}}(y(T)) \right.$$

$$\left.\text{subject to (2.115)}, y(t) = y^*(t)\right\} = \Phi_{\frac{3}{4}}(y(t)).$$

Then, by the maximum principle we have (D^* is the adjoint of D)

$$v^*(t) = D^* p_T(t), \quad \text{a.e., } t \in (0, T), \tag{2.116}$$

where p_T is the solution to the dual system

$$\begin{aligned} &p_T' - Ap_T - (F_0'(y_e))^* p_T + \gamma p_T = A^{\frac{3}{2}} y^* \quad \text{on } (0, T), \\ &p_T(T) = -Ry^*(T). \end{aligned} \tag{2.117}$$

Moreover, we have also

$$p_T(t) = -Ry^*(t), \quad \forall t \in [0, T]. \tag{2.118}$$

Now, if $y_0 \in D(A^{\frac{1}{2}})$, then as easily follows by (2.115) we have $Ay^* \in L^2(0, \infty; H)$ and so $A^{\frac{3}{2}} y^* \in L^2(0, \infty; (D(A^{\frac{1}{2}}))')$.

Next, one multiplies (2.117) by $A^{-\frac{1}{2}} p_T$ and integrate on (t, T). We obtain that

$$\begin{aligned} &\frac{1}{2}|A^{-\frac{1}{4}} p_T(t)|^2 + \int_t^T |A^{\frac{1}{4}} p_T(s)|^2 ds \\ &\quad \le \frac{1}{2} |A^{-\frac{1}{4}} p_T(T)|^2 + C \int_t^T (|Ay^*(s)|^2 + |p_T(s)|^2) ds, \quad \forall t \in (0, T), \end{aligned}$$

because $|((F_0'(y_e))^* p, A^{-\frac{1}{2}} p)| \le C|p|$.

Now, invoking the interpolating inequality $|p| \le |A^{-\frac{1}{4}} p|^{\frac{1}{2}} |A^{\frac{1}{4}} p|^{\frac{1}{2}}$, we obtain via Gronwall's lemma that

$$A^{-\frac{1}{4}} |p_T(t)|^2 + \int_t^T |A^{\frac{1}{4}} p_T(s)|^2 ds \le C, \quad \forall t \in [0, T].$$

If we multiply (2.117) by $(T - t) p_T(t)$ and integrate on (t, T), we get

$$(T - t)|p_T(t)|^2 + \int_t^T (T - s)\|p_T(s)\|^2 ds \le \int_t^T |p_T(s)|^2 ds + C \le C_1.$$

Hence, $p_T(t) \in H$ for all $t \in [0, T)$ and so $p_T(0) = -Ry_0 \in H$. Hence, $R(D(A^{\frac{1}{2}})) \subset H$ and $R \in L(D(A^{\frac{1}{2}}), H)$.

Now, if $y_0 \in D(A^{\frac{3}{4}})$, it follows (by (2.115)) that $A^{\frac{5}{4}} y^* \in L^2(0, T; H)$ and so, by (2.117), we get as above that $|A^{\frac{1}{4}} p_T(t)| \in C[0, T - \delta; H)$ and, therefore, $p_T(0) = -Ry_0 \in D(A^{\frac{1}{4}})$. Hence, $R \in L(D(A^{\frac{3}{4}}), D(A^{\frac{1}{4}}))$, as claimed.

Now, to find the Riccati equation (2.113), we start with the equation

$$\Phi_{\frac{3}{4}}(y^*(t)) = \frac{1}{2} \int_t^\infty \left| |A^{\frac{3}{4}} y^*(s)|^2 + |v^*(s)|^2_{M^*} \right| ds, \quad \forall t \ge 0, \tag{2.119}$$

and recall that (see (2.116) and (2.118)),

$$v^*(t) = -D^* R y^*(t), \quad \forall t \geq 0, \tag{2.120}$$

where D^* is the adjoint of D, that is,

$$D^* p = \{(B\psi_j, p)\}_{j=1}^{M^*}, \quad \forall p \in H.$$

Taking into account that

$$\begin{aligned}\frac{d}{dt}\Phi_{\frac{3}{4}}(y^*(t)) &= \left(Ry^*(t), \frac{dy^*}{dt}(t)\right) \\ &= -(Ry^*(t), \mathscr{A} y^*(t) - \gamma y^*(t) + DD^* R y^*(t)), \quad \forall t \geq 0,\end{aligned}$$

we obtain by (2.119) that

$$\frac{1}{2}(|A^{\frac{3}{2}} R y^*(t)|^2 + |D^* R y^*(t)|^2) = (\mathscr{A} y^*(t) - \gamma y^*(t), R y^*(t)) + |D^* R y^*(t)|^2, \\ \forall t \geq 0,$$

and for $t = 0$ we get (2.113), as claimed.

As regards (2.114), it follows immediately by (2.115). □

2.5.2 Low-gain Riccati-based Stabilizable Feedback

Proposition 2.3 *Let $\alpha = 0$. Then there is a linear self-adjoint positively semidefinite operator $R_0 \in L(H, H)$ such that $R_0 \in L(H, D(A))$ and*

$$\frac{1}{2}(R y_0, y_0) = \Phi_0(y_0), \quad \forall y_0 \in H, \tag{2.121}$$

$$(\mathscr{A} y_0 - \gamma y_0, R_0 y_0) + \frac{1}{2}\sum_{i=1}^{M^*}(B\psi_i, R_0 y_0)^2 = \frac{1}{2}\,|y_0|^2, \quad \forall y_0 \in D(A). \tag{2.122}$$

Moreover, the feedback law

$$u = -\sum_{j=1}^{M^*}(B\psi_i, R_0 y)\psi_i$$

stabilizes exponentially with decaying rate $-\gamma$ System (2.25), *that is,*

$$|y(t)| \leq C e^{-\gamma t}|y_0|, \quad \forall y_0 \in H.$$

The proof is standard and similar to that of Proposition 2.2.

We notice that also in this case we have (see (2.118))

$$\widetilde{p}_T(t) = -R_0 y^*(t), \quad \forall t \in [0, T],$$

where $\widetilde{p}_T$ is the solution to (2.117) with the right-hand side y^*. This implies an additional regularity for R_0, namely that $R_0 y_0 = -p(0) \in D(A)$ and, therefore, $R_0 \in L(H, D(A))$.

2.5.3 Internal Stabilization of Nonlinear System via High-gain Riccati-based Feedback

We assume here, besides (j) and (jj), that the following hypothesis holds.

(jjj) *G is locally Lipschitz from $V = D(A^{\frac{1}{2}})$ to $V' = (D(A^{\frac{1}{2}}))'$ and*

$$|(Gy - Gz, y - z)| \le \|y - z\|^2 + C_\varepsilon |y - z|^2 \tag{2.123}$$

for all $\|y\| + \|z\| \le \frac{1}{\varepsilon}$ and $\varepsilon > 0$. Moreover,

$$|Gy| \le \eta(\|y\|_W)|A^{\frac{3}{4}} y|^{\frac{3}{2}}, \quad \forall y \in D(A^{\frac{3}{4}}), \tag{2.124}$$

where $\eta : R \to R^+$ is continuous, increasing and $\eta(0) = 0$.

Here, $\|y\| = |A^{\frac{1}{2}} y|^2$ and $W = D(A^{\frac{1}{4}})$ with the norm $\|\cdot\|_W = |\cdot|_{D(A^{\frac{1}{4}})}$.

Theorem 2.9 *Under Assumptions* (j), (jj) *and* (jjj) *there is a neighborhood $\mathscr{U}_\rho = \{y \in W; \|y\|_W < \rho\}$ of the origin such that for all $y_0 \in \mathscr{U}_\rho$ the Cauchy problem*

$$\begin{aligned} &\frac{dy}{dt} + \mathscr{A} y + G(y) = -\sum_{j=1}^{M^*} (B\psi_j, Ry) B\psi_j, \quad \forall t \ge 0, \\ &y(0) = y_0, \end{aligned} \tag{2.125}$$

has a unique solution

$$y \in C([0, \infty); H) \cap L^2(0, \infty; D(A^{\frac{3}{4}})). \tag{2.126}$$

Moreover,

$$\begin{aligned} &\int_0^\infty e^{2\gamma t} |A^{\frac{3}{4}} y(t)|^2 dt \le C\|y_0\|_W^2, \\ &\|y(t)\|_W \le C e^{-\gamma t} \|y_0\|_W, \quad \forall t \ge 0,\ y_0 \in \mathscr{U}_\rho. \end{aligned} \tag{2.127}$$

Here $R \in L(W, W')$ is provided by Proposition 2.2.

Theorem 2.9 amounts to saying that

$$Ky = \sum_{i=1}^{M^*} (B\psi_j, Ry)\psi_j \tag{2.128}$$

is an exponentially stabilizable feedback for System (2.100) (see (2.105)).

We get therefore the following stabilization result for the equilibrium solution y_e to System (2.6), that is,

$$\begin{aligned} &\frac{dy}{dt} + Ay + F_0(y) = 0, \quad t \geq 0, \\ &y(0) = y_0. \end{aligned} \tag{2.129}$$

Corollary 2.4 *Assume that A, F_0 and $G(y) \equiv F_0(y) - F_0(y_e)$ satisfy Assumptions (j)–(jjj). Then the feedback controller $u = -K(y - y_e)$ stabilizes exponentially System (2.129) in a neighborhood of y_e. More precisely, there is $\rho > 0$ such that the closed-loop system*

$$\begin{aligned} &\frac{dy}{dt} + Ay + F_0 y = -BK(y - y_e), \quad t \geq 0, \\ &y(0) = y_0 \end{aligned} \tag{2.130}$$

has a unique solution $y \in C([0, \infty; H) \cap L^2_{loc}(0, \infty; D(A^{\frac{3}{4}}))$ which satisfies

$$\int_0^\infty e^{2\gamma t} |A^{\frac{3}{4}}(y(t) - y_e)|^2 dt < C\|y_0 - y_e\|_W^2, \tag{2.131}$$

$$\|y(t) - y_e\|_W \leq Ce^{-\gamma t}\|y_0 - y_e\|_W, \tag{2.132}$$

for all $t \geq 0$ and $\|y_0 - y_e\|_W < \rho$.

Proof of Theorem 2.9 First, we prove that the Cauchy problem (2.125) is well-posed for $y_0 \in \mathscr{U}_\rho$, where ρ is sufficiently small. To this end, we consider the truncation G_ε of the operator G, that is,

$$G_\varepsilon(y) = \begin{cases} G(y) & \text{for } \|y\| \leq \frac{1}{\varepsilon}, \\ G(\frac{y}{\varepsilon\|y\|}) & \text{for } \|y\| > \frac{1}{\varepsilon}. \end{cases} \tag{2.133}$$

Clearly, G_ε is Lipschitz from V to V' and by (2.123) we see also that

$$(G_\varepsilon y - G_\varepsilon z, y - z) \leq \|y - z\|^2 + C_\varepsilon|y - z|^2, \quad \forall y, z \in V.$$

Then, recalling (2.112), by Theorem 1.15 we conclude that for each $y_0 \in H$ there is a unique solution $y_\varepsilon \in C([0, \infty; H)) \cap L^2_{loc}(0, \infty; D(A))$, $\frac{dy_\varepsilon}{dt} \in W^1_{loc}(0, \infty; H)$ to the equation

$$\begin{aligned}&\frac{dy_\varepsilon}{dt}+\mathscr{A}y_\varepsilon+G_\varepsilon(y_\varepsilon)+BKy_\varepsilon=0,\quad \text{a.e., } t>0,\\&y_\varepsilon(0)=y_0.\end{aligned}\tag{2.134}$$

If we multiply (2.134) by Ry_ε (scalarly in H) and recall that by (2.113)

$$(\mathscr{A}y-\gamma y,Ry)+\frac{1}{2}\,(BKy,Ry)=\frac{1}{2}\,|A^{\frac{3}{4}}|^2,\quad \forall y\in D(A),$$

we obtain by (2.112), (2.124) and (2.134) that

$$\begin{aligned}&\frac{1}{2}\frac{d}{dt}(Ry_\varepsilon(t),y_\varepsilon(t))+\gamma(Ry_\varepsilon(t),y_\varepsilon(t))+\frac{1}{2}(BKy_\varepsilon(t),Ry_\varepsilon(t))+\frac{1}{2}\,|A^{\frac{3}{4}}y_\varepsilon(t)|^2\\&\quad\le|G_\varepsilon(y_\varepsilon(t))\,|Ry_\varepsilon(t)|\le C|A^{\frac{3}{4}}y(t)|^{\frac{3}{2}}\|y_\varepsilon(t)\|\eta(\|y_\varepsilon(t)\|_W),\quad \text{a.e., } t>0.\end{aligned}$$

Taking into account the interpolation inequality

$$\|y\|\le|A^{\frac{3}{4}}y|^{\frac{1}{2}}\|y\|_W^{\frac{1}{2}},\quad \forall y\in D(A^{\frac{3}{4}}),$$

we have, for $\|y_0\|_W<\rho$,

$$\frac{d}{dt}\,(Ry_\varepsilon(t),y_\varepsilon(t))+2\gamma(Ry_\varepsilon(t),y_\varepsilon(t))+(BKy_\varepsilon(t),Ry_\varepsilon(t))+\frac{1}{2}\,|A^{\frac{3}{4}}y_\varepsilon(t)|^2\le 0$$

a.e., on $(0,T^*(y_0))$, where $T^*(y_0)=\sup\{t>0;\ \|y_\varepsilon(t)\|_W\le\rho\}$ and $\rho>0$ is chosen by the condition

$$C\eta(\rho)\le\frac{1}{2}\cdot$$

This yields $T^*(y_0)=\infty$ and

$$(Ry_\varepsilon(t),y_\varepsilon(t))\le e^{-2\gamma t}(Ry_0,y_0)\le Ce^{-2\gamma t}\|y_0\|_W^2,\quad \forall t\ge 0,\tag{2.135}$$

$$\int_0^\infty e^{2\gamma t}|A^{\frac{3}{4}}y_\varepsilon(t)|^2dt\le C\|y_0\|_W^2.\tag{2.136}$$

Taking into account that $G_\varepsilon(y)=G(y)$ for $\|y\|\le\frac{1}{\varepsilon}$, it follows by (2.133), (2.135) that for each $y_0\in V\cap\mathscr{W}_\rho$ there is a solution (obviously unique by virtue of Assumption (2.123)) $y=y_\varepsilon$ to (2.125) satisfying Estimate (2.127).

The stabilizable feedback law (2.128) has the unpleasant feature that the operator R is computed from a high $D(A^{\frac{3}{4}})$-gain Riccati equation (2.113) which involves some computational problem. An alternative is to use the feedback law given in Proposition 2.3. □

2.5.4 Internal Stabilization of Nonlinear System via Low-gain Riccati-based Feedback

We study here the effect of the linear feedback

$$u(t) = -\sum_{i=1}^{M^*} (B\psi_j, R_0 y)\psi_j = -Ly(t) \tag{2.137}$$

in the system

$$\begin{aligned} &\frac{dy}{dt} + \mathscr{A}y + Gy = Bu, \quad t \geq 0, \\ &y(0) = y_0, \end{aligned} \tag{2.138}$$

where R_0 is the solution to Riccati equation (2.122) given by Proposition 2.3.

Denote by $\Gamma : D(\Gamma) \subset H \to H$ the operator

$$\Gamma y = \mathscr{A}y + BLy, \qquad D(\Gamma) = D(A). \tag{2.139}$$

By (j) it is easily seen that $-\Gamma$ generates a C_0-analytic semigroup $e^{-\Gamma t}$ on H and, by Proposition 2.3, $e^{-\Gamma t}$ is exponentially stable, that is,

$$|e^{-\Gamma t} y_0| \leq Ce^{-\gamma t}|y_0|, \quad \forall t \geq 0, \ y_0 \in H.$$

Further estimates on $z(t) = e^{-\Gamma t} y_0$ are given below. If we multiply the equation

$$\frac{dz}{dt} + \mathscr{A}z + BLz = 0, \quad t \geq 0,$$

by $A^{\frac{1}{2}}z$, we get

$$\frac{d}{dt}\|z(t)\|_W^2 + |A^{\frac{3}{4}}z(t)|^2 \leq C_1|z(t)|^2, \quad \text{a.e.}, t > 0$$

and, therefore,

$$\begin{aligned} &\frac{d}{dt}(\|z(t)\|_W^2 e^{2\gamma t}) + e^{2\gamma t}|A^{\frac{3}{4}}z(t)|^2 \\ &\leq C_2 e^{2\gamma t}(|z(t)|^2 + \|z(t)\|_W^2) \\ &\leq C_3 e^{2\gamma t}|z(t)|^2 + \frac{1}{2}e^{2\gamma t}|A^{\frac{3}{4}}z(t)|^2. \end{aligned}$$

Finally,

$$\|z(t)\|_W^2 e^{2\gamma t} + \int_0^t e^{2\gamma s}|A^{\frac{3}{4}}z(s)|^2 ds \leq C_4\|y_0\|_W^2, \quad \forall t \geq 0. \tag{2.140}$$

Now, we rewrite (2.138) with the controller u given by (2.137) as

$$y(t)=e^{-\Gamma t}y_0-\int_0^t e^{-\Gamma(t-s)}Gy(s)ds,\quad t\geq 0. \tag{2.141}$$

We assume here the following hypothesis on G.

(jv) $\|Gy-Gz\|_W\leq C|y-z|_{\frac{3}{4}}(|y|_{\frac{3}{4}}+|z|_{\frac{3}{4}})$, $\forall y,z\in D(A^{\frac{3}{4}})$.

Theorem 2.10 *Under Assumptions* (j), (jj) *and* (jv) *for each* $y_0\in\mathscr{U}_\rho$ *and* ρ *sufficiently small there is a unique solution to* (2.141)

$$y\in C([0,\infty);W)\cap L^2(0,\infty;D(A^{\frac{3}{4}})). \tag{2.142}$$

Moreover, one has

$$\|y(t)\|_W\leq Ce^{-\gamma t}\|y_0\|_W,\quad \forall t\geq 0,\ y_0\in\mathscr{U}_\rho. \tag{2.143}$$

Proof The proof will be sketched only. We are going to apply the contraction principle to the operator defined by the right-hand side $\Lambda(y)$ of (2.141),

$$y\to\Lambda(y):L^2(0,\infty;D(A^{\frac{3}{4}}))\to L^2(0,\infty;D(A^{\frac{3}{4}}))$$

defined on the set

$$\mathscr{K}_r=\left\{y\in L^2(0,\infty;D(A^{\frac{3}{4}}));\int_0^\infty|A^{\frac{3}{4}}y(t)|^2dt\leq r\right\}$$

where r will be suitable chosen. By (2.140) we have, for $y\in\mathscr{K}_r$ and $y_0\in\mathscr{U}_\rho$, via Young inequality and Hypothesis (jv) that

$$\begin{aligned}\|\Lambda(t)\|^2_{L^2(0,\infty;D(A^{\frac{3}{4}}))}&\leq C(\|y_0\|^2_W+\|G(y)\|^2_{L^1(0,\infty;W)})\\&\leq C\|y_0\|^2_W+C_1\|y\|^4_{L^2(0,\infty;D(A^{\frac{3}{4}}))})\leq C\rho^2+C_1r^2.\end{aligned} \tag{2.144}$$

Here, we have used the obvious estimates

$$|A^{\frac{3}{4}}e^{-\mathscr{A}t}y_0|\leq|A^{\frac{1}{2}}e^{-\mathscr{A}t}A^{\frac{1}{4}}y_0|,\quad\forall t>0,\ y_0\in W,$$

and the fact that, as easily follows by Hypothesis (j) and (jj), we have

$$\|A^{\frac{1}{2}}e^{-\mathscr{A}t}z_0\|_{L^2(0,\infty;H)}\leq C|z_0|,\quad\forall z_0\in H.$$

By (2.144) we see that, for $0<r\leq\mu(\rho)$ sufficiently small, we have

$$C\rho^2+C_1r^4\leq r^2$$

and so, the operator Λ leaves invariant the set $\mathscr{K}_r$.

On the other hand, we see in a similar way by (jv) that

$$\begin{aligned}
&\|\Lambda(y_1)-\Lambda(y_2)\|^2_{L^2(0,\infty;D(A^{\frac{3}{4}}))}\\
&\le C_1\|G(y_1)-G(y_2)\|^2_{L^1(0,\infty;W)})\\
&\le C_1\left(\int_0^\infty |A^{\frac{3}{4}}(y_1-y_2)|(|A^{\frac{3}{4}}y_1|+|A^{\frac{3}{4}}y_2|)dt\right)^2\\
&\le C_2\int_0^\infty |A^{\frac{3}{4}}(y_1-y_2)|^2dt\int_0^\infty (|A^{\frac{3}{4}}y_1|^2+|A^{\frac{3}{4}}y_2|^2)dt\\
&\le C_2r^2\|y_1-y_2\|^2_{L^2(0,\infty;D(A^{\frac{3}{4}}))},\qquad \forall y_1,y_2\in\mathscr{K}_r.
\end{aligned}$$

Hence, choosing r sufficiently small ($r<\frac{1}{\sqrt{C_2}}$), we have that Λ is a contraction on $\mathscr{K}_r$ and, therefore, (2.141) has a unique solu tion y satisfying (2.142).

In order to prove (2.143), we write (2.141) as

$$\frac{dy}{dt}+\mathscr{A}y+BLy+Gy=0,\qquad y(0)=y_0,$$

and repeat the previous estimates (2.140).

We get as above that, for $y_0\in\mathscr{U}_\rho$,

$$\|e^{\gamma t}y(t)\|_W\le C\|y_0\|_W,\quad \forall t\ge 0.\qquad\square$$

Theorems 2.9 and 2.10 can be applied to Example (2.17) if one assumes that β and g are C^2-functions with polynomial growth and $y_e\in L^\infty(\mathscr{O})$.

One might prove also that the linear feedback controller provided by Theorem 2.4, that is,

$$u(t)=-\eta\sum_{j=1}^N\langle y,\widetilde{\varphi}_j\rangle\widetilde{\phi}_j$$

inserted into (2.138) stabilizes exponentially the system in a neighborhood $\mathscr{U}_\rho$ of the origin. The proof is identical with that of Theorem 2.9 or 2.10, but once again the details are omitted.

2.5.5 *High-gain Feedback Controller Versus Low-gain Controller and Robustness*

Roughly speaking, Theorems 2.9 and 2.10 provide the same type of stability for the control system associated with (2.100). One might suspect, however, that the radius of stability of $\mathscr{U}_\rho$ established via Lyapunov function (Ry,y) is more exact and bigger in the first case, but this does not seem to be the principal advantage of the first approach. As a matter of fact, in both situations the stabilizing feedback controller is obtained from a linear quadratic control problem (so-called LQG design method from automatic control theory) but with different quadratic cost criteria and here arises the major difference between them because, as we show in Chap. 5, the high-gain feedback controller used in Theorem 2.1 is more robust than (2.137). The robustness of a control feedback is a central problem in automatic control and roughly speaking it is the property of the system to remain insensitive to disturbances or model imperfections. If we consider System (2.100), where $\mathscr{A}$ and G are imperfectly known but remain in a certain "neighborhood" $\mathscr{V}(\mathscr{A}^*, G^*)$ of a given state-system $\frac{dy}{dt} + \mathscr{A}^* y + G^*(y) = 0$, we say that the stabilizing feedback controller is robust in this class if $u = -Ky$ is a stabilizing feedback for all $(\mathscr{A}, G) \in \mathscr{V}(\mathscr{A}^*, G^*)$. It is well-known that a feedback controller obtained from LQG is always robust in a certain sense if all the output variables are measurable but it is also clear that the robustness performance is dependent (at least in infinite-dimensional setting) of the cost functional. One principal tool to evaluate and improve the robustness in this case is the H^∞-theory we shall speak about in Chap. 5. For the nonlinear system it is more difficult to evaluate or compare the robustness performances but one can see that the given feedback controller is more robust than another if its invariant stability class $\mathscr{V}(\mathscr{A}^*, G^*)$ is larger (measured in the same topology) than another. From this point of view, we show below that the high-gain feedback controller designed here is more robust than the low-gain feedback controller designed in Theorem 2.10.

Theorem 2.11 *Under the assumptions of Theorem* 2.9, *the feedback controller $u = -Ky$ given by* (2.128) *is still stabilizable with the rate γ for all the systems of the form*

$$\frac{dy}{dt} + \widetilde{\mathscr{A}} y + \widetilde{G} y = -BKy, \quad t \geq 0, \tag{2.145}$$

where $(\widetilde{\mathscr{A}}, \widetilde{G})$ satisfy Assumptions (j), (jj) *and* (jjj) *and*

$$|\widetilde{\mathscr{A}} y - \mathscr{A} y| \leq C\varepsilon |Ay|, \quad |\widetilde{G} y - Gy| \leq \varepsilon |Gy|, \quad \forall y \in \mathscr{U}_\rho, \tag{2.146}$$

and $\varepsilon > 0$ is sufficiently small.

Proof If we multiply (2.145) by Ry and use (2.113), we obtain by (2.146) that

$$\begin{aligned} &\frac{1}{2}\frac{d}{dt}(Ry(t), y(t)) + \frac{1}{2}\,|A^{\frac{3}{4}} y(t)|^2 + \gamma(Ry(t), y(t)) + \frac{1}{2}\,(BKy(t), Ry(t)) \\ &\quad \leq C\varepsilon |(Ay, Ry)| + (1+\varepsilon)|Gy|\,|Ry|, \quad \text{a.e.}, t \in (0, T^*), \end{aligned}$$

where $T^* = \sup\{t;\ y(t) \in \mathscr{W}_\rho\}$.

On the other hand, as seen in Proposition 2.2, we have

$$|(Ay, Ry)| \leq C|A^{\frac{3}{4}} y|^2\, \|y\|_W, \quad \forall y \in D(A)$$

and this yields, for all ε sufficiently small,

$$\frac{1}{2}\frac{d}{dt}(Ry(t),y(t))+\left(\frac{1}{2}-\varepsilon\right)|A^{\frac{3}{4}}y(t)|^2+\gamma(Ry(t),y(t))$$
$$+\frac{1}{2}(BKy(t),Ry(t))\le C\eta(\rho)|A^{\frac{3}{4}}y(t)|^2$$

and this implies, as seen earlier,

$$a_1\|y(t)\|_W^2\le(Ry(t),y(t))\le e^{-\gamma t}(Ry_0,y_0),\quad \forall t\ge 0.$$

Then, arguing as above, we find that

$$\|y(t)\|_W\le Ce^{-\gamma t}\|y_0\|_W,\quad \forall t\ge 0,$$

for all y_0, with $\|y_0\|_W\le\rho$ suitable chosen. This completes the proof. □

Theorem 2.11 amounts to saying that the feedback controller found by the high-gain Riccati equation keeps unaltered its stabilizing property for small but sharp deviations of the system. For instance, in case of the parabolic system (2.17), it turns out that it still operates with the same stabilizing rate γ on perturbed parabolic systems of the form

$$\frac{\partial y}{\partial t}-\sum_{i=1}^{N}\frac{\partial}{\partial x_i}\left(a_{ij}^{\varepsilon}(x)\,\frac{\partial y}{\partial x_i}\right)+\beta_\varepsilon(y)+g_\varepsilon(\nabla y)=-BKu\quad\text{in }(0,T)\times\mathcal{O},$$
$$y=0\quad\text{on }\partial\mathcal{O},\qquad y(0,x)=y_0(x)\quad\text{in }\mathcal{O},$$

where $|a_{ij}^{\varepsilon}-a_{ij}|\le C\varepsilon$, for $i,j=1,\dots,n$ and

$$\|\beta_\varepsilon-\beta\|_{C^2(R)}+\|g_\varepsilon-g\|_{C^2(R)}\le C\varepsilon.$$

In other words, it remains stabilizable to small structural perturbation of the system.

In particular, it follows by Theorem 2.11 that, if

$$\frac{dy}{dt}+\mathcal{A}_h y+G_h y=Bu,\quad t\ge 0 \tag{2.147}$$

is a finite element approximation of (2.100), then, if $|(\mathcal{A}_h-\mathcal{A})y|$, $|(G_h-G)y|\to 0$ as $h\to 0$ uniformly on $D(A)$ respectively, on $D(A^{\frac{3}{4}})$ (and this usually happens), then the stabilizing high-gain feedback law $u=-K_h y$ for (2.147) is, for h small, still stabilizable for System (2.100). This fact allows to stabilize the state-system (2.100) using approximating feedback laws provided by the finite-dimensional Riccati equation (2.113), that is,

$$(\mathcal{A}_h y-\gamma y,R_h y)+\frac{1}{2}\sum_{i=1}^{M^*}(B\psi_i,R_h y)^2=\frac{1}{2}\,|A_h^{\frac{3}{4}}y|^2.$$

(We notice that by the stability of the spectrum $\sigma(\mathscr{A})$ (see [59]), the spectral index M^* is invariant to small perturbations of $\mathscr{A}$.)

Now, analyzing the stability performances of low-gain Riccati-based feedback (2.137), it is easily seen that, in general, it is not robust to structural sharp perturbations of the form mentioned above or, more precisely, its robustness region is smaller than that of high-gain Riccati-based feedback discussed above.

For instance, it is not stabilizable for the linear system

$$\frac{dy}{dt} + \mathscr{A}_\varepsilon y = Bu, \quad t \ge 0,$$

where $\mathscr{A}_\varepsilon = \mathscr{A} - \varepsilon A$. Indeed, in this case we have by (2.122) (we take $\gamma = 0$),

$$\frac{1}{2}\frac{d}{dt}(R_0 y, y) + \frac{1}{2}|y|^2 + \frac{1}{2}(BKy, R_0 y) = \varepsilon(Ay, R_0 y)$$

and, obviously, this does not imply $\lim_{t\to 0}(R_0 y(t), y(t)) = 0$, as desired.

2.6 Stabilization of Time-periodic Flows

2.6.1 The Functional Setting

We consider here the controlled evolution system

$$\frac{dy}{dt}(t) + Ay(t) + B(t, y(t)) = Du(t), \quad t \in R, \tag{2.148}$$

in a Hilbert space H with the norm $|\cdot|$ and scalar product denoted $(\cdot,\cdot)$.

The following assumptions will be in effect throughout this section.

(k) A is a linear, self adjoint positive definite operator in H with domain $D(A)$. A^{-1} is completely continuous.

For $0 < \alpha < 1$ we denote, as usually, by A^α the fractional power of order α of A and by $|x|_\alpha = |A^\alpha x|$ the norm of $D(A^\alpha)$.

(kk) $B : R \times D(A^\alpha) \to H$, *where* $\frac{1}{4} \le \alpha < 1$, *satisfies the conditions*

$$B(t+T, y) = B(t, y), \quad \forall (t, y) \in R \times D(A^\alpha); \tag{2.149}$$

$$|B(t,0) - B(s,0)| \le C_1|t-s|, \quad \forall s, t \in R; \tag{2.150}$$

$$|B_y(t, y) - B_y(s, z)|_{L(D(A^\alpha),H)} \le C_2(|y|_{\frac{1}{4}} + |z|_{\frac{1}{4}})(|t-s| + |y-z|_\alpha),$$
$$\forall y, z \in D(A^\alpha);\ t, s \in R; \tag{2.151}$$

$$|B_y(t, y)|_{L(D(A^{\frac{1}{2}}),H)} \le C(1 + |Ay|), \quad \forall y \in D(A). \tag{2.152}$$

Here, $B_y(t,\cdot) \in L(D(A^\alpha), H)$ is the (*Fréchet*) *derivative of* $B(t,\cdot)$. We note that by (2.151) it follows that

$$|B(t,y) - B(t,z)| \le C_3(|y|_{\frac{1}{4}} + |z|_{\frac{1}{4}})|y-z|_\alpha, \; y,z \in D(A^\alpha), \quad t \in R. \quad (2.153)$$

(kkk) *$D \in L(U,H)$ where U is a Hilbert space with the norm $|\cdot|_U$ and the scalar product $\langle\cdot,\cdot\rangle_U$.*

Now, let $y_\pi \in C^1(R, D(A))$ be a T-periodic solution to (2.148), that is,

$$\begin{aligned} &\frac{d}{dt} y_\pi(t) + Ay_\pi(t) + B(t, y_\pi(t)) = 0, \quad t \in R,\\ &y_\pi(t) = y_\pi(t+T), \qquad \forall t \in R. \end{aligned} \quad (2.154)$$

Let $\mathscr{A}(t) \equiv A + B_y(t, y_\pi(t))$. By Assumptions (k) and (kk), and the fact that $y_\pi \in C^1(R; D(A))$, we see that the resolvent $R(\lambda; \mathscr{A}(t)) = (\lambda I + \mathscr{A}(t))^{-1}, t \in R^+$, exists for all complex $\lambda \in \Sigma$, where $\Sigma = \{\lambda;\ |\arg(\lambda - a)| \le \phi\}$ for some $a > 0$ and $\phi > \frac{\pi}{2}$. Moreover, there is a positive constant C such that $\|R(\lambda; \mathscr{A}(t))\| \le \frac{C}{|\lambda - a|}$ for all $\lambda \in \Sigma$, $t \in R^+$, and there exists a constant $C_1 > 0$ such that

$$\|(\mathscr{A}(t) - \mathscr{A}(s))(aI - \mathscr{A})^{-1}(\tau)\| \le C_1|t-s|, \quad \text{for all } s,t,\tau \in R^+.$$

Then (see, e.g., [54, 66]), there is a unique evolution operator $S(t,s)$, $0 \le s \le t < \infty$, such that

$$\begin{aligned} &\frac{d}{dt} S(t,s)x + \mathscr{A}(t)S(t,s)x = 0, \quad 0 \le s < t < \infty,\\ &S(s,s)x = x \in H. \end{aligned}$$

Moreover, $S(t,s)$ is strongly continuous in (t,s) with values in $L(D(A^\beta), D(A^\beta))$ for any $0 \le \beta < 1$ and (see [54], p. 191)

$$\begin{aligned} &|S(t,s)x|_\beta \le C(t-s)^{\gamma-\beta}|x|_\gamma, \quad 0 \le \gamma \le \beta < 1,\ t > s,\\ &\left|\frac{d}{dt} S(t,s)x\right|_\beta \le C(t-s)^{\gamma-\beta-1}|x|_\gamma. \end{aligned} \quad (2.155)$$

If we set $y(t) = S(t,0)y_0$, then $y(t) \in C(R^+; H)$ is the solution to the system

$$y'(t) + \mathscr{A}(t)y(t) = 0, \quad y(0) = y_0.$$

Now, we let $U(t) = S(T+t, t)$, $t \in R^+$, be the periodic map (Poincaré map) and recall that (see, e.g., [54], p. 198), $U(T+t) = U(t)$ and the spectrum $\sigma(U(t))$ is independent of t. Since A^{-1} is completely continuous, $U(t)$ is completely continuous as well. Moreover, $\sigma(U(t)) \setminus \{0\}$ consists entirely of eigenvalues $\{\lambda_j\}_{j=1}^\infty$, $|\lambda_j| \to 0$ as $j \to \infty$. Each eigenvalue λ_j is repeated accor ding to its algebraic multiplicity m_j. Let $U^*(t)$ and D^* be the adjoint of $U(t)$ and D, respectively. Then

$\sigma(U^*(t)) \setminus \{0\} = \{\overline{\lambda}_j\}_{j=1}^{\infty}$. We denote by X_m^* the space spanned by $\{\psi_i^*\}_{i=1}^m$, where $\psi_i^*, i = 1, \dots, m$, are eigenvectors of $U^*(T)$ corresponding to eigenvalues $\{\overline{\lambda}_j\}_{j=1}^m$. We assume that the following hypothesis holds.

(A1) $\ker\{D^*|_{X_m^*}\} = \{0\}$, $\forall m$.

In particular, Assumption (A1) implies the following unique continuation property: *If* $U^*(T)\varphi^* = \lambda\varphi^*$, *where* $\lambda \in \sigma(U^*(T)) \setminus \{0\}$, *and* $D^*\varphi^* = 0$, *then* $\varphi^* = 0$.

Assumption (A1) is a consequence of the following one.

(A1)′ *If* z *satisfies for* $\psi \in X_m^*$ *the equation*

$$\begin{aligned} &z' - \mathscr{A}^*(t)z = 0 \quad in\ (0, T), \\ &z(0) = \lambda z(T) + \psi, \end{aligned} \tag{2.156}$$

and $D^*z(T) = 0$, *then* $z \equiv 0$.

Indeed, if ψ_1^*, ψ_2^* are linearly independent eigenvectors, $U^*(T)\psi_i^* = \lambda_i\psi_i^*, i = 1, 2$ and $D^*\psi_1^* = \mu D^*\psi_2^*$, then $z(t) = S^*(T, t)\psi_1^* - \mu S^*(T, t)\psi_2^*$ satisfies (2.156) for $\psi = \psi_1^* - \mu\psi_2^* \in X_m^*$ and $D^*z(T) = 0$. Hence $z \equiv 0$ (that is, $\psi_2^* = C\psi_1^*$). The case of m eigenfunctions $\{\psi_i^*\}_{i=1}^m$ follows by induction from the previous one. A more delicate situation is that when system $\{\psi_i^*\}_{i=1}^m$ contains generalized eigenvectors ψ^*, that is, $(U^*(T) - \lambda_j I)^q\psi^* = 0$ for some $1 < q < m_j$, but we omit the proof.

In the classical Floquet theory, the eigenvalues λ of $U(t)$ are the characteristic multipliers of the linear system and $\gamma = -(\frac{1}{T})\log\lambda$ are the Floquet exponents. One knows that, if there is a characteristic multiplier with modulus greater than one, then the periodic solution y_π is unstable.

The main result of Sect. 2.6, Theorem 2.12 below, amounts to saying that under Assumptions (k)–(kkk), and (A1), (A1)′ there is a feedback controller u which stabilizes exponentially the solution y_π. Moreover, the controller u has a finite-dimensional structure $u(t) = \sum_{i=1}^N u_i(t)w_i$, where $\{w_i\}$ is a given system in U and N is the number of characteristic multipliers (repeated according to their algebraic multiplicity) with modulus greater than or equal to one.

2.6.2 Stabilization of the Linearized Time-periodic System

Let $\mathscr{A}(t) = A + B_y(t, y_\pi(t))$ with the domain $D(\mathscr{A}(t)) = D(A)$. We consider the linear system

$$\begin{cases} y'(t) + \mathscr{A}(t)y(t) = Du(t), & t \in R^+, \\ y(0) = x, \end{cases} \tag{2.157}$$

where $x \in H$. System (2.157) is just the linearization of (2.148) in $y = y_\pi(t)$.

Unless stated explicitly, by solution y to (2.157) we mean "mild" solution, that is,

$$y(t) = S(t, 0)x + \int_0^t S(t, s)Du(s)ds, \quad t \geq 0, \tag{2.158}$$

where $S(t,s)$ is the evolution generated by $\mathscr{A}(t)$. We notice, however, that such a solution is a strong solution. More precisely, y is absolutely continuous on every compact interval (δ, T) and satisfies, a.e., (2.157). It suffices to check this for $x=0$ because, as noticed earlier, $\left|\frac{d}{dt} S(t,0)x\right| \le Ct^{-1}$, for all $t>0$. By (2.155) and (2.158), we have

$$|y|_{L^2(0,T;D(A^\alpha))} \le C|Du|_{L^2(0,T;H)}.$$

Then, by Assumption (2.151) we have that

$$|B_y(t, y_\pi(t))y|_{L^2(0,T;H)} \le C_1|u|_{L^2(0,T;H)}$$

and since $-A$ generates an analytic C_0-semigroup, we see that (see Sect. 1.3)

$$\frac{d}{dt} y,\ Ay \in L^2(\delta, T; H), \quad \forall \delta > 0.$$

Lemma 2.2 *There is a controller u of the form*

$$u(t) = \sum_{i=1}^{N} u_i(t) w_i, \quad t \ge 0, \tag{2.159}$$

where $\{w_i\}_{i=1}^N \subset U$ *is a linearly independent system and* $u_i \in L^2(R^+)$, $i = 1, \dots, N$, *are such that*

$$u_i(t) = 0 \quad \text{for } t \ge T,\ i = 1, \dots, N, \tag{2.160}$$

$$\int_0^T \sum_{i=1}^N |u_i(t)|^2 dt \le C|x|^2, \tag{2.161}$$

$$|y(t)| \le Ce^{-\delta t}|x|, \quad \forall t \ge 0, \text{ where } \delta > 0. \tag{2.162}$$

Here, $y \in C(R^+; H)$ is the solution to (2.157).

Proof As in the previous cases, we can replace H by its complexified space, again denoted by $\widetilde{H}$. Similarly, we replace U by $\widetilde{U} = U + iU$. As noticed earlier, the periodic map $U(t) = S(T+t, t)$ has the property that

$$\sigma(U(t)) \setminus \{0\} = \{\lambda\}_{j=1}^\infty, \quad \lambda_j \to 0, \forall t \in R.$$

Let $\eta > 0$ be arbitrarily small but fixed. Then, outside the disk

$$\Sigma = \{\lambda \in \mathbb{C};\ |\lambda| < 1 - \eta\},$$

there remains a finite number of eigenvalues $\{\lambda_j\}_{j=1}^N$ only. (Recall the eigenvalues λ_j are repeated according to their algebraic multiplicity m_j and so $N = m_1 + m_2 +$

$\cdots + m_k$, where k is the number of distinct eigenvalues in Σ.) Then, for each $t \in R^+$, the space $\widetilde{H}$ can be decomposed as

$$\widetilde{H} = H_1(t) \oplus H_2(t), \quad \forall t \geq 0,$$

where $H_1(t) = P_1(t)H$, $H_2(t) = (I - P_1(t))H$, and $P_1(t)$ is defined by

$$P_1(t) = \frac{1}{2\pi i} \int_\Gamma (\lambda I - U(t))^{-1} d\lambda, \quad t \in R^+,$$

where Γ is a contour surrounding $\{\lambda_j\}_{j=1}^N$ but not other eigenvalues.

It is clear that $P_1(t) = P_1(t+T)$, for all $t \in R^+$, and

$$\sigma(U(t)|_{H_1(t)}) = \{\lambda_j\}_{j=1}^N, \qquad \sigma(U(t)|_{H_2(t)}) = \{\lambda_j\}_{j=N+1}^\infty.$$

It follows that (see, e.g., [54], p. 198)

$$\begin{aligned} &\dim H_1(t) = N \quad \text{and} \quad H_1(t+T) = H_1(t), \quad \forall t \geq 0; \\ &S(t,s) : H_i(s) \to H_i(t), \quad i = 1, 2, \ 0 \leq s \leq t < \infty, \\ &\quad \text{that is,} \quad S(t,s)P_i(s) = P_i(t)S(t,s), \quad i = 1, 2; \end{aligned} \tag{2.163}$$

$$|S(t,s)x| \leq Ce^{-\delta(t-s)}|x|, \quad \forall x \in H_2(s), \ t \geq s, \tag{2.164}$$

where C and δ are positive constants independent of t, s and x.

Let $U^*(t)$, $S^*(t,s)$ be the adjoints of $U(t)$ and $S(t,s)$ respectively, and let

$$H_1^*(t) = P_1^*(t)H, \quad H_2^*(t) = (I - P_1^*(t))H, \quad t \geq 0,$$

where P_1^* is the adjoint of P_1, that is,

$$P_1^*(t) = \frac{1}{2\pi i} \int_{\Gamma^*} (\lambda I - U^*(t))^{-1} d\lambda.$$

Here, Γ^* is a contour surrounding $\{\overline{\lambda}_j\}_{j=1}^N$ (the eigenvalues of $U^*(t)$). We have

$$\dim H_1^*(t) = N, \quad \forall t \in R^+,$$

and

$$\begin{aligned} &S(t+T, s+T) = S(t,s), \quad 0 < s \leq t < \infty, \\ &S^*(t+T, s+T) = S^*(t,s), \\ &S^*(t,s) : H_i^*(t) \to H_i^*(s), \quad 0 \leq s \leq t < \infty, \ i = 1, 2. \end{aligned}$$

Now, let

$$u = \sum_{i=1}^N u_i w_i,$$

where $w_i \in U$ is specified later and let $u_i \in L^2(R^+)$, $i = 1, \dots, N$. We represent the mild solution $y = y^u$ to (2.25), where $u = \sum_{i=1}^N u_i w_i$, that is,

$$y(t) = S(t,0)x + \int_0^t S(t,s)\left(\sum_{i=1}^N u_i(s)Dw_i\right) ds, \tag{2.165}$$

as $y(t) = y_1(t) + y_2(t)$, and $y_i(t) \in H_i(t)$, $i = 1, 2$, are given by

$$y_1(t) = S(t,0)x^1 + \int_0^t S(t,s)P_1(s)\left(\sum_{i=1}^N u_i(s)Dw_i\right) ds, \tag{2.166}$$

$$y_2(t) = S(t,0)x^2 + \int_0^t S(t,s)(I - P_1(s))\left(\sum_{i=1}^N u_i(s)Dw_i\right) ds. \tag{2.167}$$

Here, $x^1 = P_1(0)x$ and $x^2 = (I - P_1(0))x$.

By (2.164), it follows that

$$|y_2(t)| \le C\left(e^{-\delta t}|x^2| + \int_0^t e^{-\delta(t-s)}\left|\sum_{i=1}^N u_i(s)Dw_i\right|_U ds\right), \quad t \ge 0. \tag{2.168}$$

Next, we are going to show that there are $u_i(s)$, $i = 1, \dots, N$, such that $y_1(T) = 0$ (that is, (2.166) is exactly null controllable). To this end, we note first that for each $\xi \in H_1^*(T)$, $S^*(T,s)\xi = q(s) \in H_1^*(s)$, $0 \le s \le T$, is the so lution to the backward adjoint equation

$$q_t - \mathscr{A}^*(t)q = 0, \quad t \in (0,T), \quad q(T) = \xi, \tag{2.169}$$

where $\mathscr{A}^*(t)$ is the adjoint of $\mathscr{A}(t)$. To this purpose, we recall that the exact null controllability of (2.166) is equivalent to the following observability inequality

$$|S^*(T,0)\xi|^2 \le C \int_0^T \sum_{i=1}^N \left|\langle w_i, D^*S^*(T,s)\xi\rangle_U\right|^2 ds, \quad \forall \xi \in H_1^*(T). \tag{2.170}$$

As the space $H_1^*(T)$ is finite-dimensional, Inequality (2.170) is equivalent to the following one:

if $\langle w_i, D^*S^*(T,s)\xi\rangle_U = 0$, $\forall i = 1, \dots, N$, for all $s \in [0,T]$, then $\xi = 0$.

Inasmuch as $\xi \in H_1^*(T)$, we may write it as

$$\xi = \sum_{j=1}^N c_j \psi_j^*,$$

where $\{\psi_j^*,\ j = 1, \dots, N\}$ is a basis of $H_1^*(T)$ formed by eigenvectors of $U^*(T)$ corresponding to $\overline{\lambda}_j$, $j = 1, \dots, N$. By Assumption (A1), $\{D^*\psi_j^*\}_{j=1}^N$ is a linearly independent system in U. Thus, we may find a system $\{w_i\}_{i=1}^N \subset U$ such that

$$\det \|\langle w_i, D^*\psi_j^*\rangle_U\| \neq 0.$$

For instance, one might choose $\{w_i\}_{i=1}^N$ as the solution to the algebraic system

$$\langle w_i, D^*\psi_j^*\rangle_U = \delta_{ij}, \quad i, j = 1, \dots, N. \tag{2.171}$$

Thus, if $\langle w_i, D^*S^*(T,s)\xi\rangle = 0, i = 1, \dots, N, s \in [0, T]$, then $\xi = 0$. Hence, (2.166) is exactly null controllable. Thus, there exist $\{w_i\}_{i=1}^N \subset U$ and $u_i \in L^2(R^+)$, $i = 1, \dots, N$, such that $y_1(t), u_i(t) = 0$ for $t \geq T$, and $\int_0^T |u_i(t)|^2 dt \leq C|x^1|^2$. Then, by (2.168) we have

$$|y_1(t)| \leq C_{\gamma_0} e^{-\gamma_0 t}|x|, \quad \forall t \geq 0,$$

for any $\gamma_0 > 0$. This implies that, for some $\delta > 0$,

$$|y(t)| \leq Ce^{-\delta t}|x| \quad \text{for } t \geq 0,$$

as claimed. □

Now, if we represent $\psi_j^* = (\psi_j^1)^* + i(\psi_j^2)^*$, where $(\psi_j^1)^*, (\psi_j^2)^* \in H$, we may assume that $w_i \in U, i = 1, \dots, N$, and that the controller u is real-valued.

It is clear that Lemma 2.2 remains true on any interval $[s, T+s]$. However, the dependence of s of constants C arising in the above estimates is crucial for latter development and must be analyzed. Thus, we are lead to consider the system

$$\begin{aligned} &y'(t) + \mathscr{A}(t)y(t) = Du(t), \quad t \geq s, \\ &y(s) = x. \end{aligned} \tag{2.172}$$

Lemma 2.3 *For each $s \in [0, \infty)$, there is a controller $u_s(t) = \sum_{i=1}^N u_i^s(t)w_i$, where $\{w_i\}_{i=1}^N \subset H$, was given by Lemma 2.2, such that $u_s(t) = 0$ for $t \geq s + T$ and the solution y_s to (2.172) satisfies*

$$|y_s(t)| \leq Ce^{-\delta(t-s)}|x|, \qquad \int_0^T |u_s(t)|^2 dt \leq C|x|^2, \tag{2.173}$$

for some positive constants C and δ independent of s and x.

Proof We show first that there exists a controller $u_s(t)$ of the form $\sum_{i=1}^N u_i^s(t)w_i$, where $\{w_i\}_{i=1}^N$ are as in Lemma 2.2, such that (2.173) holds. (As before, we work in the complexified space H.) After that, we prove that $C(s)$ and $\delta(s)$ are independent of s.

As seen in the proof of Lemma 2.2, the existence of such a $u_s(t)$ is equivalent to the following observability inequality

$$|S^*(T+s,s)\xi|^2 \leq C\int_s^{T+S} \left|\sum_{i=1}^N \langle w_i, D^*S^*(T-s,\sigma)\xi\rangle_U\right|^2 d\sigma,$$

for all $\xi \in H_1^*(T+s)$, which is equivalent to the following unique continuation property:

if $\langle w_i, D^*S^*(T+s,\sigma)\xi\rangle_U$, $i=1,\dots,N$, $\sigma \in [s, T+s]$, then $\xi = 0$.

Assume that

$$\langle w_i, D^*S^*(T+s,T)\xi\rangle_U = 0 \quad \text{for all } i=1,\dots,N.$$

Since $S^*(T+s,T) : H_1^*(T+s) \to H_1^*(T)$ is one to one (see [54], p. 198), we have

$$S^*(T+s,T)\xi \in H_1^*(T) \quad \text{for } \xi \in H_1^*(T+s).$$

Hence, we may write

$$S^*(T+s,T)\xi = \sum_{j=1}^{N} \eta_j \psi_j^*,$$

where $\{\psi_j^*\}_{j=1}^N$ is the basis of $H_1^*(T)$ formed by the eigenvectors of $U^*(T)$ corresponding to $\{\overline{\lambda}_j\}_{j=1}^N$. Then, by the same argument as the used in the proof of Lemma 2.2, we obtain that $\xi = 0$, as desired.

Now, we turn to prove the independence of C and δ in (2.173) as functions of s. By the substitution $t \to t+s$, we rewrite (2.172) as

$$y_s'(t) + \mathscr{A}_s(t)y_s(t) = Du_s(t), \quad t \geq 0,\ y_s(0) = x, \tag{2.174}$$

where $\mathscr{A}_s(t) = \mathscr{A}(t+s)$. We denote by $S_s(t,\sigma)$ the evolution generated by $\mathscr{A}_s(t)$. It is clear that $S_s(t,\sigma) = S(t+s,\sigma)$. We have, of course, $\mathscr{A}_T(t) = \mathscr{A}(t)$ and $S_T(t,\sigma) = S(t,\sigma)$. By the previous discussion, the solution y_s to (2.174) may be written as $y_s(t) = y_s^1(t) + y_s^2(t)$, where

$$y_s^1(t) = S_s(t,0)P_1(s)x + \int_0^t S_s(t,\eta)P_1(\eta+s)Du_s(\eta+s)d\eta.$$

By periodicity, it suffices to assume that $s \in [0,T]$. Let $u_s^* \in L^2(0,T)$ be such that $(y_s^*)^1(T) = 0$, where y_s^* is given as above with $u_s = u_s^*$. It turns out that u_s^* can be determined by

$$u_s^*(t) = \lim_{\varepsilon \to 0} u_s^\varepsilon(t) \quad \text{strongly in } L^2(0,T), \tag{2.175}$$

where

$$u_s^\varepsilon = \arg\min \left\{ \int_0^T |u(t)|_U^2 dt + \frac{1}{\varepsilon}\,|y_s(T)|^2;\ u \in L^2(0,T;U), \right.$$
$$\left. y_s^1(t) = S_s(t,0)P_1(s)x + \int_0^t S_s(t,\eta)P_1(\eta+s)Du(\eta)d\eta \right\}. \tag{2.176}$$

Let $(\widetilde{y}_T, \widetilde{u}_T)$ be such that

$$(\widetilde{y}_T(t))' + \mathscr{A}(t)\widetilde{y}_T(t) = P_1(t)D\widetilde{u}_T(t), \quad t \in [0,T],$$
$$\widetilde{y}_T(0) = P_1(T)x, \qquad \widetilde{y}_T(T) = 0. \tag{2.177}$$

By (2.176), we have that

$$\int_0^T |u_s^\varepsilon(t)|_U^2 dt + \frac{1}{\varepsilon}|y_s(T)|^2 \le \int_0^T |u_T^*(t)|^2 dt, \tag{2.178}$$

and by (2.175), we conclude that

$$\int_0^T |u_s^*(t)|_U^2 dt \le \int_0^T |u_T^*(t)|_U^2 dt. \tag{2.179}$$

On the other hand, by (2.161) it follows that u_T^* can be chosen in such a way that

$$\int_0^T |u_T^*(t)|_U^2 dt \le C|x|^2, \tag{2.180}$$

where C is a positive constant independent of x. Thus, by (2.179), we infer that

$$\int_0^T |u_s^*(t)|_U^2 dt \le C|x|^2, \quad \forall x \in [0,T],\ s \in [0,T], \tag{2.181}$$

where C is independent of s.

Then, arguing as in the proof of Lemma 2.2, we obtain Estimate (2.173) independent of s, as claimed. □

Remark 2.5 As seen in the proof of Lemma 2.2, the dimension N of basis $\{w_j\}_{j=1}^N$ arising in construction of stabilizing controller is equal to the number of Floquet exponents for $\mathscr{A}(t)$ with nonnegative real parts.

2.6.3 The Stabilizing Riccati Equation

Throughout this sequel, we assume that Assumptions (k), (kk) hold with $\frac{1}{4} \le \alpha \le \frac{5}{8}$.

Consider the infinite horizon optimal control problem

$$\varphi(s,x) = \text{Min}\left\{\frac{1}{2}\int_s^\infty \left(|A^{\frac{3}{4}}y(t)|^2 + \sum_{i=1}^N |u_i(t)|^2\right)dt\right\} \tag{2.182}$$

subject to $u_i \in L^2(s,\infty)$, $i = 1,\dots,N$, and

$$y'(t) + Ay(t) + A_0(t)y(t) = \sum_{i=1}^N u_i(t)Dw_i, \quad t \ge 0,$$
$$y(s) = x, \tag{2.183}$$

where $\{w_i\}_{i=1}^N \subset U$ are as in Lemma 2.2 and $A_0(t) = B_y(t, y_\pi(t))$. We set $W = D(A^{\frac{1}{4}})$ with the norm $|\cdot|_W = |\cdot|_{\frac{1}{4}}$ and $\widetilde{D} \in L(R^N, H)$ given by $\widetilde{D}u = \sum_{i=1}^N u_i D w_i$, $u = \{u_i\}_{i=1}^N$. By $|u|$ we denote here the Euclidean norm of u. We also denote $\widetilde{D}^*$ the dual of $\widetilde{D}$ (that is, $\widetilde{D}^* y = \{(Dw_i, y)\}_{i=1}^N$). We note that, since A is self-adjoint, the mild solution y to (2.183) is strong solution and $y \in W^{1,2}(\delta, T; H)$ for all $0 < \delta < T$.

Lemma 2.4 *For each $s \geq 0$ there is a symmetric and positive operator $R(s) \in L(W, W')$ such that*

$$\varphi(s, x) = \frac{1}{2}(R(s)x, x), \quad \forall x \in W,\ s \geq 0.$$

There exist positive constants $\gamma_1, \gamma_2, \gamma_3$ independent of s, such that

$$\gamma_1 |x|_W^2 \leq (R(s)x, x) \leq \gamma_2 |x|_W^2, \quad \forall x \in W,\ s \geq 0, \tag{2.184}$$

and

$$|R(s)x| \leq \gamma_3 |A^{\frac{1}{2}} x|, \quad \forall x \in D(A^{\frac{1}{2}}),\ s \geq 0. \tag{2.185}$$

Moreover, $R(s)$ satisfies the Riccati equation

$$\begin{cases} (R'(s)x, x) - 2(R(s)x, (A + A_0(s))x) - \sum_{i=1}^M (R(s)x, Dw_i)^2 + |A^{\frac{3}{4}} x|^2 = 0, \\ \forall x \in D(A),\ s \geq 0, \\ R(t+T) = R(t), \quad \forall t \in (0, \infty). \end{cases} \tag{2.186}$$

Here, $R'(s) \in L(D(A), (D(A))')$ is the weak derivative of $R(s)$, that is,

$$(R'(t)x, y) = \frac{d}{dt}(R(t)x, y), \quad \forall x, y \in D(A).$$

We denote by the same symbol $(\cdot, \cdot)$ the scalar product of H and the pairing between W and its dual space W', $W \subset H \subset W'$.

Proof For any $s \geq 0$, it follows from Lemma 2.3 that there exist $u_i \in L^2(0, \infty)$ with $u_i(t) = 0$ for $t \geq s + T$, $i = 1, \ldots, N$, such that

$$|y(t)| \leq Ce^{-\delta t}|x| \quad \text{for } t \geq s, \tag{2.187}$$

and

$$\int_s^{T+s} \sum_{i=1}^N |u_i(t)|^2 dt \leq C|x|^2, \quad \forall s > 0, \tag{2.188}$$

for some positive constants C and δ independent of s. Here, y is the solution to (2.183). Moreover, it is readily seen that for each x the function $\varphi(s, x)$ is T-periodic.

Now, we fix $x \in W$. Multiplying (2.183) by $A^{\frac{1}{2}}y$, we get

$$\frac{d}{dt}|A^{\frac{1}{4}}y(t)|^2 + 2|A^{\frac{3}{4}}y(t)|^2$$
$$\leq 2|A_0(t)y(t)|\,|A^{\frac{1}{2}}y(t)| + |\widetilde{D}u|\,|A^{\frac{1}{2}}y(t)|, \quad \text{a.e.}, t > s. \qquad (2.189)$$

Since $y_\pi \in C^1(R; D(A))$, by Assumption (kk) we obtain, via the interpolation inequality, that

$$|A_0(t)y(t)||A^{\frac{1}{2}}y(t)| \leq C|y(t)|^2_{\frac{1}{2}} \leq C|y(t)|^{\frac{4}{3}}_{\frac{3}{4}}|y(t)|^{\frac{2}{3}}$$
$$\leq \frac{1}{4}\,|y(t)|^2_{\frac{3}{4}} + C|y(t)|^2, \quad \text{a.e.}, t > 0. \qquad (2.190)$$

(Here and throughout the proof of this lemma, we denote by C several positive constants independent of s, t and x.)

Integrating (2.189) over (s, ∞) and using (2.187), (2.188) and (2.190), we obtain that

$$\int_s^\infty |A^{\frac{3}{4}}y(t)|^2 dt \leq C\|x\|^2_W,$$

which implies that $\varphi(s, x) \leq C\|x\|^2_W$ for some $C > 0$ independent of s and x.

On the other hand, it is readily seen that, for each $x \in W$, Problem (2.182) has a unique pair (u^*, y^*), $u^* = \{u_i^*\}_{i=1}^N \in (L^2(s, \infty))^N$. Multiplying (2.183), where $(y, u) = (y^*, u^*)$, by $A^{\frac{1}{2}}y^*$ and integrating on (s, ∞), we obtain that

$$\frac{1}{2}|x|^2_W \leq \int_s^\infty \left(|A^{\frac{3}{4}}y^*(t)|^2 + \left|(A_0(t)y^*(t), A^{\frac{1}{2}}y^*(t))\right| + \left|(\widetilde{D}u^*(t), A^{\frac{1}{2}}y^*(t))\right|\right) dt$$
$$\leq C\int_s^\infty (|A^{\frac{3}{4}}y^*|^2 + |u^*|^2)dt = C\varphi(s, x).$$

Hence, there is a constant $C > 0$ independent of s such that

$$C\|x\|^2_W \leq \varphi(s, x).$$

In other words, $D(\varphi(s, \cdot)) = W$, for all $s \geq 0$, where $D(\varphi(s, \cdot))$ is the domain of $\varphi(s, \cdot)$. This implies that, for each $s \geq 0$, there is a linear positive and symmetric operator $R(s) : H \to H$ with the domain $D(R(s)) \subset W$ such that

$$\varphi(s, x) = \frac{1}{2}\,(R(s)x, x), \quad \forall x \in D(R(s)).$$

Moreover, $R(s)$ extends to all of W and $R(s) \in L(W, W')$.

We now turn to prove (2.184) and (2.185). To this end, we consider the optimization problem

$$\varphi_n(s, x) = \text{Min}\left\{\frac{1}{2}\int_s^n \left(|A^{\frac{3}{4}}y(t)|^2 + |u(t)|^2\right) dt\right\}, \qquad (2.191)$$

subject to

$$y'(t) + Ay(t) + A_0(t)y(t) = \widetilde{D}u(t), \quad t \in (s,n), \ y(s) = x, \tag{2.192}$$

where $u = \{u_i\}_{i=1}^N \in L^2(s,n)^N$.

By the previous discussion, for each n there is a linear symmetric operator $R_n(s) \in L(W, W')$, $R_n(s) : D(R_n(s)) \subset H \to H$, such that

$$\frac{1}{2}\,(R_n(s)x, x) = \varphi_n(s,x), \quad \forall x \in W, \ s \geq 0.$$

It is readily seen that, for $n \to \infty$,

$$(R_n(s)x, x) \to (R(s)x, x), \quad \forall x \in W, \ s \geq 0,$$

and, therefore,

$$R_n(s)x \to R(s)x \quad \text{weakly in } W', \forall x \in W, \ s \geq 0.$$

Hence, it suffices to prove Estimates (2.185) and the right-hand side part of (2.184) for R_n only.

Let $x \in D(A^{\frac{1}{2}})$ and let (y^n, u^n) be optimal for Problem (2.191). Then, by the maximum principle, we see that

$$u^n(t) = \widetilde{D}^* q^n(t) = \{\langle Dw_i, q^n(t)\rangle\}_{i=1}^N, \tag{2.193}$$

for all $t \in [x,n]$, where q^n is the solution to the Hamiltonian system

$$\begin{aligned} &y_t^n(t) + Ay^n(t) + A_0(t)y^n(t) = \widetilde{D}\widetilde{D}^* q^n(t), \quad s < t < n, \\ &q_t^n(t) - Aq^n(t) - A_0^* q^n(t) = A^{\frac{3}{2}} y^n(t), \quad s < t < n, \\ &y^n(s) = x, \qquad q^n(n) = 0. \end{aligned} \tag{2.194}$$

Moreover, one has

$$R_n(s)x = -q^n(s), \quad s \in [0,n]. \tag{2.195}$$

On the other hand, if (y_s^*, u_s^*) is an optimal pair for Problem (2.182), then we have

$$\begin{aligned} \int_s^n \left(|A^{\frac{3}{4}} y^n(t)|^2 + |u^n(t)|^2\right) dt &\leq \int_s^n \left(|A^{\frac{3}{4}} y_s^*(t)|^2 + |u_s^*(t)|^2\right) dt \\ &\leq 2\varphi(s,x) \leq C|x|_W^2. \end{aligned} \tag{2.196}$$

Now, we multiply the first equation of (2.194) by Ay^n to get

$$\begin{aligned} \frac{1}{2}\frac{d}{dt}|y^n(t)|_{\frac{1}{2}}^2 + |Ay^n(t)|^2 &\leq |(A_0(t)y^n(t), Ay^n(t))| \\ &\quad + \left|(\widetilde{D}\widetilde{D}^* q^n(t), Ay^n(t))\right|. \end{aligned} \tag{2.197}$$

We have

$$\begin{aligned}\left|(A_0(t)y^n(t), Ay^n(t))\right| &\le C\left|y^n(t)\right|_{\frac{1}{2}}|Ay^n(t)| \\ &\le \frac{1}{4}|Ay^n(t)|^2 + C|y^n(t)|^2_{\frac{1}{2}},\end{aligned} \tag{2.198}$$

and

$$\begin{aligned}\left|(\widetilde{D}\widetilde{D}^* q^n(t), Ay^n(t))\right| &\le \left|\widetilde{D}u^n(t)\right| |Ay^n(t)| \\ &\le |Ay^n(t)|^2 + C|\widetilde{D}u(t)|^2.\end{aligned} \tag{2.199}$$

Combining (2.197)–(2.199), we obtain that

$$\begin{aligned}\frac{d}{dt}|y^n(t)|^2_{\frac{1}{2}} + |Ay^n(t)|^2 &\le C\left(|\widetilde{D}u^n(t)|^2 + |y^n(t)|^2_{\frac{1}{2}}\right) \\ &\le \frac{1}{2}|Ay^n(t)|^2 + C|\widetilde{D}u^n(t)|^2\end{aligned}$$

and, integrating above on (s,n) and using (2.196), we get the estimate

$$\int_s^n |Ay^n(t)|^2 dt \le C|x|^2_{\frac{1}{2}}. \tag{2.200}$$

Multiplying the second equation of (2.194) by q^n, integrating over (s,n) and using (2.200), we see that

$$|q^n(t)| \le C|x|_{\frac{1}{2}}, \quad \forall t \in (s,n),$$

for some $C > 0$ independent of n and s, which together with (2.195) implies

$$|R_n(s)x| \le C|x|_{\frac{1}{2}}, \quad \forall s > 0,\ x \in D(A^{\frac{1}{2}}),$$

as claimed.

Finally, by a standard argument involving (2.194) and (2.195), we obtain that

$$\begin{cases} (R_n'(s)x, x) - 2(R_n(s)x, (A + A_0(s))x) - |\widetilde{D}^* R_n(s)x|^2 + |A^{\frac{3}{4}}x|^2 = 0, \\ \forall x \in D(A),\ s \ge 0,\ n \in \mathbb{N}, \\ R_n(n) = 0. \end{cases}$$

By passing to the limit for $n \to \infty$ in the latter equality, we obtain (2.186), as desired. □

2.6.4 Stabilization of Nonlinear System (2.148)

In (2.148) we insert the (feedback) controller $u(t)=\sum_{i=1}^N u_i(t)w_i$, where

$$u_i(t)=-\langle w_i, D^*R(t)(y(t)-y_\pi(t))\rangle_U,\quad i=1,\dots,N. \tag{2.201}$$

Here, $\{w_i\}_{i=1}^N$ are as in Lemma 2.2 and $R(t)\in L(W,W')$ is given by Lemma 2.4. Consider the corresponding closed-loop system

$$\begin{cases} y'(t)+Ay(t)+B(t,y(t))+\sum_{i=1}^N\langle w_i, D^*R(t)(y(t)-y_\pi(t))\rangle_U Dw_i=0, \\ \quad t\geq 0, \\ y(0)=y_0. \end{cases} \tag{2.202}$$

Lemma 2.5 *Let $y_0\in D(A^\alpha)$. Then there is $0<T_0=T_0(|y_0|_\alpha)$ such that (2.202) has a unique mild solution $y\in C([0,T];D(A^\beta))$, $\beta=\max(\alpha,\frac{1}{2})$ on the interval $[0,T_0)$. Moreover, y is absolutely continuous on each compact interval of $(0,T_0)$, satisfies, a.e., on $(0,T_0)$ (2.202) and*

$$y\in W^{1,2}(\delta,T_0;H),\quad Ay, B(t,y)\in L^2(\delta,T_0;H),\quad \forall 0<\delta<T_0. \tag{2.203}$$

Proof Since the proof is standard, we only sketch it. We write (2.202) as the integral equation

$$\begin{aligned} y(t)&=e^{-At}y_0\\ &\quad -\int_0^t e^{-A(t-s)}\left(B(s;y(s))+\sum_{i=1}^N\langle w_i, D^*R(s)(y(s)-y_\pi(s))\rangle_U Dw_i\right)ds \end{aligned} \tag{2.204}$$

and apply the Banach fixed-point theorem in the space

$$X=\{y\in C([0,T_0];D(A^\beta));\ |y(t)|_\beta\leq\mu,\ t\in[0,T_0]\},$$

where $\beta=\alpha$ if $\alpha\geq\frac{1}{2}$ and $\beta=\frac{1}{2}$ if $\alpha<\frac{1}{2}\cdot$ By (2.153) and (2.182), it follows that the operator Γ, defined by the right-hand side of (2.204), maps X into itself if $|y_0|_\beta\leq\frac{M}{2}$ and $0<T_0<\delta(\mu)$ is sufficiently small. Moreover, Γ is a contraction on X. This means that (2.204) has a unique solution $y\in C([0,T_0];D(A^\beta))$, as desired. Since $|B(s,y)|\leq C(1+|y|_\alpha|y|_{\frac{1}{4}})$ and $|R(s)y|\leq C|y|_{\frac{1}{2}}$, we conclude that $y\in W^{1,2}(\delta,T_0;H)$, that is,

$$\frac{dy}{dt}, Ay\in L^2(\delta,T_0;H),\quad \forall 0<\delta<T_0,$$

and, therefore, y is a strong solution to (2.202). This completes the proof. □

Theorem 2.12 *Assume that Hypotheses* (k)–(kkk) *and* (A1) *hold with* $\alpha = \frac{5}{8}$. *Then there is* $\rho > 0$ *such that, for* $y_0 \in D(A^\alpha)$, $\|y_0 - y_\pi(0)\|_W < \rho$, (2.201) *has a unique strong solution* $y \in C([0,\infty); H)$, *such that* $\frac{dy}{dt} \in L^2_{\text{loc}}(0,\infty; H)$, $Ay \in L^2_{\text{loc}}(0,\infty; H)$, $B(t,y) \in L^2_{\text{loc}}(0,\infty; H)$ *and*

$$\int_0^\infty |A^{\frac{3}{4}}(y(t) - y_\pi(t)|^2 dt \le C\|y_0 - y_\pi\|_W^2, \tag{2.205}$$

$$\|y(t) - y_\pi(t)\|_W \le Ce^{-\delta t}\|y_0 - y_\pi(0)\|_W, \quad \forall t \ge 0, \tag{2.206}$$

for some $\delta, C > 0$.

Proof Let $z(t) = y(t) - y_\pi(t)$, then we have

$$\begin{cases} z'(t) + Az(t) + B(t, z(t) + y_\pi(t)) - B(t, y_\pi(t)) \\ \quad + \sum_{i=1}^N \langle w_i, D^*R(t)z(t)\rangle_U Dw_i = 0, \quad t > 0, \\ z(0) = z_0 \equiv y_0 - y_\pi(0). \end{cases} \tag{2.207}$$

Let $F(t,z(t)) = B(t, z(t) + y_\pi(t)) - B(t, y_\pi(t)) - A_0(t)z(t)$. It follows from (2.151) that

$$|F(t,z)| \le C(1 + |z|_{\frac{1}{4}})|z|_\alpha^2, \quad \forall t \ge 0,\ \forall z \in D(A^\alpha), \tag{2.208}$$

where $C > 0$ is independent of t. Now, we rewrite (2.207) as

$$\begin{aligned} &z'(t) + Az(t) + A_0(t)z(t) + F(t,z(t)) + \sum_{i=1}^M \langle w_i, D^*R(t)z(t)\rangle_U Dw_i = 0, \\ &z(0) = z_0 \equiv y_0 - y_\pi(0). \end{aligned} \tag{2.209}$$

Multiplying (2.209) by Rz, we obtain

$$\begin{aligned} &(z'(t), R(t)z(t)) + (Az(t), R(t)z(t)) + (A_0(t)z(t), R(t)z(t)) \\ &\quad + \sum_{i=1}^N \langle w_i, D^*R(t)z(t)\rangle_U^2 = -(F(t,z(t), R(t)z(t))). \end{aligned} \tag{2.210}$$

Note that

$$\frac{d}{dt}(R(t)z(t), z(t)) = (R'(t)z(t), z(t)) + 2(R(t)z(t), z'(t)), \quad \text{a.e., } t > 0.$$

Then, by (2.186), (2.210), we see that

$$\begin{aligned} &\frac{d}{dt}(R(t)z(t), z(t)) + |A^{\frac{3}{4}}z(t)|^2 + \sum_{i=1}^N \langle w_i, D^*R(t)z(t)\rangle_U^2 \\ &\quad = -2(F(t,z(t)), R(t)z(t)), \end{aligned} \tag{2.211}$$

for all t in the interval of existence $(0, T_0)$ of $z(t)$.

By (2.150), (2.151), (2.152), (2.184), (2.185) and (2.208), we get via interpolation that, for $\alpha = \frac{5}{8}$,

$$\begin{aligned}
2|(F(t,z), R(t)z)| &\le 2|F(t,z)|\,|R(t)z| \le C|z|_\alpha^2(1+|z|_{\frac{1}{4}})|z|_{\frac{1}{2}} \\
&\le C|z|_{\frac{3}{4}}^{4\alpha-\frac{1}{2}}(1+|z|_{\frac{1}{4}})|z|_{\frac{1}{4}}^{\frac{7}{2}-4\alpha} = C|z|_{\frac{3}{4}}^2|z|_{\frac{1}{4}}(1+|z|_{\frac{1}{4}}) \\
&\le C\gamma_1^{-\frac{1}{2}}(R(t)z,z)^{\frac{1}{2}}(1+C\gamma_1^{-\frac{1}{2}}(R(t)z,z)^{\frac{1}{2}})|z|_{\frac{3}{4}}^2, \\
&\quad \forall t > 0,\ z \in D(A^{\frac{3}{4}}).
\end{aligned} \tag{2.212}$$

We set

$$\mathscr{U}_t = \left\{ z \in W;\ (R(t)z,z)^{\frac{1}{2}}(1+C\gamma_1^{-\frac{1}{2}}(R(t)z,z)^{\frac{1}{2}}) < \frac{\gamma_1^{\frac{1}{2}}}{2C} \right\}.$$

Equivalently,

$$\mathscr{U}_t = \{z \in W;\ (R(t)z,z) \le \eta^2(\gamma_1)\},$$

where $\eta(\gamma_1)$ is the real positive solution to equa tion $2C\lambda(1+C\gamma_1^{-\frac{1}{2}}\lambda) = \gamma_1^{\frac{1}{2}}$.

We see that, for all $t \ge 0$, we have for θ_1, θ_2 appropriately chosen,

$$\{z \in Q;\ \|z\|_W < \theta_1\} \subset \mathscr{U}_t \subset \{z \in Q;\ \|z\|_W < \theta_2\}. \tag{2.213}$$

Choose $\|z_0\|_W < \theta_1$ and consider the maximal interval $(0, T_1)$ with the property that $z(t) \in \mathscr{U}_t$, $\forall t \in (0, T_1)$. By (2.212) and (2.184), we see that $T_1 = +\infty$, that is, the solution $z(t)$ exists globally and $z(t) \in \mathscr{U}_t$, $\forall t \ge 0$. Moreover, it follows that

$$\|z(t)\|_W \le Ce^{-\delta t}\|z_0\|_W, \quad \forall t \ge 0$$

and

$$\int_0^\infty |A^{\frac{3}{4}}z(t)|^2 dt \le C\|z_0\|_W^2.$$

This completes the proof of Theorem 2.12. □

We shall briefly discuss some semilinear time-periodic parabolic equations which can be treated as special cases of Theorem 2.12. Throughout in the sequel, $\mathscr{O}$ is a bounded, open domain of R^d with smooth boundary $\partial\mathscr{O}$.

Example 2.1 Consider the controlled semilinear parabolic equation:

$$\begin{cases}
y_t(x,t) - \Delta y(x,t) + f_1(x,t,y(x,t)) + f_2(x,t)\cdot\nabla y(x,t) \\
\quad = m(x)u(x,t), \quad x \in \mathscr{O},\ t \in R^+, \\
y(x,t) = 0, \quad \forall (x,t) \in \partial\mathscr{O} \times R^+.
\end{cases} \tag{2.214}$$

Here $m = \mathbf{1}_{\mathcal{O}_0}$ is as above the characteristic function of an open domain $\mathcal{O}_0 \subset \mathcal{O} \subset R^3$ and $f_1 : \mathcal{O} \times R \times R \to R$, $f_2 : \mathcal{O} \times R \to R^3$ are given continuous functions which are T-periodic in t. More precisely, we assume that

(ℓ) *f_i, $i = 1, 2$, are analytic in (t, y) and*

$$|(f_1)_y(x,t,y) - (f_1)_y(x,s,z)| \le C_1(|y-z| + |t-s|)(|y|^p + |z|^p), \quad (2.215)$$

where $0 \le p \le \frac{5}{8}$.

Indeed, taking into account that, by the Sobolev imbedding theorem, $D(A^\alpha) \subset L^q(\mathcal{O})$ for $q < 2d/(d-4\alpha)$, $d \ge 2$, we see that Assumption (kk) holds with $\alpha = \frac{5}{8}$, $A = -\Delta$, $D(A) = H_0^1(\mathcal{O}) \cap H^2(\mathcal{O})$ and

$$B(t,y)(x) = f_1(x,t,y(x)) + f_2(x,t) \cdot \nabla y(x),$$
$$Du = mu, \quad \forall u \in U = L^2(\mathcal{O}).$$

Let y_π be a T-periodic solution to (2.214), that is,

$$\begin{aligned} &(y_\pi)_t - \Delta y_\pi + f_1(x,t,y_\pi) + f_2(x,t) \cdot \nabla y_\pi = 0 \quad \text{in } \mathcal{O} \times R, \\ &y_\pi = 0 \quad \text{on } \partial\mathcal{O} \times R, \\ &y_\pi(x,t+T) = y_\pi(x,t), \quad \forall (x,t \in \mathcal{O} \times R. \end{aligned} \quad (2.216)$$

Assuming that $t \to y_\pi(t)$ is analytic as function with values in $H_0^1(\mathcal{O}) \cap H^2(\mathcal{O})$, it follows that Assumption (A1)$'$ is satisfied. Indeed, if z and ψ_i, $i = 1, \dots, N$, are solutions to the equation

$$\begin{aligned} &z_t + \Delta z + (f_1)_y z + \operatorname{div}(f_2 z) = 0, \quad && \text{in } (0,T) \times \mathcal{O}, \\ &z = 0 && \text{on } (0,T) \times \partial\mathcal{O}, \end{aligned} \quad (2.217)$$

and satisfy, for some $\lambda, \lambda_i^* \in \mathbb{C}$,

$$\begin{aligned} &z(0,x) = \lambda z(T,x) + \sum_{i=1}^N \psi_i(0,x), \\ &\psi_i(0,x) = \lambda_i^* \psi_i(T,x), \quad i = 1, \dots, N, \end{aligned} \quad (2.218)$$

one must show that $z(T,x) = 0$ on $\mathcal{O}_0$ implies $z \equiv 0$.

We set $\psi = \sum_{i=1}^N \psi_i$ and $\zeta = z - \psi$. By periodicity, we extend ζ as solution to (2.217) on $(-T,0) \times \mathcal{O}$. By Assumption ($\ell$), it follows that ζ is analytic in t on some interval $(-\delta,\delta)$ and so

$$\zeta(t,x) = \sum_{k=0}^\infty \frac{1}{k!} \zeta_x^{(k)}(0,x) t^k, \quad \forall x \in \mathcal{O}, \ -\delta < t < \delta.$$

Since $\zeta(0,x) = 0$ on $\mathcal{O}_0$, we infer that $\zeta(t,x) = 0$ on $(-\delta,\delta) \times \mathcal{O}_0$ and so, by the unique continuation property of solutions to linear parabolic equations, we have that $\zeta \equiv 0$. Then, by (2.218), we see that $z(T,x)=0$, $\forall x \in \mathcal{O}$, and by (2.217) we conclude that $z \equiv 0$, as claimed.

Then we may apply Theorem 2.12 and conclude as follows.

Corollary 2.5 *There is a system of functions* $\{w_j\}_{j=1}^N \subset L^2(\mathcal{O})$ *such that the feedback controller*

$$u(x,t) = -\sum_{i=1}^N w_i(x) \int_{\mathcal{O}} R(t)(y(x,t) - y_\pi(x,t))dx$$

exponentially stabilizes the periodic solution y_π *in a neighborhood*

$$\mathcal{U} = \{y_0 \in W;\ \|y_0 - y_\pi(0)\|_W < \rho\}.$$

Here $R(t) : D(A^{\frac{1}{4}}) \to D((A^{\frac{1}{4}})')$ is the periodic solution to Riccati equation (2.186).

We recall that (see, e.g., [60], p. 186)

$$W = D(A^{\frac{1}{4}}) = H_{00}^{\frac{1}{2}}(\mathcal{O}) = \{y \in H^{\frac{1}{2}}(\mathcal{O});\ f(x)(\text{dist}(x,\partial\mathcal{O}))^{-\frac{1}{2}} \in L^2(\mathcal{O})\}.$$

Example 2.2 The reaction-diffusion controlled system ("Belousov–Zhabotinski" system)

$$\begin{cases} y_t - \Delta y - y(1-y-az) - bz = m(x)u + f_1(t) & \text{in } \mathcal{O} \times R, \\ z_t - \Delta z + cyz + dz = m(x)v + f_2(t) & \text{on } \mathcal{O} \times R, \\ \frac{\partial y}{\partial n} = 0, \quad \frac{\partial z}{\partial n} = 0 & \text{on } \partial\mathcal{O} \times R, \end{cases} \tag{2.219}$$

where a,b,c and d are positive constants, f_1, f_2 and T periodic functions, and $m = \mathbf{1}_{\mathcal{O}_0}$ is relevant in the theory of chemical reactions.

If (y_π, z_π) is a nontrivial T-periodic solution to the ordinary differential system

$$\begin{cases} y_\pi' - y_\pi(1 - y_\pi - az_\pi) - bz_\pi = f_1(t), & t \in R, \\ z_\pi' + cy_\pi z_\pi + dz_\pi = f_2(t), \\ y_\pi(0) = y_\pi(T), \qquad z_\pi(0) = z_\pi(T), \end{cases} \tag{2.220}$$

then (2.219) has $y \equiv y_\pi$, $z = z_\pi$ as a periodic solution with period T.

We consider the matrix $C(t)$ associated with the linearization of System (2.220) around $\{y_\pi, z_\pi\}$ and recall that this solution is asymptotically stable if the Floquet exponents associated with the monodromy matrix

$$\Phi = Y(T), \qquad Y'(t) = C(t)Y(t), \qquad Y(0) = I,$$

are in the left complex half plane, otherwise it might be asymptotically unstable. However, by Theorem 2.12 this periodic solution to (2.219) is stabilizable by internal controllers with support in $\mathcal{O}_0$.

Indeed, we may apply here Theorem 2.12, where $H=(L^2(\mathcal{O}))^2$ and

$$A(y,z)=\begin{pmatrix}-\Delta y\\ -\Delta z\end{pmatrix},$$

$$D(A)=\left\{(y,z)\in(H^2(\mathcal{O}))^2;\ \frac{\partial y}{\partial n}=0,\ \frac{\partial z}{\partial n}=0 \text{ on } \partial\mathcal{O}\right\},$$

$$B(y,z)=\begin{pmatrix}-y(1-y-az)-bz\\ cyz+dz\end{pmatrix},$$

$$D\begin{pmatrix}u\\ v\end{pmatrix}=\begin{pmatrix}mu\\ mv\end{pmatrix}.$$

Since Assumptions (k)∼(kkk) are obviously satisfied, we check (A1)′ only.

Let $(\bar{y},\bar{z})$ and $(\bar{y}_1,\bar{z}_1)$ satisfy the system

$$\begin{aligned}
&\bar{z}_t+\Delta\bar{y}+(1-y_\pi-az_\pi)\bar{y}-(cy_\pi+d)\bar{z}=0 \quad \text{in } \mathcal{O}\times(0,T),\\
&\bar{z}_t+\Delta\bar{z}-cz_\pi\bar{y}+(b-ay_\pi)\bar{z}=0,\\
&\frac{\partial}{\partial n}\bar{y}=0,\quad \frac{\partial\bar{z}}{\partial n}=0 \quad \text{on } \partial\mathcal{O}\times(0,T),\\
&\bar{y}(x,0)=\lambda\bar{y}(x,T)+\bar{y}_1(x,0),\quad \bar{z}(x,0)=\lambda\bar{z}(x,T)+\bar{z}_1(x),\quad \forall x\in\mathcal{O}.
\end{aligned}$$

Assuming that (y_π,z_π) are analytic, then, arguing as above, if $\bar{y}(x,T)=\bar{z}(x,T)\equiv 0$, it follows via the unique continuation property of solutions to linear parabolic systems that $\bar{y}\equiv 0$, $\bar{z}\equiv 0$ on $\mathcal{O}\times(0,T)$, as desired.

Then, by Theorem 2.12 there is a feedback controller

$$\begin{aligned}
u(x,t)&=\sum_{i=1}^N w_i(x)\int_{\mathcal{O}_0}(R_{11}(t)(y(x,t)-y_\pi(t))\\
&\quad+R_{12}(t)(z(x,t)-z_\pi(x,t)))w_i(x)dx,\\
v(x,t)&=\sum_{i=1}^N w_i(x)\int_{\mathcal{O}_0}(R_{12}(t)(y(x,t)-y_\pi(t))\\
&\quad+R_{22}(t)(z(x,t)-z_\pi(x,t)))w_i(x)dx,
\end{aligned}$$

which exponentially stabilizes the periodic solution (y_π,z_π).

Here,

$$R(t)=\begin{Vmatrix}R_{11}(t) & R_{12}(t)\\ R_{12}(t) & R_{22}(t)\end{Vmatrix}$$

is the T-periodic solution to the corresponding Riccati equation (2.186).

2.7 Comments to Chap. 2

Most of the results in this chapter are new and appear for the first time in this general form. However, in some particular cases these results were established earlier. For instance, Theorems 2.1 and 2.2 were previously established in the special case of Navier–Stokes equations in [26]. (See also [12].) The results of Sect. 2.4 and, in particular, Theorem 2.7 were first established for the linearized Navier–Stokes equations in [14], but the treatment extended *mutatis mutandis* to the present general case. The results of Sect. 2.5 are new in this general framework, but are straightforward extensions to similar results established firstly for Navier–Stokes equations [26, 27] or for nonlinear parabolic equations [28]. The results of Sect. 2.6 are taken from [29]. As regards the stabilization by noise, which is new in this context, it is developed in Chap. 4 for systems governed by Navier–Stokes equations.

Chapter 3
Stabilization of Navier–Stokes Flows

In this chapter we discuss the feedback stabilization of stationary (equilibrium) solutions to Navier–Stokes equations. The design of a robust stabilizing feedback control is the principal way to suppress instabilities and turbulence occurring in the dynamics of the fluid and we treat this problem in the case of internal and boundary controllers. The first case, already presented in an abstract setting in Chap. 2, is that in which the controller is distributed in a spatial domain $\mathcal{O}$ and has compact support taken arbitrarily small. The second case is that where the controller is concentrated on the boundary $\partial\mathcal{O}$. In both cases, we design a stabilizable feedback linear controller which is robust and has a finite-dimensional structure, that is, it is a linear combination of eigenfunctions for the corresponding linearized systems. From the control theory point of view, this means that the actuation, though infinite-dimensional, is confined to an arbitrary subdomain $\mathcal{O}_0$ or to the boundary.

3.1 The Navier–Stokes Equations of Incompressible Fluid Flows

The dynamics of an incompressible and homogeneous fluid in a bounded domain $\mathcal{O} \subset R^d$, $d = 2, 3$, is governed by the dimensionless Navier–Stokes equations

$$
\begin{aligned}
&\frac{\partial y}{\partial t}(t,x) - \nu\Delta y(t,x) + (y\cdot\nabla)y(t,x) = \nabla p(t,x) + f(t,x), \quad t \geq 0,\ x \in \mathcal{O},\\
&(\nabla\cdot y)(t,x) \equiv 0, \qquad\qquad (3.1)\\
&y(t,x) = 0 \quad \text{in } (0,\infty)\times\partial\mathcal{O},\\
&y(0,x) = y_0(x) \quad \text{in } \mathcal{O}.
\end{aligned}
$$

Here, $y = (y_1, y_2, \dots, y_d)$ is the velocity field, $p = p(t,x)$ is the pressure, $f = (f_1, \dots, f_d)$ is the external force, ν is the kinematic viscosity and y_0 is the initial

V. Barbu, *Stabilization of Navier–Stokes Flows*,
Communications and Control Engineering,
DOI 10.1007/978-0-85729-043-4_3, © Springer-Verlag London Limited 2011

distribution of velocity field. The boundary $\partial\mathscr{O}$ is assumed smooth (of class C^2, for instance). Everywhere in this chapter, $f \equiv f_e(x)$, where $f_e \in (L^2(\mathscr{O}))^d$.

The notation is that introduced in Sect. 1.5 and recall (see (1.57)) that we may rewrite (3.1) as

$$\begin{aligned} &\frac{dy}{dt} + \nu A y + S y = f_e, \quad t \geq 0, \\ &y(0) = y_0 \end{aligned} \tag{3.2}$$

in the space

$$H = \{y \in (L^2(\mathscr{O}))^d;\ \nabla \cdot y = 0,\ y \cdot n = 0 \text{ on } \partial\mathscr{O}\}, \tag{3.3}$$

where

$$Ay = -P\Delta y, \quad \forall y \in D(A) = (H_0^1(\mathscr{O}) \cap H^2(\mathscr{O}))^d \cap H. \tag{3.4}$$

P is the Leray projector and $S = P(y \cdot \nabla)y$ (see (1.51), (1.53), (1.54)).

For $0 < \alpha < 1$, we denote by A^α the fractional power of order α of the operator A. The space $D(A^\alpha)$ is endowed with the Hilbertian norm

$$|y|_\alpha = |A^\alpha y|, \quad \forall y \in D(A^\alpha).$$

In particular, $D(A^{\frac{1}{2}}) = V = (H_0^1(\mathscr{O}))^d \cap H$ and (see [22], p. 10)

$$\begin{aligned} &D(A^s) = (H_0^{2s}(\mathscr{O}))^d \cap H, \quad \text{for } 0 < s < \frac{1}{2}, \\ &D(A^s) = (H^{2s}(\mathscr{O}))^d \cap V, \quad \frac{1}{2} \leq s \leq 1. \end{aligned} \tag{3.5}$$

We denote by $|\cdot|$ the norm of H, and by $(\cdot,\cdot)$ its scalar product.

By definition, an *equilibrium* (stationary) solution to (3.1) is a solution $y_e \in D(A)$ to the steady-state equation

$$\begin{aligned} &-\nu\Delta y_e + (y_e \cdot \nabla)y_e = \nabla p + f_e && \text{in } \mathscr{O}, \\ &\nabla \cdot y_e = 0 && \text{in } \mathscr{O}, \\ &y_e = 0 && \text{on } \partial\mathscr{O}. \end{aligned}$$

Equivalently,

$$\nu A y_e + S y_e = f_e. \tag{3.6}$$

The existence, regularity and uniqueness of y_e as well as the dimension of the set $\{y_e\}$ was discussed in detail in literature. (See, e.g., [41, 73, 74].) Everywhere in the following, we assume that y_e is sufficiently smooth, for instance, $y_e \in (W^{2,\infty}(\mathscr{O}))^d$. The problem we address here is the stabilization of y_e by a feedback controller with support in an arbitrary open subset $\mathcal{O}_0 \subset \mathcal{O}$ or concentrated on the boundary $\partial\mathcal{O}$. In fact, it is well-known that for large Reynolds numbers $Re = \frac{1}{\nu}$ such a solution is unstable and so, its stabilization is a major problem of fluid dynamics. It is convenient to reduce the stabilization problem for y_e to that of zero solution by setting $y - y_e \Rightarrow y$ and so, to transform (3.1) into

$$\begin{aligned}
&\frac{\partial y}{\partial t} - \nu\Delta y + (y_e \cdot \nabla)y + (y \cdot \nabla)y_e + (y \cdot \nabla)y = \nabla p && \text{in } (0,\infty) \times \mathcal{O},\\
&\nabla \cdot y = 0 && \text{in } (0,\infty) \times \mathcal{O},\\
&y = 0 && \text{on } (0,\infty) \times \partial\mathcal{O},\\
&y(0,x) = y_0(x) - y_e(x), && x \in \mathcal{O}.
\end{aligned} \tag{3.7}$$

Equivalently,

$$\begin{aligned}
&\frac{dy}{dt} + \mathcal{A}y + Sy = 0 \quad \text{in } (0,\infty),\\
&y(0) = y_0 - y_e = y^0,
\end{aligned} \tag{3.8}$$

where

$$\begin{aligned}
&\mathcal{A}y = \nu Ay + A_0 y, \quad \forall y \in D(\mathcal{A}),\\
&D(\mathcal{A}) = D(A), \ A_0 y = P((y_e \cdot \nabla)y + (y \cdot \nabla)y_e).
\end{aligned} \tag{3.9}$$

The operator $\mathcal{A}$ is called the *Stokes–Oseen operator* associated with the equilibrium solution y_e.

As mentioned earlier, the main technique used to suppress instability in System (3.7) and, implicitly, the turbulence, is to design a stabilizing feedback controller u of the form

$$u(t) = Fy(t), \quad t \geq 0,$$

and insert it into the controlled system (3.7), that is,

$$\begin{aligned}
&\frac{\partial y}{\partial t} - \nu\Delta y + (y_e \cdot \nabla)y + (y \cdot \nabla)y_e + (y \cdot \nabla)y = mu + \nabla p && \text{in } (0,\infty) \times \mathcal{O},\\
&\nabla \cdot y = 0 && \text{in } (0,\infty) \times \mathcal{O},\\
&y = 0 && \text{on } (0,\infty) \times \partial\mathcal{O},\\
&y(0,x) = y^0(x) && \text{in } \mathcal{O}.
\end{aligned} \tag{3.10}$$

Here, $m = 1_{\mathcal{O}_0}$ is the characteristic function of an open domain $\mathcal{O}_0 \subset \mathcal{O}$. This means that the input $mu = mFy$ has the support in $(0,\infty) \times \mathcal{O}_0$. Such a controller is called an *internal controller* and the design of a stabilizable internal controller mu with support in an arbitrary subset $\mathcal{O}_0$ of $\mathcal{O}$ is a major objective in the following and is treated in Sect. 3.3.

A related problem, called *boundary stabilization* is that of designing a stabilizable feedback controller u with support on the boundary $\partial\mathscr{O}$, that is for the controlled system

$$\begin{aligned}
&\frac{\partial y}{\partial t} - \nu\Delta y + (y_e\cdot\nabla)y + (y\cdot\nabla)y_e + (y\cdot\nabla)y = \nabla p && \text{in } (0,\infty)\times\mathscr{O},\\
&\nabla\cdot y = 0 && \text{in } (0,\infty)\times\mathscr{O},\\
&y = u && \text{on } (0,\infty)\times\partial\mathscr{O},\\
&y\cdot n = 0 && \text{on } (0,\infty)\times\partial\mathscr{O},\\
&y(0,x) = y^0(x) && \forall x\in\mathscr{O}.
\end{aligned} \tag{3.11}$$

We have chosen Dirichlet boundary conditions here because it is a general practice in the literature on boundary stabilization of PDEs to use Dirichlet boundary values for actuation. However, in principle, one might consider other boundary value conditions as well.

Apparently, the internal stabilization which implies internal actuation is of little interest and unrealistic for practical implementation of stabilizing controllers. However, it is an important step toward a better understanding of stabilizing mechanisms and, in particular, for that of boundary stabilization. In fact, as seen in Sect. 2.3, at the level of mathematical formalism, internal and boundary control systems are represented by the same equation and so, at least formally, internal stabilizations are equivalent with boundary stabilization. On the other hand, as shown below, the boundary stabilization can be reduced to internal stabilization with support in a neighborhood of the boundary in a larger domain. The stabilizing problem for (3.10), (3.11) can be treated in the functional framework developed in Chap. 2. These topics are discussed in detail in Sects. 3.3 and 3.4.

It should be mentioned that instead of Dirichlet boundary conditions we can take in Navier–Stokes System (3.1) periodic boundary conditions in R^d of the form

$$y(t, x+\ell e) \equiv y(t,x), \quad \forall x\in R^d,$$

where $\ell\in\mathbb{N}$, $x=(x_1,\dots,x_d)$ and $e=(e_1,e_2,\dots,e_d)$ is the unitary vector. As noticed earlier, in this case System (3.1) can be also written as (3.8), where H is the space of periodic free divergence vectors. In this context, the periodic Navier–Stokes flow in a 2-D channel is of special interest and is briefly presented below. More specifically, we consider a laminar flow in a two-dimensional channel with the walls located at $y=0,1$ and assume that the velocity field $(u(t,x,y), v(t,x,y))$ and the pressure $p(t,x,y)$ are 2π periodic in $x\in(-\infty,\infty)$.

The dynamics of flow is governed by the incompressible 2-D Navier–Stokes equations

$$
\begin{aligned}
&u_t - \nu\Delta u + uu_x + vu_y = p_x, \quad x \in R,\ y \in (0,1),\ t \geq 0,\\
&v_t - \nu\Delta v + uv_x + vv_y = p_y, \quad x \in R,\ y \in (0,1),\ t \geq 0,\\
&u_x + v_y = 0,\\
&u(t,x,0) = u(t,x,1) = 0, \quad v(t,x,0) = \psi(t,x), \quad v(t,x,1) = \varphi(t,x),\\
&\quad \forall x \in R,\ t \geq 0,\\
&u(t,x+2\pi,y) \equiv u(t,x,y), \quad v(t,x+2\pi,y) \equiv v(t,x,y), \quad y \in (0,1),\\
&\quad x \in R,\ t \geq 0,\\
&u(0,x,y) = u_0(x,y), \quad v(0,x,y) = v_0(x,y), \quad x \in R,\ y \in (0,1).
\end{aligned}
\tag{3.12}
$$

Consider a steady-state flow with zero vertical velocity component, that is, $(u(x,y),0)$. (This is the stationary laminary flow sustained by a pressure gradient in the x direction.) Since the flow is freely divergent, we have $U_x \equiv 0$ and so, $U(x,y) \equiv U(y)$. Alternatively, substituting into (3.12), gives

$$-\nu U'''(y) = p_x(x,y), \qquad p_y(x,y) \equiv 0.$$

Hence, $p \equiv p(x)$ and $U'' \equiv 0$. This yields

$$U(y) \equiv C(y^2 - y), \quad \forall y \in (0,1).$$

The linearization around steady-state flow $(U(y),0)$ leads to the following system

$$
\begin{aligned}
&u_t - \nu\Delta u + u_x U + vU' = p_x, \quad y \in (0,1),\ x,t \in R,\\
&v_t - \nu\Delta v + v_x U = p_y,\\
&u_x + v_y = 0,\\
&u(t,x,0) = u(t,x,1) = 0, \qquad v(t,x,0) = \psi(t,x), \qquad v(t,x,1) = \varphi(t,x),\\
&u(t,x+2\pi,y) = u(t,x,y), \qquad v(t,x+2\pi,y) = v(t,x,y).
\end{aligned}
\tag{3.13}
$$

The stabilization of (3.12) and (3.13) by normal boundary controllers $\{\psi(t,x), \varphi(t,x)\}$ is treated in Sect. 3.5.

3.2 The Spectral Properties of the Stokes–Oseen Operator

Consider the operator $\mathscr{A}$ defined by (3.9) and denote again by $\mathscr{A}$ its extension on the complexified space $\widetilde{H} = H + iH$. We denote by $|\cdot|_{\widetilde{H}}$ (or sometimes, simply, by $|\cdot|$) the norm in $\widetilde{H}$ and by the same symbol $(\cdot,\cdot)$ the scalar product in H and $\widetilde{H}$.

We begin with the following simple proposition on the spectrum of $\mathscr{A}$.

Proposition 3.1 *For* $\mathrm{Re}\,\lambda \leq -\alpha_0$, *where* α_0 *is sufficiently large, the operator* $\mathscr{A}$ *has compact resolvent* $(\lambda I - \mathscr{A})^{-1}$ *and* $-\mathscr{A}$ *is the infinitesimal generator of a* C_0-*analytic semigroup* $e^{-\mathscr{A}t}$ *on* $\widetilde{H}$.

Proof By (3.9) we have $\mathscr{A} = \nu A + A_0$, where $A_0 = P((y_e \cdot \nabla)y + (y \cdot \nabla)y_e)$. Since νA is self-adjoint and positive definite, while $|A_0 y| \le C|A^{\frac{1}{2}} y|$, $\forall y \in D(A)$, we infer that $-\mathscr{A}$ generates a C_0-analytic semigroup and (see Theorem 1.13)

$$\|(\lambda I - \mathscr{A})^{-1} f\|_V \le C|f|/|\lambda + \alpha_0|, \qquad \operatorname{Re}\lambda < -\alpha_0.$$

Taking into account that $(\lambda I - \mathscr{A})^{-1} f = z$ is the solution to the equation

$$\lambda z - \nu A z - A_0 z = f$$

and so, for $\operatorname{Re}\lambda \le -\alpha_0$ where α_0 is sufficiently large,

$$\|z\|^2_{(H^2(\mathscr{O})\cap H^1_0(\mathscr{O}))^d} \le C|f|^2_{\widetilde{H}},$$

we infer that the operator $\mathscr{A}$ has a compact resolvent $(\lambda I - \mathscr{A})^{-1}$. Consequently, $\mathscr{A}$ has a countable number of eigenvalues $\{\lambda_j\}_{j=1}^\infty$ in the complex half-space $\{\lambda \in \mathbb{C};\ \operatorname{Re}\lambda > -\alpha_0\}$ with corresponding eigenfunctions φ_j each with finite algebraic multiplicity m_j. In the following, each eigenvalue λ_j is repeated according to its algebraic multiplicity m_j.

Note also that for each γ there is a finite number of eigenvalues $\{\lambda_j\}_{j=1}^N$ with $\operatorname{Re}\lambda_j \le \gamma$ and that the spaces $X_u = \operatorname{lin\,span}\{\varphi_j\}_{j=1}^N = P_N\widetilde{H}$, $X_s = (I - P_N)\widetilde{H}$ are invariant with respect to $\mathscr{A}$. (Here, P_N is defined as in (1.2), that is

$$P_N = \frac{1}{2\pi i}\int_\Gamma (\lambda I - \mathscr{A})^{-1} d\lambda,$$

where Γ is a closed curve which contains in interior the eigenvalues $\{\lambda_j\}_{j=1}^N$.)

If set $\mathscr{A}_u = \mathscr{A}|_{X_u}$, $\mathscr{A}_s = \mathscr{A}|_{X_s}$, then we have

$$\sigma(\mathscr{A}_u) = \{\lambda_j;\ \operatorname{Re}\lambda_j \le \gamma\}, \qquad \sigma(\mathscr{A}_s) = \{\lambda_j;\ \operatorname{Re}\lambda_j > \gamma\}. \tag{3.14}$$

(Here, φ_j are the corresponding eigenfunctions of $\mathscr{A}$, that is, $\mathscr{A}\varphi_j = \lambda_j\varphi_j$ or $(\mathscr{A} - \lambda_j)^k\varphi_j = 0$, $k = 1, \dots, m_j$, if λ_j is not semisimple.) □

We recall that the eigenvalue λ_j is called semisimple if its algebraic multiplicity m_j coincides with its geometric multiplicity m_j^g (see Sect. 1.1). In particular, this happens if λ_j is simple. It turns out that the property of eigenvalues λ_j to be simple is generic. More precisely, we have (see Theorem 3.16).

Proposition 3.2 *The set*

$$\mathscr{M} = \{y_e \in D(A);\ \textit{all the eigenvalues } \lambda_j \textit{ are simple}\}$$

is a residual in the space $D(A)$, that is a countable intersection of open and dense sets.

Roughly speaking, this means that for "almost all" $y_e \in \mathscr{W}$ the eigenvalues λ_j of the Stokes–Oseen operator $\mathscr{A}$ are simple.

We denote by $\{\varphi_j^*\}$ the eigenfunctions to the dual operator $\mathscr{A}^*$ which is given by

$$\{\mathscr{A}^*\psi\}_k = P(-\nu\Delta\psi_k - (y_e)_i D_i\psi_k + \psi_i D_k(y_e)_i), \quad k = 1, 2, \dots, d.$$

Of course, the eigenvalues λ_j and $\overline{\lambda}_j$ have the same multiplicity and the corresponding eigenfunctions have the same properties.

We denote by P_N^* the dual of the projector P_N, that is,

$$P_N^* = \frac{1}{2\pi i}\int_\Gamma (\lambda I - \mathscr{A}^*)^{-1} d\lambda,$$

where Γ is a rectifiable contour which encircles the eigenvalues $\{\overline{\lambda}_j\}_{j=1}^N$ of $\mathscr{A}^*$.

Proposition 3.3 *Let $\mathscr{O}_0 \subset \mathscr{O}$ be an open subset of $\mathscr{O}$ and let $\{\varphi_j\}_{j=1}^N$, $\{\varphi_j^*\}_{j=1}^N$ be eigenfunction systems for $\mathscr{A}$ and $\mathscr{A}^*$. Then $\{\varphi_j\}_{j=1}^N$, $\{\varphi_j^*\}_{j=1}^N$ are linearly independent on $\mathscr{O}_0$. In particular, if $\varphi_j \equiv 0$ on $\mathscr{O}_0$ or $\varphi_j^* \equiv 0$ on $\mathscr{O}_0$, then $\varphi_j \equiv 0$, respectively, $\varphi_j^* \equiv 0$ on $\mathscr{O}$.*

The proof of Proposition 3.3 is given in Sect. 3.8 (see Theorems 3.14 and 3.15).

3.3 Internal Stabilization via Spectral Decomposition

This stabilization approach was already presented in Chap. 2 for the abstract parabolic-like systems. Here, we develop it on the special case of the controlled Navier–Stokes system

$$\begin{cases} y_t - \nu\Delta y + (y\cdot\nabla)y_e + (y_e\cdot\nabla)y + (y\cdot\nabla)y = \nabla p + mu & \text{in } (0,\infty)\times\mathscr{O}, \\ \nabla\cdot y = 0 & \text{in } (0,\infty)\times\mathscr{O}, \\ y = 0 & \text{on } (0,\infty)\times\partial\mathscr{O}, \\ y(0,x) = y_0(x), \end{cases} \tag{3.15}$$

or, equivalently (see (3.7), (3.8)),

$$\begin{aligned} &\frac{dy}{dt} + \mathscr{A}y + Sy = P(mu), \quad t \ge 0, \\ &y(0) = y_0, \end{aligned} \tag{3.16}$$

where $P : (L^2(\mathscr{O}))^d \to H$ is, as usually, the Leray projector and $m = \mathbf{1}_{\mathscr{O}_0}$ is the characteristic function of an open subset $\mathscr{O}_0 \subset \mathscr{O}$.

Everywhere in the following, $\{\varphi_j\}_{j=1}^N$ is a system of eigenfunctions for the operator $\mathscr{A}$, that is, $\varphi_j = \{\varphi_{jk}\}_{k=1}^{m_j}$, where

$$(\mathscr{A} - \lambda_j I)^k \varphi_{jk} = 0 \quad \text{for } k = 1, \ldots, m_j,\ j = 1, \ldots, \ell, \tag{3.17}$$

where ℓ is the number of distinct eigenvalues and m_k is the algebraic multiplicity of eigenvalue λ_k (repeated according its algebraic multiplicity). Here, N is chosen by the condition

$$\operatorname{Re} \lambda_j \le \gamma, \quad j = 1, \ldots, N, \tag{3.18}$$

and $\gamma > 0$ is arbitrary but fixed.

In the following, we denote by M the number

$$M = \max\{m_j;\ j = 1, 2, \ldots, \ell\}, \tag{3.19}$$

which has an important role in the stabilization procedure to be described.

3.3.1 The Internal Stabilization of the Stokes–Oseen System

The exponential global stabilization of the linearized controlled system associated with (3.1), that is,

$$\frac{dy}{dt} + \mathscr{A} y = u, \quad t \ge 0, \qquad y(0) = y_0, \tag{3.20}$$

where support $u(t) \subset \mathscr{O}_0$ is the first step toward the internal stabilization of Navier–Stokes equations. In fact, the linearized system arises from the equations describing the small perturbations of the flow and it captures most of the evolution of nonlinear dynamics. Theorem 3.1 is a first result in this direction.

Theorem 3.1 *There is a controller u of the form*

$$u(t) = \sum_{i=1}^{M} P(m\phi_i) v_i(t), \quad \forall t \ge 0, \tag{3.21}$$

which stabilizes exponentially System (3.20).

More precisely, the solution $y \in C([0, \infty); \widetilde{H})$ *to System* (3.20) *with control u given by* (3.21) *satisfies*

$$|y(t)|_{\widetilde{H}} \le C e^{-\gamma t} |y_0|, \quad t \ge 0. \tag{3.22}$$

The controller $v = \{v_i\}_{i=1}^M$ *can be chosen in* $L^2(0, T; \mathbb{C}^M)$ *and such that*

$$\int_0^T |v_i(t)|_M^2 dt \le C |y_0|^2, \quad v_i(t) = 0 \ \textit{for}\ t \ge T. \tag{3.23}$$

Here, $[0, T]$ *is an arbitrary interval and* $\{\phi_i\}_{i=1}^M \subset D(A)$ *is a system of functions which is made precise below.*

The controller $v = \{v_j\}_{j=1}^M$ *can also be found as a* C^1*-function on* $[0, \infty)$ *such that*

$$|v_j(t)| + |v_j'(t)| \le Ce^{-\gamma t}|y_0|, \quad \forall t \ge 0,\ j = 1, \dots, M. \tag{3.24}$$

We note that, in terms of the controller $v \in L^2(0, T; \mathbb{C}^M)$, the control system (3.20) can be, equivalently, written as

$$\begin{aligned} &\frac{dy}{dt} + \mathscr{A} y = Bv, \quad t \ge 0, \\ &y(0) = y_0, \end{aligned} \tag{3.20$'$}$$

where $B : \mathbb{C}^M \to \widetilde{H}$ is given by

$$Bv = \sum_{i=1}^M P(m\phi_i)v_i, \quad \forall v \in \mathbb{C}^M.$$

Proof First, we prove the theorem in the special case where

1° *the eigenvalues* λ_j, $j = 1, \dots, N$, *are semisimple.*

As seen earlier, in this case the systems $\{\varphi_j\}_{j=1}^N$ and $\{\varphi_j^*\}_{j=1}^N$ can be chosen biorthogonal, that is,

$$(\varphi_i, \varphi_j^*) = \delta_{ij}, \quad i, j = 1, \dots, N. \tag{3.25}$$

We take in (3.21)

$$\phi_i = \varphi_i^*, \quad i = 1, \dots, N, \tag{3.26}$$

and we are decoupling the system

$$\begin{aligned} &\frac{dy}{dt} + \mathscr{A} y = \sum_{i=1}^M P(m\varphi_i^*)v_i, \quad t \ge 0, \\ &y(0) = y_0, \end{aligned}$$

in the finite-dimensional part

$$\begin{aligned} &\frac{dy_u}{dt} + \mathscr{A}_u y_u = P_N \sum_{i=1}^M P(m\varphi_i^*)v_i, \\ &y_u(0) = P_N y_0, \end{aligned} \tag{3.27}$$

and the infinite-dimensional γ-stable part

$$\begin{aligned}
&\frac{dy_s}{dt} + \mathscr{A}_s y_s = (I - P_N) \sum_{i=1}^{M} P(m\varphi_i^*) v_i, \\
&y_s(0) = (I - P_N) y_0.
\end{aligned} \tag{3.28}$$

Here, $y_u = \sum_{i=1}^{N} y_i \varphi_i$ and $\mathscr{A}_u = \mathscr{A}|_{X_u}$, $\mathscr{A}_s = \mathscr{A}|_{X_s}$, $X_u = P_N(\widetilde{H}) = \text{lin span}\{\varphi_j\}_{j=1}^{N}$, $X_s = (I - P_N)\widetilde{H}$.

As a matter of fact, X_s can be, equivalently, defined as lin span $\{\varphi_j\}_{j=N+1}^{\infty}$. (See [22], Remark 1.1.)

By (3.25), System (3.27) can be rewritten as

$$\begin{aligned}
&\frac{dy_j}{dt} + \lambda_j y_j = \sum_{i=1}^{M} (\varphi_i^*, \varphi_j^*)_0 v_i, \quad j = 1, \dots, N, \\
&y_j(0) = y_j^0 = (P_N y_0, \varphi_j^*),
\end{aligned} \tag{3.29}$$

where $(\cdot, \cdot)_0$ is the scalar product in $(L^2(\mathscr{O}_0))^d$. This yields

$$\begin{aligned}
&z'(t) + \Lambda z(t) = \mathscr{B} v(t), \quad t \ge 0, \\
&z(0) = z_0,
\end{aligned} \tag{3.30}$$

where

$$z(t) = \{y_j(t)\}_{j=1}^{N}, \qquad z_0 = \{(y)_j^0\}_{j=1}^{N} \quad \text{and} \quad \mathscr{B} = \|(\varphi_i^*, \varphi_j^*)_0\|_{i,j=1}^{M,N}$$

while Λ is the diagonal $N \times N$ matrix

$$\Lambda = \begin{Vmatrix} J_1 & & & \\ & J_2 & & 0 \\ & 0 & \ddots & \\ & & & J_\ell \end{Vmatrix}$$

where J_j is the $m_j \times m_j$ diagonal matrix

$$J_j = \begin{Vmatrix} \lambda_j & & & \\ & \lambda_j & & 0 \\ 0 & & \ddots & \\ & & & \lambda_j \end{Vmatrix}, \quad j = 1, \dots, \ell.$$

(Recall that ℓ is the number of distinct eigenvalues λ_j with $\text{Re}\,\lambda_j \le \gamma$.)

Then, as argued in Lemma 2.1, it follows that System (3.30) is exactly null controllable on $(0, T)$, where $T > 0$ is arbitrary but fixed. Indeed, the equation

$$\mathscr{B}^* e^{-\Lambda t} x = 0, \quad t \ge 0,$$

implies that

$$\sum_{i=1}^{m_1} b_{ij}x_i = 0, \qquad \sum_{i=m_1+1}^{m_2} b_{ij}x_i = 0, \quad \ldots, \quad \sum_{i=m_{\ell-1}+1}^{m_\ell} b_{ij}x_i = 0, \tag{3.31}$$

where $x = \{x_i\}_{i=1}^N$, $b_{ij} = (\varphi_j^*, \varphi_i^*)_0$, $i = 1, \ldots, N$, $j = 1, \ldots, M$.

Since the system $\{\varphi_j^*\}_{j=1}^M$ is linearly independent on $\mathcal{O}_0$ (Proposition 2.1), we have that $\operatorname{rank}\|b_{ij}\|_{i,j=1}^{N,M} = M$ and so, we conclude by (3.31) that $x = 0$, as claimed. Hence, by the Kalman controllability theorem there is a control input $\{v_j\}_{j=1}^M \subset C([0,T]; \mathbb{C}^M)$ such that $y_u(T) = 0$. Moreover, by the linear finite-dimensional controllability theory, we know that $\{v_j\}_{j=1}^M$ can be chosen in such a way that

$$\int_0^T |v_j(t)|^2 dt \le C|z_0|^2, \quad \forall t \in [0,T], \ j = 1, \ldots, M.$$

As seen earlier in Sect. 2.2 (Proof of Theorem 2.1), from the exact null controllability of (3.30) on $[0,T]$ it follows also, via the linear quadratic stabilization technique, the existence of a stabilizable controller $v_j \in C^1([0,\infty))$, $j = 1, \ldots, M$, satisfying (3.24).

Now, recalling that $\sigma(\mathcal{A}_s) = \{\lambda_j;\ \operatorname{Re}\lambda_j > \gamma\}$ and that $-\mathcal{A}_s$ generates a C_0-analytic semigroup, we have by Theorem 1.14 that

$$\|e^{-\mathcal{A}_s t}\|_{L(\tilde{H},\tilde{H})} \le Ce^{-\gamma t}, \quad \forall t \ge 0,$$

and so, by (3.28) we have

$$|y(t,y_0)_{\tilde{H}} \le C\left(e^{-\gamma t}|y_0| + \int_0^t \sum_{j=1}^M |e^{-\mathcal{A}_s(t-s)}P(m\varphi_j^*)v_j(s)|ds\right) \le Ce^{-\gamma t}|y_0|, \tag{3.32}$$

as claimed. This completes the proof. □

2° We consider now the general case of non-semisimple eigenvalues λ_j, $j = 1, \ldots, N$. By the Gram–Schmidt orthogonalization algorithm, we may replace $\{\varphi_j\}_{j=1}^N$ by an orthonormal system again denoted by $\{\varphi_j\}_{j=1}^N$. Then, we take in (3.20) the controller u again of the form (3.21), where $\{\phi_i\}_{i=1}^M$ is specified later on.

Setting as above $y_u = \sum_{i=1}^N y_i\varphi_i$, we have also in this case $y = y_u + y_s$, where

$$\begin{aligned} &\frac{dy_i}{dt} + \sum_{j=1}^M a_{ij}y_i = \sum_{j=1}^N (P(m\phi_j), P_N^*\varphi_i)v_j, \quad i = 1, \ldots, N, \\ &y_i(0) = y_i^0 = (y_0, \varphi_i), \end{aligned} \tag{3.33}$$

$$\frac{dy_s}{dt} + \mathscr{A}_s y_s = (I - P_N)\sum_{j=1}^{M} P(m\phi_j)v_j,$$
$$y_s(0) = (I - P_N)y_0. \tag{3.34}$$

As in the previous case, it suffices to show that the finite-dimensional system (3.33) is exactly null controllable on some interval $[0, T]$. Though in this case the matrix $A^0 = \|a_{ij}\|_{i,j=1}^N$ is not the Jordan matrix associated with $\{\lambda_j\}_{j=1}^N$ there is, however, a nonsingular matrix $\Lambda^0 = \|\gamma_{ij}\|_{i,j=1}^N$ such that $J = \Lambda^0 A^0 (\Lambda^0)^{-1}$ is. Then, System (3.33) can be written as

$$\frac{dz}{dt} + Jz = Dv, \quad t \geq 0,$$
$$z(0) = z_0, \tag{3.35}$$

where $z = \Lambda^0 y_u$ and $D = \Lambda^0 B$, $B = \|b_{ij}\|_{i,j=1}^{N,M}$, $b_{ij} = (P(m\phi_j), P_N^*\varphi_i^*) = (\phi_j, P_N^*\varphi_i^*)_0$.

We have the following lemma.

Lemma 3.1 *There is a system* $\{\phi_j\}_{j=1}^M$ *of the form*

$$\phi_j = \sum_{k=1}^{N} \alpha_{jk} P_N^* \varphi_k, \quad j = 1, \ldots, M, \tag{3.36}$$

such that the finite-dimensional system (3.33) (*equivalently,* (3.35)) *is exactly null controllable on each interval* $[0, T]$.

Proof If $M = N$, it suffices to take $\phi_j = P_N^*\varphi_j$, $j = 1, \ldots, N$. Indeed, since the system $\{P_N^*\varphi_j\}_{j=1}^N$ is linearly independent on $\mathcal{O}_0$ we have in this case that $\det B \neq 0$ and so $\det D \neq 0$, too. This, clearly, implies that System (3.35) is exactly null controllable on $[0, T]$.

Assume now that M given by (3.19) is less than N. The Jordan matrix J has, therefore, the following form

$$J = \{\lambda_1 E_{\tilde{m}_1} + H_{\tilde{m}_1}, \ldots, \lambda_\ell E_{\tilde{m}_\ell} + H_{\tilde{m}_\ell}\},$$

where $E_{\tilde{m}_i}$ is a unitary matrix of order $\tilde{m}_i \leq m_i$ and $H_{\tilde{m}_i}$ is a $\tilde{m}_i \times \tilde{m}_i$ matrix of the form

$$H_{\tilde{m}_i} = \left\| \begin{matrix} 0 & 1 & 0 & \cdots & 0 \\ 0 & 0 & 1 & \cdots & 0 \\ 0 & 0 & 0 & \cdots & 1 \\ 0 & 0 & 0 & \cdots & 0 \end{matrix} \right\|.$$

(Some of λ_j might be repeated.)

If we set $J_i = \lambda_i E_{\tilde{m}_i} + H_{\tilde{m}_i}$, then J is the matrix

$$J = \begin{Vmatrix} J_1 & & & \\ & J_2 & & 0 \\ & & & \\ 0 & & \ddots & \\ & & & J_\ell \end{Vmatrix}. \tag{3.37}$$

For simplicity, we assume that the blocks J_ℓ are distinct, that is, $\lambda_j \neq \lambda_i$ and $\tilde{m}_i = m_i$. (The general case follows in a similar way.)

We start with the equation

$$D^* e^{-J^* t} x = 0, \quad \forall t \geq 0, \tag{3.38}$$

where $*$ stands for adjoint operation.

Taking into account that

$$e^{-J^* t} x = \begin{Vmatrix} e^{-J_1^* t} x^1 \\ \cdots \\ e^{-J_\ell^* t} x^\ell \end{Vmatrix}^{\mathsf{T}},$$

where

$$x^1 = \begin{pmatrix} x_1 \\ \vdots \\ x_{m_1} \end{pmatrix}, \quad \ldots, \quad x^\ell = \begin{pmatrix} x_{m_\ell - 1} \\ \vdots \\ x_{m_\ell} \end{pmatrix}$$

and

$$D^* = B^* \Lambda^* = \left\| \sum_{k=1}^N b_{jk} \gamma_{ki} \right\|_{i,j=1}^{M,N} = \| d_{ij}^* \|_{i=1,j=1}^{M,N}. \tag{3.39}$$

We have

$$e^{-J_k^* t} x^k = \begin{Vmatrix} x_{m_{k-1}} \\ t x_{m_{k-1}} + x_{m_{k-2}} \\ \cdots \\ \frac{t^{m_{k-1}}}{(m_{k-1})} x_{m_{k-1}} + \cdots + x_{(m_{k-1}+m_k)} \end{Vmatrix}$$

for all $k = 1, 2, \ldots, \ell$. Then, by (3.38) we have that

$$\sum_{j=1}^{m_1} d_{kj}^* x_j = 0, \quad \sum_{j=m_1+1}^{m_1+m_2} d_{kj}^* x_j = 0, \quad \ldots, \quad \sum_{j=m_{\ell-1}}^{N} d_{kj}^* x_j = 0, \quad k = 1, \ldots, M. \tag{3.40}$$

Consider the matrices

$$D_i^* = \|d_{kj}^*\|_{k=1,j=m_i+1}^{M,m_i+m_{i+1}}, \quad i=1,2,\dots,\ell,$$

where (see (3.39))

$$d_{kj}^* = \sum_{r=1}^{N} b_{jr}\gamma_{ri}.$$

Then, in order to conclude by (3.40) that $x=0$, that is, System (3.35) is exactly null controllable, it suffices to take $\{\phi_j\}_{j=1}^M$ of the form (3.36) in such a way that

$$\text{rank } D_i^* = m_i \quad \text{for all } i=1,\dots,\ell. \tag{3.41}$$

□

Taking into account that $\Lambda^0 = \|\gamma_{ik}\|$ is not singular while $b_{ij} = (\phi_j, P_N^*\varphi_i)_0$, it is easily seen that, if α_{jk} are suitable chosen, for the system ϕ_j defined by (3.36), then Condition (3.41) holds. (Here, we use once again the fact that system $\{P_N^*\varphi_i\}_{i=1}^N$ is independent on $\mathscr{O}_0$.) This completes the proof of Theorem 3.1.

It should be noticed that Theorem 3.1 provides a minimal finite-dimensional controller u of the form (3.21) which stabilizes the Stokes–Oseen operator $\mathscr{A}$. This controller takes values in a finite-dimensional subspace of $\widetilde{H}$ related to eigenfunctions of $\mathscr{A}$ or $\mathscr{A}^*$ and, more precisely, in lin span$\{\varphi_j^*\}_{j=1}^N$ and, respectively, in lin span$\{P_N^*\varphi_j\}_{j=1}^N$.

3.3.2 *The Stabilization of Stokes–Oseen System by Proportional Feedback Controller*

This subsection reproduces the procedure presented in Sect. 2.2 to stabilize System (3.20) by linear feedback controller u of the form (see (2.56))

$$u(t) = -\eta\sum_{i=1}^{N}(y(t),\varphi_j^*)P(m\phi_j), \tag{3.42}$$

where $\{\phi_j\}_{j=1}^N \subset \widetilde{H}$ is a system of functions of the form

$$\phi_j = \sum_{k=1}\alpha_{kj}\varphi_k^*, \quad j=1,\dots,N, \tag{3.43}$$

satisfying the conditions

$$(\phi_i,\varphi_j^*)_0 = \delta_{ij}, \quad i,j=1,\dots,N. \tag{3.44}$$

(Here we are working under the hypothesis that all λ_j, $j = 1, \dots, N$, are semisimple.)

Since the system $\{\varphi_j^*\}$ is linearly independent on $\mathcal{O}_0$, System (3.43), (3.44) has a solution $\{\alpha_{kj}\}$ and so such $\{\phi_j\}_{j=1}^N$ exists. If plug u in System (3.20) and write the latter under the form (3.27), (3.28), we obtain as above that $y = y_u + y_s$, where $y_u = \sum_{i=1}^N y_i \varphi_i$, $y_s = (I - P_N)y$ and

$$\frac{dy_i}{dt} + (\lambda_i + \eta)y_i = 0, \quad i = 1, \dots, N,$$

$$\frac{dy_s}{dt} + \mathcal{A}_s y_s = -\eta \sum_{j=1}^N y_j (I - P_N) P(m\phi_j).$$

Then, arguing as in the proof of Theorem 2.3 (see also Theorem 3.1), we obtain the following proposition.

Proposition 3.4 *For* $\eta > \lambda_0 + \gamma$ *the feedback controller* (3.42) *stabilizes exponentially with exponent decay* $-\gamma$ *System* (3.20).

This result extends *mutatis mutandis* to the general case of non semisimple eigenvalues by the argument indicated in the proof of Theorem 2.4, but the details are omitted.

For future convenience, we reformulate Theorem 3.1 in the real space H. Taking into account that

$$y(t) = \operatorname{Re} y(t) + i \operatorname{Im} y(t), \qquad \phi_j = \operatorname{Re} \phi_j + i \operatorname{Im} \phi_j, \quad v_j = \operatorname{Re} v_j + i \operatorname{Im} v_j,$$
$$j = 1, \dots, M^*,$$

and recalling the discussion in Sect. 2.2, we infer by Theorem 3.1 that there is a controller $u^* : [0, \infty) \to H$ of the form

$$u^*(t) = \sum_{j=1}^{M^*} P(m \operatorname{Re} \phi_j) \operatorname{Re} v_j(t) - P(m \operatorname{Im} \phi_j) \operatorname{Im} v_j(t), \tag{3.45}$$

which exponentially stabilizes the real system

$$\begin{aligned} &\frac{dy}{dt} + \mathcal{A} y = u^*, \quad t \geq 0, \\ &y(0) = y_0. \end{aligned} \tag{3.46}$$

Here, $1 < M^* \leq N$ is given by (2.55) and

$$\phi_j = \varphi_j^*, \quad j = 1, \dots, M^*,$$

if the spectrum is semisimple, while

$$\phi_j \in \text{lin span}\{P_N^*\varphi_i\}_{i=1}^N, \quad j=1,\dots,M^*,$$

in the general case. (See (3.36).)

In particular, if all the eigenvalues λ_j, $j=1,\dots,N$, are real, then $M^*=M$, while, if all the eigenvalues λ_j are simple, then $M^*=2$. (It follows also that $M^*=1$ if all the eigenvalues are real and simple.)

Therefore, by Theorem 3.1 we have the following result.

Theorem 3.2 *There is a real controller* u^* *of the form*

$$u^*(t)=\sum_{j=1}^{M^*} P(m\psi_j)v_j^*(t) \tag{3.47}$$

such that the corresponding solution

$$y^* \in C([0,T;H)\cap L^2(0,T;V)\cap L^2_{\text{loc}}(0,T;D(A)), \quad \forall T>0, \tag{3.48}$$

to (3.46) *satisfies the estimate*

$$|y^*(t)|\le Ce^{-\gamma t}|y_0|, \quad \forall t\ge 0, \tag{3.49}$$

and

$$v_j\in C^1[0,\infty), \quad |v_j(t)|+|v_j'(t)|\le Ce^{-\gamma t}|y_0|, \quad j=1,\dots,M^*,\ t>0.$$

Here,

$$\{\psi_j\}_{j=1}^{M^*}\in (H^2(\mathcal{O}))^d\cap (H_0^1(\mathcal{O}))^d\cap H, \qquad \psi_j\in \text{lin span}\{P_N^*\varphi_k\}_{k=1}^N,$$
$$j=1,\dots,N.$$

If all λ_j *are semisimple, then*

$$\psi_j=\operatorname{Re}\varphi_j^* \ \ \text{for } 1\le j\le \frac{M^*}{2}, \qquad \psi_j=\operatorname{Im}\varphi_j^* \ \ \text{for } \frac{M^*}{2}<j<M^*.$$

Remark 3.1 We see that the structure of the stabilizing controller is quite simple if the spectrum $\{\lambda_j\}$ is semisimple and if all the eigenvalues $\{\lambda_j\}_{j=1}^N$ are simple, then the controller u^* is two-dimensional, that is,

$$u^*(t)=P(m\operatorname{Re}\varphi_1^*)v_1^*(t)+P(m\operatorname{Im}\varphi_1^*)v_2^*(t).$$

Taking into account that, by Proposition 1.1, the property of eigenvalues λ_j to be simple is a generic one, we may conclude that for "almost all" steady-state solutions y_e the stabilizable controller u^* is of the above form and this fact allows us to simplify the numerical construction of the controller.

3.3.3 Internal Stabilization via Feedback Controller; High-gain Riccati-based Feedback

The stabilizable controller found in Theorem 3.1 is an open-loop controller only, while that found in Theorem 3.2, though it is a feedback-stabilizable controller, it is not robust, however (see Remark 2.2).

Roughly speaking, Theorem 3.1 implies only that the linear controlled equation (3.20) is stabilizable by a finite-dimensional controller u with support in $[0, T]$ or by $C^1[0, \infty)$-exponential decaying controller. Starting from this preliminary result, our aim here is to obtain a stabilizable feedback controller via an infinite-dimensional algebraic Riccati equation associated with the linearized system (3.20). The model we follow here is essentially the same as that developed in Sect. 2.5 with some specific differences.

Everywhere in the following, $(\cdot,\cdot)$ is the scalar product in the space H or in a duality pair with H as pivot space. We set $V = D(A^{\frac{1}{2}})$, $W = D(A^{\frac{1}{4}})$ and denote by $\|\cdot\|$ the norm in V. By $(\cdot,\cdot)_0$ denote the scalar product in $(L^2(\mathcal{O}_0))^d$.

First, we prove a stabilization result for the linearized Stokes–Oseen system (3.20).

Theorem 3.3 *Let $\gamma > 0$ and M^* and N as in Theorems 3.1 or 3.2. Then there is a linear self-adjoint operator $R : D(R) \subset H \to H$ such that for some constants $0 < a_1 < a_2 < \infty$ and $C_1 > 0$,*

$$a_1|A^{\frac{1}{4}}y|^2 \le (Ry, y) \le a_2|A^{\frac{1}{4}}y|^2, \quad \forall y \in D(A^{\frac{1}{4}}); \tag{3.50}$$

$$|Ry| \le C_1\|y\|, \quad \forall y \in V; \tag{3.51}$$

$$(\nu Ay + A_0y - \gamma y, Ry) + \frac{1}{2}\sum_{i=1}^{M^*}(\psi_i, Ry)_0^2 = \frac{1}{2}|A^{\frac{3}{4}}y|^2, \quad \forall y \in D(A). \tag{3.52}$$

Moreover, the feedback controller

$$u^*(t) = -P\left(m\sum_{j=1}^{M^*}(Ry(t), \psi_i)_0\psi_i\right) \tag{3.53}$$

exponentially stabilizes the linear system (3.20), *that is, the solution y to the corresponding closed-loop system satisfies*

$$\|y(t)\|_W \le e^{-\gamma t}\|y_0\|_W, \quad \forall y_0 \in W, \tag{3.54}$$

$$\int_0^\infty e^{2\gamma t}|A^{\frac{3}{4}}y(t)|^2dt \le C\|y_0\|_W^2. \tag{3.55}$$

Here, $\{\psi_i\}_{i=1}^{M^*}$ are as in Theorem 3.2.

Proof The proof is similar to that of Proposition 2.2. We consider the optimization problem

$$\varphi(y_0) = \text{Min}\,\frac{1}{2}\int_0^\infty (|A^{\frac{3}{4}}y(t)|^2 + |u(t)|^2_{M^*})dt, \tag{3.56}$$

subject to $u \in L^2(0,\infty; R^{M^*})$ and

$$y' + \nu Ay + A_0 y - \gamma y = P\left(m\sum_{i=1}^{M^*}\psi_i u_i\right), \qquad y(0) = y_0. \tag{3.57}$$

Let us show first that $\varphi(y_0) < \infty$, $\forall y_0 \in D(A^{\frac{1}{4}})$. We set

$$Du = P\left(m\sum_{i=1}^{M^*}\psi_i u_i\right), \qquad u \in R^{M^*}.$$

(Here, $|\cdot|_{M^*}$ is the Euclidean norm in the space R^{M^*}.) By Theorem 3.2 there is an admissible pair (y,u) such that $y \in L^2(0,\infty; H) \cap L^2_{\text{loc}}(0,\infty; D(A))$. For such a pair, we have, by virtue of (3.57) (recall that, by (3.9), $(A_0 y, z) = b(y, y_e, z) + b(y_e, y, z)$, $\forall z \in V$)

$$\begin{aligned}\frac{1}{2}\frac{d}{dt}|y(t)|^2 + \nu\|y(t)\|^2 &\le |b(y, y_e, y)(t)| + |Du|\,|y(t)| + \gamma|y|^2\\ &\le C(|y(t)|\,\|y(t)\| + |y(t)|^2 + |u(t)|^2_{M^*})\\ &\le \frac{\nu}{2}\|y(t)\|^2 + C_1(|y(t)|^2 + |u(t)|^2_{M^*}), \quad \forall t > 0.\end{aligned}$$

Hence, $y \in L^2(0,\infty; V)$. Next, we multiply (3.57) by $A^{\frac{1}{2}}y(t)$ to obtain by Proposition 1.7 that

$$\begin{aligned}&\frac{1}{2}\frac{d}{dt}|A^{\frac{1}{4}}y(t)|^2 + \nu|A^{\frac{3}{4}}y(t)|^2\\ &\le |b(y(t), y_e, A^{\frac{1}{2}}y(t))| + |b(y_e, y(t), A^{\frac{1}{2}}y(t))| + |Du(t)|\,|A^{\frac{1}{2}}y(t)|\\ &\quad + \gamma|A^{\frac{1}{2}}y(t)|\,|y(t)|\\ &\le C(\|y(t)\|\,\|y_e\|_{\frac{3}{2}}|A^{\frac{1}{2}}y(t)| + \|y_e\|_2\|y(t)\|\,|A^{\frac{1}{2}}y(t)| + |u(t)|_{M^*}|A^{\frac{1}{2}}y(t)|\\ &\quad + \gamma|A^{\frac{1}{2}}y(t)|\,|y(t)|)\\ &\le C(\|y(t)\|^2 + |u(t)|_{M^*}\|y(t)\|), \quad t > 0.\end{aligned}$$

Integrating on $(0,\infty)$, we see that $\varphi(y_0) < \infty$, $\forall y_0 \in D(A^{\frac{1}{4}})$. Moreover, it follows from the previous equality that

$$\alpha_1|A^{\frac{1}{4}}y_0|^2 \le \varphi(y_0) \le \alpha_2|A^{\frac{1}{4}}y_0|^2, \quad \forall y_0 \in D(A^{\frac{1}{4}}),$$

where $\alpha_i > 0,\ i = 1, 2$. Thus, there is a linear self-adjoint operator $R : D(R) \subset H \to H$ such that $R \in L(W, W')$ and

$$\varphi(y_0) = \frac{1}{2}(Ry_0, y_0), \quad \forall y_0 \in W = D(A^{\frac{1}{4}}).$$

In other words, $\nabla\varphi = R$ and (3.50) follows.

Let us prove (3.51) and (3.52). By the dynamic programming principle, for each $T > 0$, the solution (u^*, y^*) to (3.56) is also the solution to the optimization problem

$$\text{Min}\left\{\frac{1}{2}\int_0^T (|A^{\frac{3}{4}}y(s)|^2 + |u(s)|^2_{M^*})ds + \varphi(y(T)), \text{ subject to (3.57)}\right\},$$

and so, by the maximum principle,

$$u^*(t) = \{(q_T(t), \psi_i)_0\}_{i=1}^{M^*}, \quad \text{a.e., } t \in (0, T), \tag{3.58}$$

where q_T is the solution to the dual backward equation

$$\begin{aligned} q_T' - (\nu A + A_0)^* q_T + \gamma q_T &= A^{\frac{3}{2}} y^*, \quad \forall t \in (0, T), \\ q_T(T) &= -Ry^*(T). \end{aligned} \tag{3.59}$$

Since T is arbitrary, we have

$$Ry^*(t) = -q_T(t), \quad \forall t > 0, \tag{3.60}$$

and, therefore,

$$u^*(t) = -\{(Ry^*(t), \psi_i)_0\}_{i=1}^{M^*}, \quad \forall t \geq 0. \tag{3.61}$$

Now, let $y_0 \in V$ be arbitrary but fixed. By (3.57), multiplying by Ay^*, we have as above that

$$\begin{aligned} &\frac{d}{dt}\|y^*(t)\|^2 + 2\nu|Ay^*(t)|^2 \\ &\quad \leq 2(|b(y_e, y^*(t), Ay^*(t))| + |b(y^*(t), y_e, Ay^*(t))|) \\ &\qquad + 2|Du^*(t)|\,|Ay^*(t)| + \gamma\|y^*(t)\|^2 \\ &\quad \leq C(\|y_e\|_2\|y^*(t)\|\,|Ay^*(t)| + |y^*(t)|_{\frac{1}{2}}\|y_e\|_2|Ay^*(t)| \\ &\qquad + |Du^*(t)|\,|Ay^*(t)| + \gamma\|y^*(t)\|^2). \end{aligned}$$

This implies that

$$\|y^*(t)\|^2 + \int_0^t |Ay^*(s)|^2 ds \leq C(1 + \|y_0\|^2), \quad \forall t \geq 0. \tag{3.62}$$

On the other hand, by (3.59), we see that $z = A^{-\frac{1}{2}} q_T$ satisfies the equation

$$\frac{dz}{dt} - \nu A z - A^{-\frac{1}{2}} A_0^* A^{\frac{1}{2}} z + \gamma z = A y^*, \quad \text{a.e., } t > 0,$$

and this yields (by multiplying with Az)

$$\begin{aligned}\frac{1}{2}\frac{d}{dt}\|z(t)\|^2 &\geq \nu|Az(t)|^2 - |b(y_e, A^{\frac{1}{2}} z(t), A^{\frac{1}{2}} z(t))| - \gamma\|z(t)\|^2\\ &\quad - |b(A^{\frac{1}{2}} z(t), y_e, A^{\frac{1}{2}} z(t))| + |Ay^*(t)|\,|Az(t)|\\ &\geq \nu|Az(t)|^2 - C|Az(t)|\,\|z(t)\|\,\|y_e\|_2 - |Ay^*(t)|\,|Az(t)| - \gamma\|z(t)\|^2.\end{aligned}$$

Integrating on $(0, t)$ and using (3.62), we see that $z \in L^\infty(0, T; V) \cap C([0, T]; H)$. Hence, $z : [0, T] \to V$ is weakly continuous and, therefore, $q_T \in C_w([0, T]; H)$. This shows that $q_T(0) \in H$ and, recalling that $Ry_0 = -q_T(0)$, we conclude that $Ry_0 \in H$, as claimed. Inequality (3.51) follows, therefore, by the closed graph theorem.

Finally, to show that R is a solution to Riccati equation (3.52), we first notice that, again by the dynamic programming principle, we have

$$\varphi(y^*(t)) = \frac{1}{2}\int_t^\infty (|A^{\frac{3}{4}} y^*(s)|^2 + |u^*(s)|^2_{M^*})ds, \quad \forall t \geq 0.$$

This yields, by virtue of (3.61) that (recall that $\frac{d}{dt}\varphi(y^*) = (Ry^*, \frac{dy^*}{dt})$)

$$\left(Ry^*(t), \frac{dy^*}{dt}(t)\right) + \frac{1}{2}|A^{\frac{3}{4}} y^*(t)|^2 + \frac{1}{2}\sum_{i=1}^{M^*}(Ry^*(t), \psi_i)_0^2 = 0, \quad \forall t \geq 0.$$

This leads, via a standard device involving (3.57), to

$$-(Ry^*(t), \nu Ay^*(t) + A_0 y^*(t) - \gamma y^*(t)) - \frac{1}{2}\sum_{i=1}^{M^*}(Ry^*(t), \psi_i)_0^2 + \frac{1}{2}|A^{\frac{3}{4}} y^*(t)|^2 = 0,$$
$$\forall t \geq 0,$$

which implies (3.52).

In order to prove (3.54), (3.55), it suffices to multiply the closed-loop equation

$$\frac{dy}{dt} + \mathscr{A} y + P\left(m\sum_{j=1}^{M^*}(Ry, \psi_i)_0 \psi_i\right) = 0, \quad t \geq 0,$$

by Ry and use (3.52). We get

$$\frac{1}{2}\frac{d}{dt}(Ry(t),y(t))+\frac{1}{2}|A^{\frac{3}{4}}y(t)|^2+\gamma(Ry(t),y(t))+\frac{1}{2}\sum_{i=1}^{M^*}(Ry(t),\psi_i)_0^2=0,$$
$$\forall t\ge 0,$$

which, by integration, yields (3.54), (3.55). This completes the proof. □

We have, incidently, proven that the solution y to the closed-loop system (3.20) with the feedback controller (3.53) is just the solution y^* to the minimization problem (3.56).

Now, we formulate the main internal stabilization result for System (3.1) (equivalently, (3.2)). The notation is as that from Theorem 3.3.

Theorem 3.4 *The feedback controller*

$$u=-\sum_{i=1}^{M^*}(R(y-y_e),\psi_i)_0\psi_i, \tag{3.63}$$

exponentially stabilizes the steady-state solution y_e to (3.2) *in a neighborhood*

$$\mathscr{U}_\rho=\{y_0\in W;\ \|(y_0-y_e)\|_W<\rho\}$$

of y_e for suitable $\rho>0$. More precisely, if $\rho>0$ is sufficiently small, then for each $y_0\in\mathscr{U}_\rho$ there exists a strong solution $y\in C([0,\infty);W)\cap C((0,\infty);V)$ to the closed-loop system

$$\frac{dy}{dt}+\nu Ay+Sy+P\left(m\sum_{i=1}^{M^*}(R(y-y_e),\psi_i)_0\psi_i\right)=Pf_e,\quad t\ge 0; \tag{3.64}$$
$$y(0)=y_0,$$

such that $\sqrt{t}\,A(y-y_e)\in L^2(0,\infty;H)$, $\sqrt{t}\,\frac{dy}{dt}\in L^2(0,\infty;H)$ and

$$\int_0^\infty e^{2\gamma t}|A^{\frac{3}{4}}(y(t)-y_e)|^2dt\le C_2\|y_0-y_e\|_W^2, \tag{3.65}$$

$$\|y(t)-y_e\|_W\le C_3e^{-\gamma t}\|y_0-y_e\|_W,\quad \forall t\ge 0, \tag{3.66}$$

where $C_2,C_3>0$.

In particular, it follows that, for all $y_0\in\mathscr{U}_\rho$, (3.64) has a unique strong solution y satisfying (3.65), (3.66). (If $d=2$, then this is, of course, implied by Theorem 1.17.) It should be noticed also that, as we see later on, the radius ρ of the stability region $\mathscr{U}_\rho$ is independent of γ. Moreover, the stabilization effect of the feedback controller

(3.63) is in force for all $\nu > 0$. In other words, no assumption on viscosity coefficient ν and, implicitly, on Reynolds number $\frac{1}{\nu}$, is assumed whatever.

Proof of Theorem 3.4 By substitution $y - y_e \to y$ and $y^0 = y_0 - y_e$, we reduce the closed-loop system (3.64) to

$$\frac{dy}{dt} + \nu Ay + A_0 y + Sy + P\sum_{i=1}^{M^*}(Ry, \psi_i)_0\psi_i = 0, \quad t > 0, \qquad (3.67)$$
$$y(0) = y^0.$$

We are going to show that $\varphi(y) = \frac{1}{2}(Ry, y)$ is a Lyapunov function for System (3.67) in a neighborhood of the origin.

As seen in Sect. 1.5 (Theorem 1.18), (3.67) has at least one weak solution $y \in C_w([0,T]; H) \cap L^2(0,T;V)$, $\forall T > 0$, given as limit of strong solutions $y_\varepsilon \in C((0,T); H) \cap L^2(\delta, T; D(A))$, $\forall \delta > 0$, to the equation

$$\frac{dy_\varepsilon}{dt} + \nu Ay_\varepsilon + A_0 y_\varepsilon + S_\varepsilon y_\varepsilon + P\left(m\sum_{i=1}^{M^*}(Ry_\varepsilon, \psi_i)_0\psi_i\right) = 0, \quad t > 0, \qquad (3.68)$$
$$y_\varepsilon(0) = y^0,$$

where S_ε is the truncated operator

$$S_\varepsilon y = \begin{cases} Sy & \text{if } \|y\| \le \frac{1}{\varepsilon}, \\ \frac{Sy}{\varepsilon^2\|y\|^2} & \text{if } \|y\| > \frac{1}{\varepsilon}. \end{cases}$$

In fact, one has

$$y_\varepsilon \to y \quad \text{strongly in } L^2(0,T;H), \text{ weakly in } L^2(0,T;V),\ \forall T > 0, \qquad (3.69)$$

and the following estimate holds

$$|y_\varepsilon(t)|^2 + \int_0^t \left(\|y_\varepsilon(s)\|^2 + \left\|\frac{dy_\varepsilon}{ds}(s)\right\|_{V'}^{\frac{4}{3}}\right) ds \le |y^0|^2 + C_T, \quad \forall \varepsilon > 0. \qquad (3.70)$$

Using Riccati equation (3.52), we obtain by (3.68) that

$$\frac{d}{dt}(Ry_\varepsilon, y_\varepsilon) + \sum_{i=1}^{M}(Ry_\varepsilon, \psi_i)_0^2 + |A^{\frac{3}{4}}y_\varepsilon|^2 + 2\gamma(Ry_\varepsilon, y_\varepsilon) = -2(S_\varepsilon y_\varepsilon, Ry_\varepsilon),$$
$$\text{a.e., } t > 0. \qquad (3.71)$$

On the other hand, we have, by standard estimate on Navier–Stokes equations in 3-D (see Proposition 1.7)

$$|(S_\varepsilon y_\varepsilon, Ry_\varepsilon)| \le |b(y_\varepsilon, y_\varepsilon, Ry_\varepsilon)| \le C\|y_\varepsilon\|\,|y_\varepsilon|_{\frac{3}{4}}|Ry_\varepsilon|$$

$$\leq C\|y_\varepsilon\|\,|A^{\frac{3}{4}}y_\varepsilon|^2|Ry_\varepsilon| \leq C|A^{\frac{3}{4}}y_\varepsilon|\,\|y_\varepsilon\|^2$$
$$\leq C|A^{\frac{3}{4}}y_\varepsilon|^2(Ry_\varepsilon, y_\varepsilon)^{\frac{1}{2}}, \tag{3.72}$$

where the various constants C are independent of ε. (Here, we have used (3.50), (3.51) and the interpolation inequality $\|y\|^2 \leq |A^{\frac{3}{4}}y|\,|A^{\frac{1}{4}}y|$.)

Substituting (3.72) into (3.71), we obtain for $(Ry^0, y^0) \leq \rho$ and ρ sufficiently small that on the maximal interval $(0, T_\varepsilon)$, where $(Ry_\varepsilon(t), y_\varepsilon(t)) \leq \rho$, we have

$$\frac{d}{dt}(Ry_\varepsilon(t), y_\varepsilon(t)) + 2\gamma(Ry_\varepsilon(t), y_\varepsilon(t)) + \frac{1}{2}|A^{\frac{3}{4}}y_\varepsilon(t)|^2 \leq 0.$$

We have, therefore, by (3.50) that

$$\int_0^{T_\varepsilon} |A^{\frac{3}{4}}y_\varepsilon(t)|^2 e^{2\gamma t}dt \leq C|A^{\frac{1}{4}}y^0|^2$$

and

$$\|y_\varepsilon(t)\|_W^2 = |A^{\frac{1}{4}}y_\varepsilon(t)|^2 \leq Ce^{-2\gamma t}|A^{\frac{1}{4}}y^0|^2, \quad \forall t \in (0, T_\varepsilon),$$

where C is independent of ε. Recalling (3.50), the latter implies that, for ρ sufficiently small, $T_\varepsilon = \infty$ for all $\varepsilon > 0$ and so, the previous estimates extend to $(0, \infty)$.

On the other hand, if we multiply (3.68) by $t\,Ay_\varepsilon(t)$ and integrate on $(0, t)$, we obtain that

$$\frac{1}{2}t(Ay_\varepsilon(t), y_\varepsilon(t)) + \int_0^t s|Ay_\varepsilon(s)|^2 ds$$
$$\leq \frac{1}{2}\int_0^t (Ay_\varepsilon(s), y_\varepsilon(s))ds + \int_0^t s(b(y_\varepsilon(s), y_e, Ay_\varepsilon(s)) + b(y_e, y_\varepsilon(s), Ay_\varepsilon(s)))ds$$
$$+ \int_0^t sb(y_\varepsilon(s), y_\varepsilon(s), Ay_\varepsilon)ds + C\int_0^t s|Ay_\varepsilon(s)|\,|Ry_\varepsilon(s)|ds, \quad \forall t \geq 0.$$

Recalling that, by Proposition 1.7,

$$|b(y_\varepsilon, y_e, Ay_\varepsilon)| + |b(y_e, y_\varepsilon, Ay_\varepsilon)| + |b(y_\varepsilon, y_\varepsilon, Ay_\varepsilon)|$$
$$\leq C|Ay_\varepsilon|(|y_e|_{\frac{1}{2}}|y_\varepsilon|_1 + |y_e|_1|y_\varepsilon|_{\frac{1}{2}} + |y_\varepsilon|_{\frac{1}{2}}|y_\varepsilon|_{\frac{3}{4}}),$$

where $|y|_\alpha = |A^\alpha y|$, we obtain by the previous estimates that

$$t\|y_\varepsilon(t)\|^2 + \int_0^t s|Ay_\varepsilon(s)|^2 ds \leq C\|y^0\|_W^2, \quad \forall \varepsilon > 0,\ t \geq 0,$$

and so, letting ε tend to zero, we obtain that

$$t\|y(t)\|^2 + \int_0^t s|Ay(s)|^2 ds \leq C\|y^0\|_W^2, \quad \forall t \geq 0,$$

$$\int_0^\infty |A^{\frac{3}{4}} y(t)|^2 e^{2\gamma t} dt \le C\|y^0\|_W^2,$$

$$\|y\|_W \le Ce^{-\gamma t}\|y^0\|_W, \quad \forall t > 0,$$

for all $y^0 \in W$ such that $(Ry^0, y^0) \le \rho_0$.

This implies also that y is strong solution to (3.67) with

$$y \in C([0,\infty); W) \cap C((0,\infty); V),$$

$$\sqrt{t}\, Ay \in L^2(0,\infty; H), \qquad \sqrt{t}\,\frac{dy}{dt} \in L^2(0,\infty; H).$$

Then $y - y_e$ satisfies Conditions and Estimates (3.65), (3.66) of Theorem 3.4. This completes the proof. □

Remark 3.2 As seen in (3.72), the optimal radius ρ of stability domain $\mathscr{U}_\rho$ might be determined by the formula

$$\max_{|y|_{\frac{1}{4}} \le \rho} \frac{2|b(y,y,Ry)|}{|y|^2_{\frac{3}{4}}} < 1,$$

and so, as easily follows by interpolation inequality $|y|_{\frac{1}{2}} \le |y|_{\frac{1}{4}} |y|_{\frac{3}{4}}$, for $d = 3$, the radius of $\mathscr{U}_\rho$ might be taken any ρ such that

$$0 < \rho < \frac{1}{2}\, \|R\|_{L(D(A^{\frac{1}{2}}),H)}.$$

This result is, of course, not optimal because it uses a linear stabilizing feedback for the linearized equation (3.20) in the nonlinear equation (3.16) and, it is well-known in control system theory and, in particular, in fluid flows control that the stabilizable feedback controllers for the linearized system can have a destabilizing effect if applied outside a certain neighborhood of the equilibrium state. One might expect, however, that a larger if not maximal domain of stabilization for System (3.16) might be found by using a nonlinear stabilization strategy, that is, via nonlinear feedback control design. Indeed, one might speculate that for $\gamma = 0$ an optimal control feedback for (3.16) with the cost functional

$$J(y,u) = \frac{1}{2}\int_0^\infty (|y|^2_{\frac{3}{4}} + |u(t)|^2_{M^*})dt$$

has a wider domain of stability and the design of such a feedback law was developed in [16]. The main result established there is that the feedback law

$$u(t) = -\sum_{i=1}^{M^*} (\nabla\varphi(y(t) - y_e), \psi_i)_0 \psi_i$$

exponentially stabilizes System (3.2), if φ is the solution to the stationary Hamilton–Jacobi equation

$$((\nu A + A_0)x + Sx, \nabla\varphi(x)) + \frac{1}{2}\sum_{i=1}^{M^*}(\nabla\varphi(x), \psi_i)_0^2 = \frac{1}{2}|A^{\frac{3}{4}}x|^2 \tag{3.73}$$

and the maximum domain $\mathscr{U}_\rho$ of stabilization for (3.16) is the set of all $x \in W$ for which (3.73) has a solution φ (in $C^1(\mathscr{U}_\rho)$ or in some generalized sense). Anyway, one might expect that this approach to the stabilization of Navier–Stokes equations leads to sharper results and estimates on the stabilization domain.

Remark 3.3 An important property of the stabilizable feedback controller (3.63) is its robustness to static perturbations in inputs of the form

$$u = (I + \varepsilon\phi)(u_{nom}) = -(I + \varepsilon\phi)\sum_{i=1}^{M^*}(R(y - y_e), \psi_i)_0\psi_i, \tag{3.74}$$

where $\phi : H \to H,\ \phi(0) = 0$ is a smooth nonlinear mapping satisfying the condition

$$|\phi(z)| \le \alpha|z|, \quad \forall z \in H. \tag{3.75}$$

Indeed, substituting the feedback input (3.74) into System (3.2), we obtain the closed-loop system

$$\frac{dy}{dt} + \mathscr{A}y + Sy + P\left(m(I + \varepsilon\phi)\sum_{i=1}^{M^*}(Ry, \psi_i)_0\psi_i\right) = 0$$

and using the Riccati equation (3.52), we find, as above (see (3.71)), that

$$\frac{d}{dt}(Ry, y) + 2\gamma(Ry, y) + \frac{1}{2}|A^{\frac{3}{4}}y|^2$$
$$\le \varepsilon\left(Ry, P\left(m\phi\left(\sum_{i=1}^{M}(Ry, \psi_i)_0\psi_i\right)\right)\right) \le C\varepsilon|Ry|^2.$$

This implies that for $0 < \varepsilon \le \varepsilon_0$ the feedback input is still exponentially stabilizable for System (3.16). It should be said that the robustness of feedback (3.63) comes from its design via the high-gain Riccati equation (3.52), that is, from a linear quadratic optimal control problem on $(0, \infty)$ with high-gain energetic cost functional $\int_0^\infty |A^{\frac{3}{4}}y|^2 dt$ and might fail for lower-gain cost functionals.

It turns out that the feedback controller (3.63) is also robust with respect to stochastic Gaussian perturbations of the form $\sqrt{\varepsilon}\,Q^{\frac{1}{2}}dW_t$, where W_t is an H-valued

Gaussian process and Q is a linear continuous self-adjoint positive operator in H with finite trace. More precisely, the solution $X = X(t)$ to the stochastic equation

$$dX + \mathscr{A}X\,dt + SX\,dt + P\left(m\sum_{i=1}^{M^*}(R(X - y_e), \psi_i)_0\psi_i\right)dt = \sqrt{\varepsilon}\,Q^{\frac{1}{2}}dW_t,$$

$$X(0) = x$$

for $\|x - y_e\|_W \le \rho$, is exponentially convergent to zero for $t \to \infty$ with high probability, that is,

$$\mathbb{P}\left[\max_{0\le t\le\infty}\|X(t)\|_W^2 e^{\mu t} \le \lambda\right] \ge 1 - \frac{\delta}{\lambda}\|x - y_e\|_W^2,$$

for all $\lambda > \delta$ and $0 < \varepsilon < \mu$ (see [16]).

3.3.4 Internal Stabilization; Low-gain Riccati-based Feedback

In this case, we plug into (3.2) the feedback controller

$$u(t) = -\sum_{i=1}^{M^*}(R_0(y - y_e), \psi_i)_0\psi_i, \tag{3.76}$$

where $R_0 \in L(H, H)$ is the solution to the corresponding Riccati equation (2.122), that is

$$(\mathscr{A}y - \gamma y, R_0 y) + \frac{1}{2}\sum_{i=1}^{M^*}(R_0 y, \psi_i)_0^2 = \frac{1}{2}|y|^2, \quad \forall y \in D(A), \tag{3.77}$$

$$\frac{1}{2}(R_0 y, y) = \Phi_0(y), \quad y \in H, \tag{3.78}$$

and Φ_0 is given by (2.109). A nice feature of R_0 already mentioned in Proposition 2.3 is that R_0 maps H into $D(A)$ which does not happen for R given by Theorem 3.3.

Theorem 3.5 *Let $W = D(A^{\frac{1}{4}})$ if $d = 2$ and $W = D(A^{\frac{1}{4}+\varepsilon})$ if $d = 3$. Let $\mathscr{U}_\rho = \{y_0 \in W;\ \|y_0 - y_e\|_W \le \rho\}$. For ρ sufficiently small there is a unique solution $y \in C([0,\infty); W) \cap L^2(0,\infty; Z)$ to* (3.2) *with the feedback controller* (3.76) *such that*

$$\|y(t) - y_e\|_W \le Ce^{-\gamma t}\|y_0 - y_e\|_W, \quad \forall t > 0. \tag{3.79}$$

Here, $Z = D(A^{\frac{3}{4}})$ if $d = 2$ and $Z = D(A^{\frac{3}{4}+\varepsilon})$ for $d = 3$ and $\varepsilon > 0$ is positive and small.

Proof By the substitution $y \to y - y_e$, we reduce the equation to the closed-loop system

$$\frac{dy}{dt} + \mathscr{A}y + Sy + P\left(m\sum_{i=1}^{M^*}(R_0 y, \psi_i)_0\psi_i\right) = 0, \tag{3.80}$$
$$y(0) = y^0 = y_0 - y_e.$$
□

In order to prove that System (3.80) is exponentially asymptotical stable, we cannot use in this case the Lyapunov function argument as in Theorem 3.4 because the function $y \to \frac{1}{2}(R_0 y, y)$ is no longer a Lyapunov function for System (3.80). In fact, the Riccati equation (3.77) is not sufficiently strong in energetic meaning to give to its solution R_0 a dissipation property which can dominate the inertial term S. So, the argument we use here (already described in Theorem 2.10 in an abstract setting) relies on a fixed-point argument. In this approach, the Riccati equation (3.77) has a secondary role only. (See Remark 3.3.)

In the following, we denote by $\Gamma : D(\Gamma) = D(A) \to H$ the operator

$$\Gamma y = \mathscr{A}y + \sum_{i=1}^{M^*} P(m\psi_i)(R_0 y, \psi_i)_0, \quad \forall y \in D(\Gamma).$$

We know that

$$\|e^{-\Gamma t}z_0\|_{L(H,H)} \le Ce^{-\gamma t}|z_0|, \quad \forall z_0 \in H. \tag{3.81}$$

Indeed, by (3.77) we see that Φ_0 is a Lyapunov function for the linear system

$$\frac{dz}{dt} + \Gamma z = 0, \quad t \ge 0,$$

and, more precisely,

$$\frac{d}{dt}\Phi_0(z(t)) + \gamma\Phi_0(z(t)) + \frac{1}{2}|z(t)|^2 \le 0, \quad \forall t \ge 0.$$

Hence

$$\frac{d}{dt}(e^{2\gamma t}\Phi_0(z(t))) + \frac{1}{2}e^{2\gamma t}|z(t)|^2 \le 0, \quad \forall t \ge 0,$$

which yields

$$\int_0^\infty e^{2\gamma t}|z(t)|^2 dt \le C|z(0)|^2,$$

and this implies (3.81), as claimed.

We also have the following key estimate.

Lemma 3.2 *We have*

$$\int_0^\infty \|e^{-(\Gamma-\gamma)t}z_0\|_Z^2 dt \le C\|z_0\|_W^2, \quad \forall z_0 \in W, \tag{3.82}$$

$$\|e^{-\Gamma t}z_0\|_W \le Ce^{-\gamma t}\|z_0\|_W, \quad \forall t \ge 0. \tag{3.83}$$

Proof The function $z(t) = e^{-(\Gamma-\gamma)t}z_0$ is the solution to the equation

$$\frac{dz}{dt} + \nu Az + A_0 z + P\left(\sum_{i=1}^{M^*} m\psi_i(R_0 z, \psi_i)_0\right) = \gamma z,$$
$$z(0) = z_0.$$

Assume first that $d = 2$. If we multiply the latter by $A^{\frac{1}{2}}z$, we obtain that

$$\frac{1}{2}\frac{d}{dt}|z(t)|^2_{\frac{1}{4}} + \nu|A^{\frac{3}{4}}z(t)|^2 \le C(|A^{\frac{1}{2}}z(t)|^2 + |A^{\frac{1}{2}}z(t)|\,|z(t)|, \quad \text{a.e., } t > 0.$$

By an interpolation inequality, we have

$$|A^{\frac{1}{2}}z|^2 \le |A^{\frac{3}{4}}z|^{\frac{4}{3}}|z|^{\frac{2}{3}} \le \frac{\nu}{2}|A^{\frac{3}{4}}z|^2 + C|z|^2$$

and this yields

$$\frac{d}{dt}|z(t)|^2_{\frac{1}{4}} + \frac{\nu}{2}|A^{\frac{3}{4}}z(t)|^2 \le C_1|z(t)|^2, \quad \forall t > 0.$$

Taking into account that, by (3.81),

$$|z(t)| \le Ce^{-\gamma t}|z_0|, \quad \forall t \ge 0,$$

we obtain (3.82), (3.83), as claimed.

If $d = 3$, by multiplying the equation by $A^{\frac{1}{2}+2\varepsilon}z$, we obtain, similarly,

$$|z(t)|^2_{\frac{1}{4}+\varepsilon} + \int_0^\infty |A^{\frac{3}{4}+\varepsilon}z(t)|^2 dt \le C\left(|y^0|^2_{\frac{1}{4}+\varepsilon} + \int_0^\infty |z(t)|^2 dt\right) \le C_1|y^0|^2_{\frac{1}{4}+\varepsilon}.$$

This completes the proof of Lemma 3.2. □

In the following, we use a few H^s estimates for the nonlinear term $y \to Sy$ given in Lemma 3.3.

Lemma 3.3 *Let $0 \le s < 1$. Then we have, for some constant $K > 0$,*

$$|A^s Sy| \le K|A^{s+\frac{1}{2}}y|^2, \quad \forall y \in D(A^{s+\frac{1}{2}}) \tag{3.84}$$

for $d = 2$ and

$$|A^{s+\varepsilon} Sy| \le K|A^{s+\varepsilon+\frac{1}{2}}y|^2, \quad \forall y \in D(A^{s+\varepsilon+\frac{1}{2}}) \tag{3.85}$$

for $d = 3$ and $\varepsilon > 0$, $\frac{1}{4} - \varepsilon < s < 1$.

Proof We recall that $D(A^{\frac{1}{4}+\varepsilon}) = (H_0^{\frac{1}{2}+2\varepsilon}(\mathcal{O}))^d \cap H$ for $0 < \varepsilon \le \frac{1}{2}$ and $D(A^{\frac{3}{4}+\varepsilon}) = (H^{\frac{3}{2}+2\varepsilon}(\Omega))^d \cap V$ for $\varepsilon > 0$. It should be said that Estimates (3.84), (3.85) do not follow directly from Proposition 1.7 and so, to get them, we need some stronger estimates in the space $H^s(\mathcal{O})$. We invoke here the following estimate (see [21], or [50], Lemma 4.7, Chap. 3)

$$\|zD_iz\|_{H^{s_3}(\mathcal{O})} \le C\|D_iz\|_{H^{s_1}(\mathcal{O})}\|z\|_{H^{s_2}(\mathcal{O})}, \tag{3.86}$$

for $z \in H^{s_2}(\mathcal{O})$, $D_iz = \frac{\partial}{\partial\xi_i} z$ and $s_3 \le s_j$, $j = 1, 2$, $s_1 + s_2 - s_3 > \frac{d}{2}$.

We apply (3.86) with $s_2 = s_1 + 1 = 2s + 1$, $s_3 = 2s$ in the case $d = 2$ and $s_2 = s_1 + 1 = 2(s+\varepsilon) + 1$, $s_3 = 2(s+\varepsilon)$ if $d = 3$. We have, for $d = 3$,

$$\begin{aligned} |A^{s+\varepsilon}Sy| &\le C\|(y\cdot\nabla)y\|_{(H^{2(s+\varepsilon)}(\mathcal{O}))^d} \le C\|\nabla y\|_{(H^{2(s+\varepsilon)}(\mathcal{O}))^d}\|y\|_{(H^{2(s+\varepsilon)+1}(\mathcal{O}))^d} \\ &\le C\|y\|^2_{(H^{2(s+\varepsilon)+1}(\mathcal{O}))^d} \le C|A^{s+\varepsilon+\frac{1}{2}}y|^2, \end{aligned}$$

while, for $d = 2$, by the same estimate (3.86), we get (3.84), as claimed. □

Proof of Theorem 3.5 (continued). By the substitution $z(t) = e^{\gamma t}y(t)$, System (3.80) reduces to

$$\begin{aligned} &\frac{dz}{dt} + \Gamma^* z + e^{-\gamma t}Sz = 0, \quad t \ge 0, \\ &z(0) = y^0, \end{aligned} \tag{3.87}$$

where $\Gamma^* = \Gamma - \gamma I$.

Equivalently,

$$z = \mathcal{N}z,$$

where the operator $\mathcal{N} : L^2(0,\infty; H) \to L^2(0,\infty; H)$ is defined by

$$(\mathcal{N}z)(t) = e^{-\Gamma^* t}y^0 - \int_0^t e^{-\Gamma^*(t-s)}e^{-\gamma s}Sz(s)ds.$$

As in the proof of Theorem 2.10, we apply the Banach fixed-point theorem to the operator $\mathcal{N}$ defined on the set

$$\mathcal{K}_r = \left\{z \in L^2(0,\infty; Z);\ \int_0^\infty \|z(t)\|_Z^2 dt \le r^2\right\}.$$

To this end, we prove first the following estimate

$$\|\mathcal{N}z\|^2_{L^2(0,\infty;Z)} = \int_0^\infty \|\mathcal{N}z(t)\|_Z^2 dt \le C\left(\|y^0\|_W^2 + \left(\int_0^\infty \|z(t)\|_Z^2 dt\right)^2\right). \tag{3.88}$$

It suffices to show that

$$\|\mathcal{N}z\|_{L^2(0,\infty;Z)} = \left(\int_0^\infty \|\mathcal{N}z(t)\|_Z^2 dt\right)^{\frac{1}{2}}$$
$$= \int_0^\infty \left(\left\| e^{-\Gamma^* t}y^0 - \int_0^t e^{-\Gamma^*(t-\tau)}e^{-\gamma\tau}Sz(\tau)d\tau\right\|_Z^2 dt\right)^{\frac{1}{2}}$$
$$\leq C\left(\|Sz\|_{L^1(0,\infty;W)} + \|y^0\|_W\right) \tag{3.89}$$

because, by Estimates (3.84), (3.85) we obtain (3.88).

We prove the latter by a duality argument. Namely, let $\zeta \in L^2(0,\infty;Z')$, where Z' is the dual of Z with respect to H as a pivot space. We have (for $\gamma = 0$)

$$\int_0^\infty ((\mathcal{N}z)(t), \zeta(t))dt$$
$$= \int_0^\infty \left(\int_0^t e^{-\Gamma^*(t-\tau)}(Sz)(\tau)e^{-\gamma\tau}d\tau, \zeta(t)\right) dt$$
$$\leq \int_0^\infty \int_0^t \|e^{-\Gamma^*(t-\tau)}(Sz)(\tau)e^{-\gamma\tau}\|_Z\|\zeta(t)\|_{Z'}\,d\tau\,dt$$
$$= \int_0^\infty \left\{\int_\tau^\infty \|e^{-\Gamma^*(t-\tau)}(Sz)(\tau)e^{-\gamma\tau}\|_Z\|\zeta(t)\|_{Z'}\,dt\right\} d\tau$$
$$\leq \int_0^\infty \left\{\left[\int_\tau^\infty \|e^{-\Gamma^*(t-\tau)}(Sz)(\tau)e^{-\gamma\tau}\|_Z^2 dt\right]^{\frac{1}{2}} \left[\int_0^\infty \|\zeta(t)\|_{Z'}^2 dt\right]^{\frac{1}{2}}\right\} d\tau$$
$$= \|\zeta\|_{L^2(0,\infty;Z')} \int_0^\infty \left[\int_0^\infty \|e^{-\Gamma^*\sigma}(Sz)(\tau)e^{-\gamma\tau}\|_Z^2 d\sigma\right]^{\frac{1}{2}} d\tau.$$

(Here, we have used the Fubini theorem several times.)

By Lemma 3.2, the latter implies (3.89) as desired because

$$\left[\int_0^\infty \|e^{-\Gamma^*\sigma}Sz(\tau)\|_Z^2 d\sigma\right]^{\frac{1}{2}} \leq C\|Sz(\tau)\|_W.$$

Now, using Estimates (3.84) (respectively, (3.85)) in (3.89), we obtain (3.88).

By (3.89), we see that if $z \in \mathcal{K}_r$, then, for r suitable chosen and $\|y^0\|_W \leq \frac{r}{4C} = \rho$, we have that $\mathcal{N}z \in \mathcal{K}_r$, that is $\mathcal{N}$ leaves invariant $\mathcal{K}_r$.

$\mathcal{N}$ is a contraction on $\mathcal{K}_r \subset L^2(0,\infty;Z)$. Indeed, we have for $z_1, z_2 \in L^2(0,\infty;Z)$

$$\|\mathcal{N}z_1 - \mathcal{N}z_2\|_{L^2(0,\infty;Z)}^2 \leq C\int_0^\infty \|(Sz_1)(t) - (Sz_2)(t)\|_W^2 dt$$
$$= C\|Sz_1 - Sz_2\|_{L^2(0,\infty;W)}^2.$$

The proof (by duality) is exactly the same as the proof of (3.89) and so, it is omitted.
Then, once again using Lemma 3.3, we obtain

$$\begin{aligned}\|Sz_1 - Sz_2\|_W &= \|P[(z_1 \cdot \nabla)z_1 - (z_2 \cdot \nabla)z_2]\|_W \\ &= \|P[((z_1 - z_2) \cdot \nabla)z_1 + z_2 \cdot \nabla)(z_1 - z_2)]\|_W \\ &\leq K\{\|z_1\|_Z + \|z_2\|_Z\}\|z_1 - z_2\|_Z.\end{aligned}$$

Thus, the Schwarz inequality yields

$$\begin{aligned}&\|\mathcal{N} z_1 - \mathcal{N} z_2\|^2_{L^2(0,\infty;Z)} \\ &\quad \leq C\left\{\int_0^\infty [\|z_1(\tau)\|_Z + \|z_2(\tau)\|_Z]\,\|z_1(\tau) - z_2(\tau)\|_Z d\tau\right\}^2 \\ &\quad \leq C\left(\int_0^\infty [\|z_1(\tau)\|^2_Z + \|z_2(\tau)\|^2_Z d] d\tau\right)\left(\int_0^\infty \|z_1(\tau) - z_2(\tau)\|^2_Z d\tau\right) \\ &\quad \leq Cr^2 \int_0^\infty \|z_1(\tau) - z_2(\tau)\|^2_Z d\tau \leq Cr^2\|z_1 - z_2\|^2_{L^2(0,\infty;Z)}.\end{aligned}$$

(We have denoted by C several positive constant independent of z.)

We may conclude, therefore, that for r sufficiently small and $\|y^0\|_W \leq \rho = \frac{r}{4C}$ there is a unique solution $z \in L^2(0, \infty; Z)$ to (3.87).

Moreover, recalling the estimate on $\|\mathcal{N} z\|_{L^2(0,\infty;Z)}$, we also have that

$$\|z(t)\|_W \leq C\|y^0\|_W, \quad \forall t \geq 0.$$

Then, recalling that $z(t) = e^{\gamma t} y(t)$, we may conclude that (3.80) has a unique solution $y \in C([0, \infty); W) \cap L^2(0, \infty; Z)$ and that

$$\|y(t)\|_W \leq Ce^{-\gamma t}\|y^0\|_W, \quad \forall t > 0,$$

for all $\|y^0\|_W \leq \rho < 0$. This concludes the proof of Theorem 3.5. □

Remark 3.4 As mentioned in Sect. 2.3 (see also Remark 3.1) the low-gain controller is less robust than the high-gain controller. On the other hand, the special structure (3.76) of the feedback control is not relevant in Theorem 3.5. It suffices to have a linear continuous feedback controller $u = Fy$, which stabilizes exponentially the linear system (3.20). In particular, it follows that Theorem 3.5 remains true for feedback controllers of the form (3.42) designed in Sect. 3.3.2. In other words, by the model of the previous proof, one can show that the solution y to the closed-loop system (see Proposition 3.4)

$$\begin{aligned}&y_t + \nu Ay + Sy + \eta P\left(m \sum_{i=1}^{N} (y(t) - y_e, \varphi_j^*)\phi_j\right) = 0, \\ &y(0) = y_0,\end{aligned}$$

where ϕ_j are chosen as in (3.43), satisfies

$$\|y(t) - y_e\|_W \le Ce^{-\gamma t}\|y_0 - y_e\|_W, \quad \forall t \ge 0,$$

for η sufficiently large and $\|y_0 - y_e\|_W \le \rho$ sufficiently small.

Without any doubt, this proportional stabilizable feedback is simpler and its stabilizing performances are comparable with that of (3.53) or (3.76). However, as mentioned earlier in Sect. 2.2.2, such a stabilizable feedback is not robust being highly sensitive to perturbation of spectrum to Stokes–Oseen operator and, consequently, to structural perturbations of the system.

Remark 3.5 There are some important features of the feedback controller designed in this section. The first is that the controller uses an arbitrarily small set $\mathcal{O}_0 \subset \mathcal{O}$ for actuation. The second is its finite-dimensional structure as a finite linear combination of eigenvectors of the dual linearized system. This is a considerable advantage over numerical implementations of this controller and has already been tested on some special problems (see Sect. 3.7). On the other hand, the robustness of the controller to structural perturbations of the system allows us to work on finite-dimensional approximations of the Stoke–Oseen system.

Remark 3.6 It is well-known (see, e.g., [74]) that, generically, for "almost all" f_e there is a finite number of equilibrium solutions y_e^i, $i = 1, 2, \dots, N_e$, to (3.6). By Theorem 3.3 (respectively, Theorem 3.4), for each y_e^i there is a feedback controller $u = F_i(y - y_e^i)$, $i = 1, \dots, n$, of the form (3.63) (respectively (3.76)), which stabilizes exponentially System (3.2) in a neighborhood $\mathcal{U}(\rho_i) = \{y;\ \|y - y_e^i\|_W \le \rho_i\}$. If $\chi_i \in C^\infty(W)$ is taken in such a way that

$$\chi_i(y) = 1 \quad \text{in } \mathcal{U}(\rho_i), \qquad \chi_i(y) = 0 \quad \text{in } \mathcal{U}^c(\rho_i + \varepsilon),$$

we see that the feedback controller

$$u = \sum_{i=1}^{N} \chi_i(y) F_i(y - y_e^i)$$

stabilizes System (3.2) for $y_0 \in \bigcup_{i=1}^{N} U(\rho_i)$. As a matter of fact, the semigroup $S(t)y_0 = y(t)$ defined by the closed-loop system

$$\frac{dy}{dt} + \nu A y + S y = \sum_{i=1}^{N_e} \chi_i(y) F_i(y - y_e^i), \quad t \ge 0,$$
$$y(0) = y_0$$

has a compact attractor in the space H and

$$\lim_{t\to\infty} S(t)y_0 = y_e^i, \quad \forall y_0 \in \mathcal{U}(\rho_i),\ i = 1, 2, \dots, N_e.$$

The dynamic and asymptotic behavior of $S(t)$ for $y_0 \overline{\in} \bigcup_{i=1}^{N_e} \mathscr{U}(\rho_i)$ is, however, still open and its description might be essential to understanding the global stabilization effect of the above feedback controller.

3.4 The Tangential Boundary Stabilization of Navier–Stokes Equations

Here, we study the boundary feedback stabilization of Navier–Stokes equations via the methods developed in the previous sections as well as that in Sect. 2.3.

It should be mentioned that by internal stabilization theorems we may find boundary stabilization results by a standard device which is briefly described below and refer to [22] for details.

We consider $\widetilde{\mathcal{O}}_0$ an exterior open neighborhood of the domain $\mathcal{O}$ and set $\widetilde{\mathcal{O}} = \overline{\mathcal{O}} \cup \widetilde{\mathcal{O}}_0$. (See Fig. 3.1.) Then, there is a controller u with the support in $\widetilde{\mathcal{O}}_0 \subset \widetilde{\mathcal{O}}$ such that the corresponding solution $\widetilde{y}$ to Navier–Stokes equation (3.1) on $\overline{\mathcal{O}}$ and Dirichlet homogeneous boundary conditions on $\partial\widetilde{\mathcal{O}}$ is exponentially stable. Then, $y = \widetilde{y}|_{\mathcal{O}}$ satisfies (3.1) on $\widetilde{\mathcal{O}}$ and $y = u$ on $\partial\mathcal{O}$, where $u = y|_{\partial\mathcal{O}}$. Clearly, this boundary controller u stabilizes exponentially System (3.1) in $(0, \infty) \times \mathcal{O}$.

It should be said, however, that this simple device does not provide a feedback controller which, from the point of view of automatic control theory, is a major limitation. However, it can be used to design an open-loop stabilizable boundary controller.

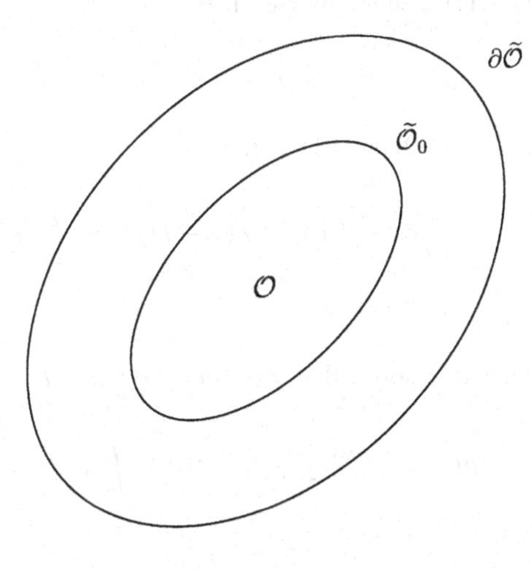

Fig. 3.1

3.4.1 The Tangential Boundary Stabilization of the Stokes–Oseen Equation

Consider here the linearized equation (3.11) with boundary controller input u, that is

$$\begin{aligned}
&\frac{\partial y}{\partial t} - \nu\Delta y + (y_e\cdot\nabla)y + (y\cdot\nabla)y_e = \nabla p && \text{in } (0,\infty)\times\mathcal{O},\\
&y(0,x) = y^0(x) = y_0(x) - y_e(x) && \text{in } \mathcal{O},\\
&\nabla\cdot y = 0 && \text{in } (0,\infty)\times\mathcal{O},\\
&y(t,x) = u(t,x) && \forall (t,x)\in(0,\infty)\times\partial\mathcal{O}.
\end{aligned}\tag{3.90}$$

We assume here that the controller $u\in L^2((0,\infty)\times\partial\mathcal{O})$, which represents the velocity field on the boundary, is *tangential*, that is $u(t)\cdot n = 0$ a.e., on $(0,\infty)\times\partial\mathcal{O}$. From the point of view of the fluid control theory, this means that the actuation is tangential and, as argued in [7], it is technologically feasible.

There is a general method to "move" the boundary controller u on $\mathcal{O}$ as a distributed controller and so, to homogenize (3.90). Namely, for k sufficiently large, the steady-state equation

$$\begin{aligned}
&k\theta - \nu\Delta\theta + (y_e\cdot\nabla)\theta + (\theta\cdot\nabla)y_e = \nabla p && \text{in } \mathcal{O},\\
&\nabla\cdot\theta = \theta && \text{in } \mathcal{O},\\
&\theta = u && \text{on } \partial\mathcal{O},
\end{aligned}\tag{3.91}$$

has a unique solution $\theta = Du$, which is in $(H^{s+1}(\mathcal{O}))^d\cap H$ if $u\in(H^s(\partial\mathcal{O}))^d$, $s\ge\frac{1}{2}$, and $u\cdot n = 0$ on $\partial\mathcal{O}$ (see Theorem A.2.1 in [22]). Moreover, D is continuous from $(H^s(\partial\mathcal{O}))^d$ to $(H^{s+\frac{1}{2}}(\mathcal{O}))^d\cap H$ for $s\ge\frac{1}{2}$. If we denote by $\mathcal{A}_k : D(\mathcal{A}) = D(A)\to\widetilde{H}$ the operator $kI+\mathcal{A}$, where $\mathcal{A}$ is defined by (3.9), and consider $\widetilde{\mathcal{A}}: H\to(D(A))'$ defined by (see 1.9))

$$_{(D(A))'}(\widetilde{\mathcal{A}}y,\psi)_{D(A)} = (y,\mathcal{A}^*\psi),\quad \forall\psi\in D(\mathcal{A}^*) = D(A),$$

we have by (3.90) and (3.91) that

$$\begin{aligned}
&\frac{d}{dt}(y-Du) + \mathcal{A}(y-Du) = -\frac{d}{dt}Du - kDu,\quad t\ge 0,\\
&y(0) = y^0.
\end{aligned}$$

This yields (we notice that $y - Du\in D(\mathcal{A}) = D(A)$)

$$y(t) - Du(t) = e^{-\mathcal{A}t}(y^0 - Du(0)) - \int_0^t e^{-\mathcal{A}(t-s)}\left(kDu(s) + \frac{d}{ds}Du(s)\right)ds.\tag{3.92}$$

Integrating by parts, we see that y is the solution to the equation

$$\begin{aligned}&\frac{d}{dt}y(t)+\widetilde{\mathscr{A}}y(t)=(\widetilde{\mathscr{A}}+k)Du(t),\quad t>0,\\&y(0)=y^0.\end{aligned}\tag{3.93}$$

Equivalently,

$$\frac{d}{dt}(y(t),\psi)+(y(t),\mathscr{A}^*\psi)=(u(t),((\mathscr{A}+k)D)^*\psi),\quad \forall\psi\in D(A).\tag{3.94}$$

It is easily to see, via Green's formula, that the dual $((\widetilde{\mathscr{A}}+k)D)^*$ of the operator $(\widetilde{\mathscr{A}}+k)D$ is given by

$$((\widetilde{\mathscr{A}}+k)D)^*\psi=-\nu\,\frac{\partial\psi}{\partial n},\quad \forall\psi\in D(A).\tag{3.95}$$

The control system (3.93) is of the form (3.20)′ where

$$B=(\widetilde{\mathscr{A}}+k)D:(L^2(\partial\mathscr{O}))^d\to(D(A))',$$

so one might expect that the stabilization technique used in the previous case applies in the present situation as well. We see that this is indeed the case, but here the operator B is unbounded and this fact leads to some serious problems. Since (3.92) (or, equivalently, (3.93)) plays a central role in the boundary stabilization analysis, we pause briefly to discuss its existence in the space $D(A^s)$, where $s=\frac14-\varepsilon$ or $\frac14+\varepsilon$. (Here, $0<\varepsilon<\frac14$.)

To this end, we mention the following lemma which, in various versions, arose several times in our analysis.

Lemma 3.4 *Let $0<s<1$ and $d=2,3$. Then, for $y^0\in D(A^{\frac{s}{2}})$ we have*

$$e^{-\mathscr{A}t}y^0\in C([0,T];D(A^{\frac{s}{2}}))\cap L^2(0,T;D(A^{\frac{1+s}{2}})),$$

for each $0<T<0$.

Proof Let $y(t)=e^{-\mathscr{A}t}y^0$. Then, we have

$$\frac{dy}{dt}+\nu Ay+A_0y=0,\quad \forall t\in[0,T]$$

and by interpolation this yields (see Proposition 1.7)

$$\begin{aligned}\frac12\frac{d}{dt}(A^sy,y)+\nu|A^{\frac{1+s}{2}}y|^2&\le|(A_0y,A^sy)|\le C(|b(y,y_e,A^sy)|+|b(y_e,y,A^sy)|)\\&\le C|A^sy|\,|A^{\frac12}y|\le C|A^{\frac{1+s}{2}}y|^{\frac{2s+1}{1+s}}|y|^{\frac{1}{1+s}}.\end{aligned}$$

Hence

$$\begin{aligned}|(A^s y(t), y(t))| &+ \int_0^t |A^{\frac{1+s}{2}} y(\tau)|^2 d\tau \\ &\le C\left((A^s y^0, y^0) + \int_0^t |y(\tau)|^2 d\tau\right) \\ &\le C(|A^{\frac{s}{2}} y^0|^2 + |y^0|^2), \quad t \in (0, T)\end{aligned}$$

(by Gronwall's inequality). This implies the desired result. □

Proposition 3.5 *Let y be the solution to* (3.92). *If $\varepsilon \in (0, \frac{1}{4})$ and*

$$y^0 \in D(A^{\frac{1}{4}-\varepsilon}), \quad u \in H^1([0,T]; (H^{\frac{1}{2}}(\partial\mathscr{O}))^d), \quad u \cdot \mathbf{n} \equiv 0 \quad on\ \partial\mathscr{O}, \tag{3.96}$$

then

$$y, y - Du \in C([0,T]; D(A^{\frac{1}{4}-\varepsilon})) \cap L^2(0,T; D(A^{\frac{3}{4}-\varepsilon})). \tag{3.97}$$

In addition, assume that

$$y^0 - Du(0) \in D(A^{\frac{1}{4}+\varepsilon}), \tag{3.98}$$

then

$$y - Du \in C([0,T]; D(A^{\frac{1}{4}+\varepsilon})) \cap L^2(0,T; D(A^{\frac{3}{4}+\varepsilon})). \tag{3.99}$$

Proof We recall that $D(\mathscr{A}) = D(A)$ and $D(A^s) = D(\mathscr{A}^s)$ for $0 \le s \le 1$. By (3.92), we have (for simplicity, we take $k = 0$)

$$\begin{aligned}&|A^{\frac{1}{4}-\varepsilon}(y(t) - Du(t))| \\ &\quad\le C|A^{\frac{1}{4}-\varepsilon}(y^0 - Du(0))| + \left|\int_0^t A^{\frac{1}{4}-\varepsilon} e^{-\mathscr{A}(t-s)} \frac{d}{ds} Du(s) ds\right| \\ &\quad\le C_1 \|y^0 - Du(0)\|_{D(A^{\frac{1}{4}-\varepsilon})} + C_2 \int_0^t \left\|\frac{d}{ds} Du(s)\right\|_{D(A^{\frac{1}{4}-\varepsilon})} ds \\ &\quad\le C_1 \|y^0 - Du(0)\|_{D(A^{\frac{1}{4}-\varepsilon})} + C_2 \|u\|_{H^1(0,T;(H^1(\partial\mathscr{O}))^2)}.\end{aligned}$$

Hence, $y - Du, y \in C([0,T]; D(A^{\frac{1}{4}-\varepsilon}))$, as claimed.

Similarly, we have

$$\begin{aligned}|A^{\frac{3}{4}-\varepsilon}(y(t) - Du(t))| &\le |A^{\frac{3}{4}-\varepsilon} e^{-\mathscr{A}t}(y^0 - Du(0))| \\ &\quad + \int_0^t |A^{\frac{3}{4}-\varepsilon} e^{-\mathscr{A}(t-s)} \frac{d}{ds} Du(s)| ds.\end{aligned}$$

Taking into account that, by (3.96),

$$\left\| \frac{d}{ds} Du(s) \right\|_{L^2(0,T;D(A^{\frac{1}{2}}))} \le C\|u\|_{L^2(0,T;(H^{\frac{1}{2}}(\partial\mathscr{O}))^d)},$$

we obtain, by Lemma 3.4,

$$\|A^{\frac{3}{4}-\varepsilon}(y - Du)\|_{L^2(0,T)} \le C\|y^0 - Du(0)\|_{D(A^{\frac{1}{4}-\varepsilon})} + C\|u\|_{L^2(0,T;(H^{\frac{1}{2}}(\partial\mathscr{O}))^d)},$$

as claimed. □

In a similar way, taking into account that, by Lemma 3.4,

$$\|e^{-\mathscr{A}t} f\|_{L^2(0,T;D(A^{\frac{3}{4}+\varepsilon}))} \le C\|f\|_{D(A^{\frac{1}{4}+\varepsilon})},$$

(3.99) follows by Assumption 3.98.

In the following, we come back to the representation of System (3.93) in the complexified space $\widetilde{H} = H + iH$. However, for simplicity, we denote by the same symbols $(\cdot,\cdot)$ and $|\cdot|$ the scalar product and norm in H and $\widetilde{H}$. (Sometimes, however, we write H instead of $\widetilde{H}$.)

As in the previous sections, φ_j are eigenfunctions to $\mathscr{A}$ and φ_j^* to the dual operator $\mathscr{A}^*$.

The first stabilization result is under Assumptions (K1), (K2) below.

(K1) *The eigenvalues* $\{\lambda_j\}_{j=1}^N$ *are semisimple.*

(K2) *The system* $\{\frac{\partial\varphi_j^*}{\partial n}\}_{j=1}^N$ *is linearly independent on* $(L^2(\partial\mathscr{O}))^d$.

Therefore, as seen earlier, we may choose an eigenfunction system such as

$$(\varphi_i, \varphi_j^*) = \delta_{ij}, \quad i, j = 1, \dots, N.$$

Here, γ is chosen in such a way that $\operatorname{Re}\lambda \le \gamma$ for $j = 1, \dots, N$.

Denote, also, by M the number

$$M = \max\{m_j;\ 1 \le j \le N\},$$

where m_j is the multiplicity of the eigenvalue λ_j. (The notation is that from Sect. 3.1.)

Theorem 3.6 *Let* $d = 2, 3$ *and* $0 < \varepsilon < \frac{1}{4}$. *Then, if* (K1) *and* (K2) *hold, there is a controller* $u \in C^1([0,\infty); (L^2(\partial\mathscr{O}))^d)$ *of the form*

$$u(t) = \sum_{j=1}^M v_j(t) \frac{\partial\varphi_j^*}{\partial n}, \quad t \ge 0, \tag{3.100}$$

such that the corresponding solution y^u *to System* (3.93) *satisfies the following conditions*

$$|y^u(t)| \le Ce^{-\gamma t}|y^0|, \quad \forall y^0 \in H,\ t \ge 0. \tag{3.101}$$

If $y^0 \in D(A^{\frac{1}{4}-\varepsilon})$, *we have*

$$\|y^u(t)\|_{D(A^{\frac{1}{4}-\varepsilon})} \le Ce^{-\gamma t}\|y^0\|_{D(A^{\frac{1}{4}-\varepsilon})}, \tag{3.102}$$

$$\int_0^\infty e^{2\gamma t}\|y^u(t)\|^2_{D(A^{\frac{3}{4}-\varepsilon})}\,dt \le C\|y^0\|^2_{D(A^{\frac{1}{4}-\varepsilon})}. \tag{3.103}$$

Moreover, $\{v_j\}_{j=1}^M$ *can be chosen in such a way that*

$$v_j \in C^1[0,\infty), \quad j = 1, 2, \ldots, M, \tag{3.104}$$

$$|v_j(t)| + |v_j'(t)| \le Ce^{-3\gamma t}|y^0|, \quad \forall t \ge 0. \tag{3.105}$$

Proof Arguing as in the previous cases, we write System (3.93) as

$$\begin{aligned} &\frac{dy_u}{dt} + \mathscr{A}_u y_u = P_N \sum_{j=1}^M v_j(t)(\widetilde{\mathscr{A}} + k)\frac{\partial \varphi_j^*}{\partial n}, \\ &y_u(0) = P_N y^0, \end{aligned} \tag{3.106}$$

$$\begin{aligned} &\frac{dy_s}{dt} + \widetilde{\mathscr{A}_s} y_s = (I - P_N) \sum_{j=1}^M v_j(t)(\widetilde{\mathscr{A}} + k)\frac{\partial \varphi_j^*}{\partial n}, \\ &y_s(0) = (I - P_N) y^0. \end{aligned} \tag{3.107}$$

Here,

$$\mathscr{A}_u = \mathscr{A}|_{X_u}, \qquad \widetilde{\mathscr{A}_s} = \widetilde{\mathscr{A}}|_{X_s},$$

where $X_u = \text{lin span}\{\varphi_j\}_{j=1}^N = P_N\widetilde{H}$, $X_s = (I - P_N)\widetilde{H}$.

If we set $y_u = \sum_{j=1}^M y_i\varphi_i$ and take into account the biorthogonality relation, we get by (3.106) and (3.95)

$$\begin{cases} \frac{dy_i}{dt} + \lambda_i y_i = -\nu \sum_{j=1}^M v_j(t)(\frac{\partial \varphi_j^*}{\partial n}, \frac{\partial \varphi_i^*}{\partial n})_{(L^2(\partial\mathscr{O}))^d}, \\ y_i(0) = y_i^0, \quad i = 1, \ldots, N. \end{cases} \tag{3.108}$$

Equivalently,

$$\begin{aligned} &\frac{dz}{dt} + \Lambda z = Bv, \quad t \ge 0, \\ &z(0) = z_0, \end{aligned} \tag{3.109}$$

where

$$\Lambda = \operatorname{diag}\|\lambda_j\|_{j=1}^{N}, \qquad B = -\nu\|b_{ij}\|_{i=1,j=1}^{N.M},$$
$$b_{ij} = \left(\frac{\partial \varphi_j^*}{\partial n}, \frac{\partial \varphi_i^*}{\partial n}\right)_{(L^2(\partial\mathscr{O}))^d} \qquad \text{and } v = \{v_j\}_{j=1}^{M}.$$

Taking into account Assumption (K2), we conclude by the same argument as that in the proof of Theorem 3.1 that, for each $T > 0$, System (3.109) is exactly null controllable on $[0, T]$. This implies, however, via finite-dimensional quadratic stabilization, that the stabilizable controller v can be taken as in (3.104), (3.105). Indeed, in this case there is a feedback controller $v(t) = Rz(t)$ which stabilizes exponentially System (3.109) that is $z' + \Lambda z = BRz$ on $(0, \infty)$ and

$$|z(t)| \le Ce^{-3\gamma t}|z_0|, \quad \forall t \ge 0.$$

Clearly, this implies that this controller $v \in C([0, \infty); \mathbb{C}^M)$ is differentiable on $[0, \infty)$ and satisfies (3.102)–(3.105), as claimed.

Now, we come back to the infinite-dimensional system (3.107) which, by virtue of (3.92), can be written as

$$\begin{aligned} y_s(t) = {} & \sum_{j=1}^{M} v_j(t) D\left(\frac{\partial \varphi_j^*}{\partial n}\right) + e^{-\mathscr{A}_s t}\left(y_s(0) - \sum_{j=1}^{M} v_j(0) D\left(\frac{\partial \varphi_j^*}{\partial n}\right)\right) \\ & - k\int_0^t e^{-\mathscr{A}_s(t-s)} \sum_{j=1}^{M} v_j(s) D\left(\frac{\partial \varphi_j^*}{\partial n}\right) ds \\ & - \int_0^t \sum_{j=1}^{M} v_j'(s) e^{-\mathscr{A}_s(t-s)} D\left(\frac{\partial \varphi_j^*}{\partial n}\right) ds. \end{aligned} \tag{3.110}$$

Recalling (3.105) and that $\sigma(\mathscr{A}_s) \subset \{\lambda;\ \operatorname{Re}\lambda > \gamma\}$, we have, for some $\delta > 0$,

$$\|e^{-\mathscr{A}_s t}\|_{L(H,H)} \le Ce^{-(\gamma+\delta)t}, \quad \forall t \ge 0, \tag{3.111}$$

while

$$\left\| D\left(\frac{\partial \varphi_j^*}{\partial n}\right)\right\|_{(H^2(\mathscr{O}))^d} \le C\left\|\frac{\partial \varphi_j^*}{\partial n}\right\|_{(H^{\frac{3}{2}}(\partial\mathscr{O}))^d}, \tag{3.112}$$

and so, (3.101) holds.

We have, also, by (3.110)

$$|A^{\frac{1}{4}-\varepsilon} y_s(t)| \le \sum_{j=1}^{M} |v_j(t)| \left|A^{\frac{1}{4}-\varepsilon} D\left(\frac{\partial \varphi_j^*}{\partial n}\right)\right|$$

$$+\left|A^{\frac{1}{4}-\varepsilon}e^{-\mathscr{A}_s t}\left(y_s(0)-\sum_{j=1}^{M}v_j(0)D\left(\frac{\partial\varphi_j^*}{\partial n}\right)\right)\right|$$

$$+C\int_0^t e^{-3\gamma s}\sum_{j=1}^{M}\left|A^{\frac{1}{4}-\varepsilon}e^{-\mathscr{A}_s(t-s)}D\left(\frac{\partial\varphi_j^*}{\partial n}\right)\right|ds. \tag{3.113}$$

Taking into account (3.111), (3.112), we obtain that

$$|A^{\frac{1}{4}-\varepsilon}y_s(t)|\le Ce^{-\gamma t}|A^{\frac{1}{4}-\varepsilon}y_s(0)| + C\int_0^t e^{-3\gamma s}\sum_{j=1}^{M}\left|A^{\frac{1}{4}-\varepsilon}e^{-\mathscr{A}_s(t-s)}D\left(\frac{\partial\varphi_j^*}{\partial n}\right)\right|ds,$$
$$\forall t\ge 0. \tag{3.114}$$

By (3.110), we have also that

$$|A^{\frac{3}{4}-\varepsilon}y_s(t)|\le Ce^{-2\gamma t}\sum_{j=1}^{M}\left|A^{\frac{3}{4}-\varepsilon}D\left(\frac{\partial\varphi_j^*}{\partial n}\right)\right|$$
$$+\left|A^{\frac{3}{4}-\varepsilon}\left(e^{-\mathscr{A}_s t}(y_s(0))-\sum_{j=1}^{M}v_j(0)D\left(\frac{\partial\varphi_j^*}{\partial n}\right)\right)\right|$$
$$+\int_0^t e^{-3\gamma s}\sum_{j=1}^{M}\left|A^{\frac{3}{4}-\varepsilon}e^{-\mathscr{A}_s(t-s)}D\left(\frac{\partial\varphi_j^*}{\partial n}\right)\right|ds. \tag{3.115}$$

On the other hand, if we denote $z(t)=e^{\gamma t}e^{-\mathscr{A}_s t}z_0$, we have

$$z'+\nu Az+A_0 z-\gamma z=0,\quad \forall t\ge 0,$$

and this yields, by multiplying with $A^{\frac{1}{2}-2\varepsilon}z$ and using the interpolation inequality,

$$\frac{1}{2}\frac{d}{dt}|z(t)|^2_{\frac{1}{4}-\varepsilon}+\nu|z(t)|^2_{\frac{3}{4}-\varepsilon}\le C|z(t)|^2,\quad \forall t\ge 0.$$

Since, by (3.111), $|z|_{L^2(0,\infty;H)}\le C|z_0|$, we have

$$|z(t)|^2_{\frac{1}{4}-\varepsilon}+\int_0^\infty |A^{\frac{3}{4}-\varepsilon}z(t)|^2dt\le C|z_0|^2_{\frac{1}{4}-\varepsilon}$$

and, therefore,

$$|e^{-\mathscr{A}_s t}z_0|^2_{\frac{1}{4}-\varepsilon}e^{2\gamma t}+\int_0^\infty e^{2\gamma t}|A^{\frac{3}{4}-\varepsilon}e^{-\mathscr{A}_s t}z_0|^2dt\le C|z_0|^2_{\frac{1}{4}-\varepsilon}.$$

By (3.112), we have also

$$\int_0^\infty \left| e^{-\mathscr{A}_s t} A^{\frac{3}{4}-\varepsilon} D\left(\frac{\partial \varphi_j^*}{\partial n}\right)\right|^2 e^{2\gamma t} dt \le C.$$

Then, by (3.114), (3.115) we obtain that

$$e^{2\gamma t}|A^{\frac{1}{4}-\varepsilon} y_s(t)|^2 + \int_0^\infty e^{2\gamma t}|A^{\frac{3}{4}-\varepsilon} y_s(t)|^2 dt \le C\|y_s(0)\|^2_{D(A^{\frac{1}{4}-\varepsilon})},$$

which implies (3.102), (3.103). This completes the proof. □

Remark 3.7 Controller (3.100) is tangential, that is, $u(t)\cdot n = 0$, a.e., on $\partial\mathscr{O}$. Indeed, by Lemma 3.3.1 in [22], we know that

$$(\nabla\varphi_j^* \cdot n)\cdot n = 0, \quad \text{a.e., on } (0,\infty)\times\partial\mathscr{O},\ j=1,\dots,N,$$

where n is the normal to $\partial\mathscr{O}$.

Remark 3.8 As seen earlier in Proposition 3.5, Hypothesis (K1) holds "almost for all y_e" in the sense of genericity. As regards Hypothesis (K2), it is not clear whether it is satisfied for all eigenfunctions systems $\{\varphi_j\}_{j=1}^N$, $\{\varphi_j^*\}_{j=1}^N$. In fact, it is equivalent with the following unique continuation property

$$\nabla\varphi_j\cdot n \not\equiv 0 \quad (\text{respectively, } \nabla\varphi_j^*\cdot n \not\equiv 0) \quad \text{on } \partial\mathscr{O},$$

which to our knowledge is still an open problem. One might suspect, however, that this property is generic too in the class of all $\{y_e\}$ and it is strong evidence that this is, indeed, the case.

Now, we design a stabilizable feedback controller u for (3.91) in the absence of Hypothesis (K2). To this purpose, we orthogonalize the system $\{\varphi_j\}_{j=1}^N$, that is, $(\varphi_j,\varphi_j)=\delta_{ij}$, $i,j=1,\dots,N$, and set as above

$$X_u == P_N(\widetilde{H}) = \text{lin span}\{\varphi_j\}_{j=1}^N, \qquad X_s = (I-P_N)\widetilde{H},$$

where $P_N : H \to X_u$ is the algebraic projection on X_u, and denote by P_N^* its dual. We assume that

$$\textit{The system } \left\{\frac{\partial\varphi_j}{\partial n}\right\}_{j=1}^N \textit{ is linearly independent in } (L^2(\partial\mathscr{O}))^d. \tag{3.116}$$

Then, as noticed earlier, there is a system $\{\Phi_j\}_{j=1}^N \subset (L^2(\partial\mathscr{O}))^d$ of the form

$$\Phi_j = \sum_{k=1}^N \alpha_{jk}\frac{\partial\varphi_k}{\partial n}, \quad j=1,\dots,N, \tag{3.117}$$

such that

$$\int_{\partial\Omega} \frac{\partial \varphi_j}{\partial n} \overline{\Phi}_i \, dx = \delta_{ij}, \quad i, j = 1, \ldots, N. \tag{3.118}$$

Consider the feedback controller

$$u(t) = -\eta \sum_{j=1}^{N} (y, \varphi_j) \Phi_j, \tag{3.119}$$

which, inserted into System (3.93), yields

$$\begin{aligned} &\frac{dy}{dt} + \widetilde{\mathscr{A}} y = -\eta \sum_{j=1}^{N} (y, \varphi_j)(\widetilde{\mathscr{A}} + k) D\Phi_j, \quad t \geq 0, \\ &y(0) = y^0. \end{aligned} \tag{3.120}$$

We have the following theorem.

Theorem 3.7 *For* $\eta \geq \eta_0 > 0$ *sufficiently large, the solution* $y \in C([0,\infty); \widetilde{H})$ *to System* (3.120) *satisfies*

$$\|y(t)\|_{D(A^{\frac{1}{4}-\varepsilon})} \leq Ce^{-\gamma t} \|y^0\|_{D(A^{\frac{1}{4}-\varepsilon})}, \quad \forall t > 0, \tag{3.121}$$

$$\int_0^\infty e^{2\gamma t} \|y(t)\|^2_{D(A^{\frac{1}{4}-\varepsilon})} \, dt \leq C \|y^0\|^2_{D(A^{\frac{1}{4}-\varepsilon})}. \tag{3.122}$$

Proof We proceed as in the previous case. Namely, we set

$$\mathscr{A}_u = \mathscr{A}|_{X_u}, \qquad \widetilde{\mathscr{A}_s} = \widetilde{\mathscr{A}}|_{X_s}.$$

Then, setting

$$y = y_u + y_s, \quad y_u = \sum_{j=1}^{N} y_j \varphi_j,$$

we rewrite System (3.120) as

$$\frac{dy_i}{dt} + \sum_{j=1}^{N} a_{ij} y_j = -\eta y_i, \quad i = 1, \ldots, N, \tag{3.123}$$

$$\begin{aligned} &\frac{dy_s}{dt} + \widetilde{\mathscr{A}_s} y_s = -\eta \sum_{j=1}^{N} y_j (\widetilde{\mathscr{A}} + k) D\Phi_j, \quad t \geq 0, \\ &y_s(0) = (I - P_N) y_0. \end{aligned} \tag{3.124}$$

Here, $a_{ij} = (\mathscr{A}\varphi_i, \varphi_j)$. Then, for $\eta \geq \eta_0$ sufficiently large, we have for solution $\{y_i\}_{i=1}^N$ to (3.123)

$$|y_i(t)| \leq e^{-\gamma t} |y_i(0)|, \quad i = 1, \ldots, N,$$

and, inserting the latter into (3.124), we obtain exactly as in the proof of Theorem 3.6 that

$$\|y_s(t)\|_{D(A^{\frac{1}{4}-\varepsilon})} \le Ce^{-\gamma t}\|y(0)\|_{D(A^{\frac{1}{4}-\varepsilon})}$$

and

$$\int_0^\infty e^{2\gamma t}\|y_s(t)\|^2_{D(A^{\frac{1}{4}-\varepsilon})}\,dt \le C\|y(0)\|^2_{D(A^{\frac{1}{4}-\varepsilon})},$$

which completes the proof. □

We remark also that Controller (3.119) is tangential too, that is $u(t)\cdot n = 0$ on $\partial\mathcal{O}$.

It should be recalled that the above results were obtained for the complexified Stokes–Oseen systems in the space $\widetilde{H} = H + iH$. For the time being, it is useful, however, to express them for the real Stokes–Oseen systems.

Corollary 3.1 *Under Assumptions* (K1), (K2), *there is a stabilizable real-valued controller u^* (in the sense of* (3.102)∼(3.103))

$$u^*(t) = \sum_{j=1}^{M^*} v_j^*(t)\,\frac{\partial\psi_j^*}{\partial n}, \tag{3.125}$$

where $\psi_j^* = \operatorname{Re}\varphi_j^*$ or $\operatorname{Im}\varphi_j^*$ for $j = 1, \dots, M^*$ and M^* is defined by (2.63).

Remark 3.9 Taking into account that $\gamma < \operatorname{Re}\lambda_{N+1}$, where λ_{N+1} is the first stable eigenvalue for the operator $\mathscr{A}$, it is clear that in Theorems 3.6 and 3.7 respectively in Estimates (3.102), (3.103) respectively (3.121), (3.122), the exponent γ can be replaced by $\gamma + \delta$, where $\delta = \operatorname{Re}\lambda_{N+1} - \gamma$.

Indeed, taking $y^1 = \operatorname{Re} y$ in System (3.93) with the controller u given by (3.100), we see that

$$\begin{aligned}\frac{dy^1}{dt} + \widetilde{\mathscr{A}}y^1 &= (\widetilde{\mathscr{A}} + kI)\sum_{j=1}^{M^*}\left(\operatorname{Re} v_j(t)\operatorname{Re}\frac{\partial\varphi_j^*}{\partial n} - \operatorname{Im} v_j(t)\operatorname{Im}\frac{\partial\varphi_j^*}{\partial n}\right)\\ &= (\widetilde{\mathscr{A}} + kI)\sum_{j=1}^{M^*} v_j^*(t)\frac{\partial\varphi_j^*}{\partial n}.\end{aligned}$$

(Since, for every complex eigenvalue λ_j the system $\{\lambda_j\}_{j=1}^N$ contains also its conjugate $\overline{\lambda}_j$ with eigenfunction $\overline{\varphi}_j$ it is clear that the dimension of the controller u^* remains M^*.)

Corollary 3.2 *Under Assumptions* (3.116), *there is a real stabilizable controller* u *of the form*

$$u^*(t) = -\eta \sum_{i=1}^{N} v_i(t)\phi_j^*(t), \quad t \geq 0, \tag{3.126}$$

where $\eta \geq \eta_0 > 0$ *and*

$$\phi_i^* \in \text{lin span}\left\{\text{Re}\,\frac{\partial \varphi_\ell}{\partial n}, \text{Im}\,\frac{\partial \varphi_\ell}{\partial n}\right\}_{\ell=1}^{N}, \quad i = 1, \dots, N.$$

Remark 3.10 Since $\gamma < \text{Re}\,\lambda_{N+1}$, it is clear that in Estimates (3.107), (3.108), (3.128) and (3.129) of Theorems 3.6 and 3.7 the exponent γ can be replaced by $\gamma + \delta$, where $\delta = \text{Re}\,\lambda_{N+1} - \gamma$.

3.4.1.1 The $(H^{\frac{1}{2}-\varepsilon}(\mathcal{O}))^d$ Topological Level Versus $(H^{\frac{1}{2}+\varepsilon}(\mathcal{O}))$ Level

By virtue of Proposition 3.5, Part (3.98), (3.99) in order to have for the state y of the control system (3.92) the regularity level $L^2(0, \infty; D(A^{\frac{3}{4}+\varepsilon}))$ it is necessary to assume that the controller u satisfies the compatibility condition (3.98) and since $y \in C([0, T]; H^{\frac{1}{2}+2\varepsilon}(\mathcal{O}) \cap H)$ we have by trace theorem that $y = u$ in $C([0, T], (H^{2\varepsilon}(\partial\mathcal{O}))^d)$ and, therefore, $y^0|_{\partial\mathcal{O}} = u(0)$. This precludes the existence of a finite-dimensional stabilizable controller u of the form (3.100) or (3.126) and so, for this purpose we must confine to the pair $((H^{\frac{1}{2}-\varepsilon}(\mathcal{O}))^d, (H^{\frac{3}{2}-\varepsilon}(\mathcal{O}))^d)$.

Of course, for the linear stabilization theory this basic space $(H^{\frac{1}{2}-\varepsilon}(\mathcal{O}))$ is quite convenient and does not impose any restriction on the dimension d. However, as seen in Sect. 3.1 (see the proof of Theorem 3.5) the analysis of nonlinear inertial term $Sy = P((y \cdot \nabla)y)$ requires in dimension $d = 3$ some estimates in $(H^{\frac{3}{2}+\varepsilon}(\mathcal{O}))^d$ (Lemma 3.3). In the boundary stabilization case (see next section), we need the same requirement which precludes the boundary stabilization analysis for $d = 3$ in the case of the Navier–Stokes equation but not for its linearization, in which case the topological level $(H^{\frac{1}{2}-\varepsilon}(\mathcal{O}))^d$ works.

3.4.2 Stabilizable Boundary Feedback Controllers via Low-gain Riccati Equation

For the time being, it is useful to have under the assumptions of Theorem 3.6 a stabilizable feedback controller of the form (3.125) or (3.126) respectively. The standard way to find such a controller is via the quadratic optimal control problem governed by the Stokes–Oseen operator with boundary control. We have seen in the case of the internal stabilization problem that two quadratic cost functionals are appropriate in this case; the high-gain cost functional

$$\int_0^\infty (|A^{\frac{3}{4}} y(t)|^2 + |v(t)|^2_{M^*})dt$$

and the low-gain cost functional

$$\int_0^\infty (|y(t)|^2 + |v(t)|^2_{M^*})dt.$$

As mentioned earlier, the advantage of high-gain cost is that it provides a robust feedback controller for the linear system and keeps this quality when inserted into the nonlinear (Navier–Stokes) equation. On the other hand, the low-gain cost observation leads to a simpler Riccati equation and robustness of the stabilizable feedback controller in a narrower class of perturbations. Here, we design a feedback stabilizable controller starting from a low-gain observation (cost). Namely, we consider the cost functional

$$J(v) = \int_0^\infty (|y_v(t)|^2 + |v(t)|^2_{M^*})dt,$$

where $|v|_{M^*}$ is the norm in the Euclidean space R^{M^*}, while y_v is the solution to the controlled system

$$\begin{aligned} &\frac{dy}{dt} + \tilde{\mathscr{A}} y - \gamma y = \sum_{i=1}^{M^*} v_i(t)(\tilde{\mathscr{A}} + k) D\left(\frac{\partial \psi_i^*}{\partial n}\right), \quad t \geq 0, \\ &y(0) = y_0. \end{aligned} \tag{3.127}$$

By Theorem 3.6, we know that, under Assumptions (K) and (KK), the minimization problem

$$\inf\{J(v);\ v \in L^2(0, \infty; R^{M^*})\} \tag{3.128}$$

has a unique solution v^* and, by standard theory of linear quadratic optimal control problems, we know that there is $R_0 \in L(H, H)$, $R_0 = R_0^*$ such that

$$(R_0 y_0, y_0) = \inf\{J(v);\ v \in L^2(0, \infty; R^{M^*}))= \int_0^\infty (|y^*(t)|^2 + |v^*(t)|^2_{M^*})dt, \tag{3.129}$$

where $y^* = y_{v^*}$.

In Theorem 3.8, we collect together the main properties of the operator R_0 and we prove that the optimal controller $v^*(t) = v(t, y_0)$ in (3.128) is a stabilizing feedback controller of the form

$$v_i^*(t) = \nu\left(\frac{\partial}{\partial n} R_0 y^*(t), \frac{\partial \psi_i^*}{\partial n}\right)_{(L^2(\partial\mathscr{O}))^d}, \quad t \geq 0. \tag{3.130}$$

Theorem 3.8 *Assume that Hypotheses* (K1) *and* (K2) *hold. Then the operator* $R_0 \in L(H,H)$ *is the unique self-adjoint and positive solution to the algebraic Riccati equation*

$$\begin{aligned}(R_0y, \mathscr{A}z - \gamma z) &+ (\mathscr{A}^*y - \gamma y, R_0z) \\ &+ \nu^2\left(\frac{\partial}{\partial n}R_0y, \frac{\partial}{\partial n}R_0z\right)_{(L^2(\partial\mathscr{O}))^2} = (y,z), \quad \forall y,z \in D(\mathscr{A}). \qquad (3.131)\end{aligned}$$

We have also

(i) $R_0y \in D(A)$, $\forall y \in H$ *and the operator* $F = \nu\frac{\partial}{\partial n}R_0$ *is continuous from* H *to* $(H^{\frac{1}{2}}(\partial\mathscr{O}))^2$.

(ii) *The operator* $A_F : D(A_F) \subset H \to H$, $-A_Fy = \mathscr{A}(y - DFy) - kDFy, \forall y \in D(A_F) = \{y \in H;\ y - DFy \in D(A)\}$ *is the infinitesimal generator of a* C_0*-analytic semigroup* e^{-A_Ft} *in* H *and*

$$\|e^{-A_Ft}\|_{L(H,H)} \le Ce^{-\gamma t}, \quad \forall t \ge 0. \qquad (3.132)$$

(iii) e^{-A_Ft} *is a* C_0*-analytic semigroup in* $W = D(A^{\frac{1}{4}-\varepsilon})$ *and*

$$\|e^{-A_Ft}\|_{L(W,W)} \le Ce^{-\gamma t},$$

$$\int_0^\infty |A^{\frac{3}{4}-\varepsilon}e^{-A_Ft}y_0|^2e^{2\gamma t}dt \le C\|y_0\|_W^2, \quad \forall y_0 \in W.$$

Moreover, the optimal controller v^* *in Problem* (3.128) *is expressed in the feedback form* (3.130) *or, equivalently,*

$$v_i^*(t,y_0) = \nu\left(\frac{\partial}{\partial n}R_0e^{-A_Ft}y_0, \frac{\partial\psi_i^*}{\partial n}\right)_{(L^2(\partial\mathscr{O}))^d}\frac{\partial\psi_i^*}{\partial n} \quad \forall t \ge 0. \qquad (3.133)$$

By Theorem 3.8, we have

Corollary 3.3 *Under Assumptions* (K1) *and* (K2), *the feedback controller*

$$u(t) = \nu\sum_{i=1}^{M^*}\left(\frac{\partial}{\partial n}(R_0y(t)), \frac{\partial\psi_i^*}{\partial n}\right)_{(L^2(\partial\mathscr{O}))^d}\frac{\partial\psi_i^*}{\partial n}, \quad t \ge 0,$$

exponentially stabilizes System (3.90) *in* H *and* W *with exponent decay* $-\gamma$. *In other words, the solution* y *to the closed-loop system*

$$\begin{aligned}&\frac{\partial y}{\partial \nu} - \nu\Delta y + (y\cdot\nabla)y_e + (y_e\cdot\nabla)y = \nabla p && \text{in } (0,\infty)\times\mathscr{O},\\ &y(0,x) = y_0(x), && \forall x \in \mathscr{O},\\ &y = \nu\sum_{i=1}^{M^*}\left(\frac{\partial}{\partial n}(R_0y), \frac{\partial\psi_i^*}{\partial n}\right)_{(L^2(\partial\mathscr{O}))^d}\frac{\partial\psi_i^*}{\partial n} && \text{on } (0,\infty)\times\partial\mathscr{O},\end{aligned}$$

satisfies

$$\|y(t)\|_W \le Ce^{-\gamma t}\|y_0\|_W, \quad \forall t \ge 0,\ y_0 \in W,$$
$$|y(t)| \le Ce^{-\gamma t}|y_0|, \quad \forall t \ge 0,\ y_0 \in H.$$

Proof of Theorem 3.8 We denote by $\Gamma : R^{M^*} \to (D(\widetilde{\mathscr{A}}))'$ the operator

$$\Gamma v = \sum_{i=1}^{M^*} v_i(\widetilde{\mathscr{A}} + k)D\left(\frac{\partial \psi_i^*}{\partial n}\right), \quad v \in R^{M^*}.$$

Then, System (3.127) can be rewritten as

$$\begin{aligned} &\frac{dy}{dt} + \widetilde{\mathscr{A}}y - \gamma y = \Gamma v, \quad t \ge 0,\\ &y(0) = y_0, \end{aligned} \tag{3.134}$$

and so, by standard maximum principle for infinite time horizon linear quadratic optimal control problems, the optimal controller v^* in Problem (3.128) is expressed as

$$v^*(t) = \Gamma^* p(t) = \left\{ -\nu \int_{\partial\mathscr{O}} \frac{\partial p}{\partial n} \cdot \frac{\partial \psi_i^*}{\partial n}\, dx \right\}_{i=1}^{M^*}, \tag{3.135}$$

where Γ^* is the dual of Γ and p is the solution to the dual backward system

$$\begin{aligned} &\frac{dp}{dt} - \mathscr{A}^* p + \gamma p = y^*, \quad t \ge 0,\\ &p(\infty) = 0. \end{aligned} \tag{3.136}$$

(Since $y^* \in L^2(0,\infty; H)$, the dual equation to (3.134) involves $\mathscr{A}^*$ only.)

By the dynamic programming principle, we have that v^* is still optimal in the problem

$$\operatorname{Min}\left\{ \int_0^t (|y_v(t)|^2 + |v(t)|^2_{M^*})dt + (R_0 y(t), y(t)) \right\}, \quad \forall t \ge 0,$$

and this yields

$$p(t) = -R_0 y^*(t), \quad \forall t \ge 0, \tag{3.137}$$

which, by virtue of (3.135), implies that v^* has the feedback representation

$$v^*(t) = -\Gamma^* R_0 y^*(t) = \nu \left\{ \int_{\partial\mathscr{O}} \frac{\partial}{\partial n} R_0 y^*(t) \frac{\partial}{\partial n} \psi_i^* dx \right\}_{i=1}^{M^*}. \tag{3.138}$$

Since, by (3.136) and the smoothing effect of the semigroup $e^{-\mathscr{A}t}$, $p(t) \in D(A)$, $\forall t \geq 0$, we have that $R_0 \in L(H, D(A))$ and so, by the trace theorem,

$$\frac{\partial}{\partial n} R_0 y^*(t) \in (H^{\frac{1}{2}}(\partial\mathscr{O}))^d, \quad \forall t \geq 0.$$

We note, also, that by (3.138) and by

$$(R_0 y^*(t), y^*(t)) = \int_t^\infty (|y^*(s)|^2 + |v^*(s)|^2) ds, \quad \forall t \geq 0,$$

it follows that

$$\tfrac{d}{dt} (R_0 y^*(t), y^*(t)) = -(|y^*(t)|^2 + |\Gamma^* R_0 y^*(t)|^2_{M^*}), \quad t \geq 0.$$

Together with (3.128) and (3.138), the latter yields

$$(\mathscr{A} y^*(t) - \gamma y^*(t), R_0 y^*(t)) + \frac{1}{2} |\Gamma^* R_0 y^*(t)|^2_{M^*} = \frac{1}{2} |y^*(t)|^2, \quad \forall t \geq 0,$$

which, clearly, implies (3.131), as claimed.

Now, to prove (ii) and (iii), consider the closed-loop system (3.134) with the feedback controller (3.138), that is,

$$\begin{aligned} &\frac{dy}{dt} + \tilde{\mathscr{A}} y - \gamma y + \Gamma \Gamma^* R_0 y = 0, \quad t \geq 0, \\ &y(0) = y_0. \end{aligned} \tag{3.139}$$

Equivalently,

$$\begin{aligned} &\frac{dy}{dt} + A_F y - \gamma y = 0, \quad t \geq 0, \\ &y(0) = y_0. \end{aligned} \tag{3.140}$$

As seen above, for each $y_0 \in H$, (3.140) has a unique solution $y^* = y(t, y_0) \in C([0, \infty); H)$ and the dynamics $t \to y^*(t)$ is a C_0-semigroup on H, that is, $y(t, y_0) = e^{-A_F t} y_0$, $\forall y_0 \in H$. Moreover, by (3.140) and (3.131), we see that

$$\frac{1}{2} \frac{d}{dt} (R_0 y(t), y(t)) + \gamma (R_0 y(t), y(t)) + \frac{1}{2} |\Gamma^* R_0 y(t)|^2_{M^*} + \frac{1}{2} |y(t)|^2 = 0,$$
$$\text{a.e., } t > 0,$$

and, therefore,

$$\begin{aligned} &(R_0 y(t), y(t)) \leq e^{-2\gamma t} (R_0 y_0, y_0), \quad \forall t > 0, \\ &\int_0^\infty |y(s)|^2 e^{2\gamma s} ds \leq C |y^0|^2, \quad \forall y_0 > H. \end{aligned} \tag{3.141}$$

The latter implies (3.132).

As regards (iii), it follows directly by Theorem 3.6 (Part (3.102), (3.103)) that the operator $A_F y = \mathscr{A}(y - DFy) - kDFy$ satisfies Conditions (ii), (iii) of Theorem 3.8. The proof is completely similar to that of Theorem 3.7 and so it is omitted. □

Remark 3.11 In (ii) and (iii), the exponent γ can be replaced by $\gamma + \delta$, where $\delta = \operatorname{Re}\lambda_{N+1} - \gamma$ (see Remark 3.9).

3.4.3 The Boundary Feedback Stabilization of Navier–Stokes Equations

One might suspect that the stabilizable feedback controllers for the Stokes–Oseen system found in the previous section would stabilize the Navier–Stokes equation in a neighborhood of the origin (respectively, of equilibrium solution). We see below that this is, indeed, the case and we prove this by a fixed-point argument similar to that used in the proof of Theorem 3.4.

We come back to the Navier–Stokes equation (3.1) with $f \equiv f_e$ and a boundary tangential controller u, that is

$$\begin{aligned} &\frac{\partial y}{\partial t} - \nu \Delta y + (y \cdot \nabla) y = f_e + \nabla p && \text{in } (0,\infty) \times \mathcal{O}, \\ &y(0) = y_0 && \text{in } \mathcal{O}, \\ &\nabla \cdot y = 0 && \text{in } (0,\infty) \times \mathcal{O}, \\ &y = u && \text{on } (0,\infty) \times \partial\mathcal{O}. \end{aligned} \tag{3.142}$$

Equivalently (see (3.7)),

$$\begin{aligned} &\frac{\partial y}{\partial t} - \nu \Delta y + (y \cdot \nabla) y + (y \cdot \nabla) y_e + (y \cdot \nabla) y = \nabla p && \text{in } (0,\infty) \times \mathcal{O}, \\ &y(0) = y_0 - y_e && \text{in } \mathcal{O}, \\ &\nabla \cdot y = 0 && \text{in } (0,\infty) \times \mathcal{O}, \\ &y = u && \text{on } (0,\infty) \times \partial\mathcal{O}. \end{aligned}$$

The latter reduces as above to (see (3.93))

$$\begin{aligned} &\frac{dy}{dt} y(t) + \widetilde{\mathscr{A}} y(t) + S y(t) = (\widetilde{\mathscr{A}} + k) D u, \quad t \ge 0, \\ &y(0) = y_0. \end{aligned} \tag{3.143}$$

(We have denoted $y_0 - y_e$ again by y_0.)

We study below the stabilization of the controlled equation (3.143) via the abstract feedback controller

$$u = Fy, \tag{3.144}$$

where $F \in L(W, (L^2(\partial\mathcal{O}))^d)$, and

$$W = D(A^{\frac{1}{4}-\varepsilon}) = H^{\frac{1}{2}-2\varepsilon}(\mathcal{O}) \cap H \quad \text{if } d=2, \quad \text{and}$$

$$W = D(A^{\frac{1}{4}+\varepsilon}) = H^{\frac{1}{2}+2\varepsilon}(\mathcal{O}) \cap H \quad \text{if } d=3.$$

We also set

$$Z = D(A^{\frac{3}{4}-\varepsilon}) = H^{\frac{3}{2}-2\varepsilon}(\mathcal{O}) \cap V \quad \text{if } d=2,$$

$$Z = D(A^{\frac{3}{4}+\varepsilon}) = H^{\frac{3}{2}+2\varepsilon}(\mathcal{O}) \cap V \quad \text{if } d=3.$$

Consider the operator

$$A_F y = \mathcal{A}(y - DFy) - kDFy, \quad \forall y \in D(A_F),$$

$$D(A_F) = \{y \in H;\ y - DFy \in D(A)\}.$$

If we take u of the form (3.144) and plug it into (3.143), we are lead to the closed-loop system

$$\begin{aligned} &\frac{dy}{dt} + A_F y + Sy = 0 \quad \text{in } (0,\infty), \\ &y(0) = y_0. \end{aligned} \tag{3.145}$$

Everywhere in the following, the solution y to (3.145) is considered in the following "mild" sense

$$y(t) = e^{-A_F t} y_0 - \int_0^t e^{-A_F(t-s)} Sy(s)ds. \tag{3.146}$$

We assume for the time being that the following assumptions hold.

(j) $-A_F$ generates a C_0-analytic semigroup $e^{-A_F t}$ in W and H. Moreover, there are $C, c, \delta > 0$, such that

$$\|e^{-A_F t} y_0\|_W \le \sqrt{c}\, e^{-(\gamma+\delta)t} \|y_0\|_W, \quad \forall t \ge 0,\ y_0 \in W. \tag{3.147}$$

(jj) $\int_0^\infty \|e^{-A_F t} y_0\|_Z^2 e^{2\gamma t} dt \le C\|y_0\|_W^2$, $\forall y_0 \in W$.

Theorem 3.9 is the main stabilization result.

Theorem 3.9 *Assume that Assumptions* (j) *and* (jj) *hold. Then there is* $\rho > 0$ *such that, for all* $\|y_0\|_W \le \rho$, (3.145) *has a unique solution* $y \in C([0,\infty); W)$ *such that*

$$\|y(t)\|_W \le Ce^{-\gamma t}\|y_0\|_W, \quad \forall t \ge 0, \tag{3.148}$$

$$\int_0^\infty e^{2\gamma t}\|y(s)\|_Z^2 ds \le C\|y_0\|_W^2. \tag{3.149}$$

Here,

$$W = D(A^{\frac{1}{4}-\varepsilon}), \quad Z = D(A^{\frac{3}{4}-\varepsilon}), \quad \text{if } d = 2 \quad \text{and}$$

$$W = D(A^{\frac{1}{4}+\varepsilon}), \quad Z = D(A^{\frac{3}{4}+\varepsilon}), \quad \text{if } d = 3.$$

Later on, we present some significant examples of such feedback laws $u = Fy$. In particular, we obtain the following stabilization result by Theorem 3.9.

Corollary 3.4 *Under Assumptions* (j) *and* (jj), *the boundary feedback controller*

$$u(t) = -F(y(t) - y_e), \quad \forall t \ge 0, \tag{3.150}$$

exponentially stabilizes the equilibrium solution y_e *to* (3.142) *for* $\|y_0 - y_e\|_W \le \rho$, *that is,*

$$\|y(t) - y_e\|_W \le Ce^{-\gamma t}\|y_0 - y_e\|_W, \quad \forall t \ge 0. \tag{3.151}$$

Proof of Theorem 3.9 For simplicity, we take $\gamma = 0$ as the general case follows by substituting into (3.145), $y(t)$ by $y(t)e^{\gamma t}$. The proof is quite similar to that of Theorem 3.5. For any $r > 0$, we introduce the ball $\mathscr{K}_r$ of radius r, centered at the origin, of the space $L^2(0,\infty;Z)$:

$$\mathscr{K}_r \equiv \left\{ z \in L^2(0,\infty;Z) : \|z\|_{L^2(0,\infty;Z)} = \left\{ \int_0^\infty \|z(t)\|_Z^2 dt \right\}^{\frac{1}{2}} \le r \right\}.$$

Next, for any $\eta_0 \in W$ and $z \in L^2(0,\infty;Z)$, we introduce the map

$$(\Lambda z)(t) \equiv e^{-A_F t} y_0 - (Nz)(t);$$

$$(Nz)(t) \equiv \int_0^t e^{-A_F(t-\tau)}(Sz)(\tau)d\tau.$$

Clearly, (3.145) reduces to $y = \Lambda y$ and, therefore, to existence of a fixed-point to operator Λ on $\mathscr{K}_r$.

First, we prove the following inequalities

$$\begin{aligned}
\|Nz\|_{L^2(0,\infty;Z)} &\equiv \left\{ \int_0^\infty \|(Nz)(t)\|_Z^2 dt \right\}^{\frac{1}{2}} \\
&= \left\{ \int_0^\infty \left\| \int_0^t e^{-A_F(t-\tau)}(Sz)(\tau)d\tau \right\|_Z^2 \right\}^{\frac{1}{2}} \\
&\le \sqrt{c}\|Sz\|_{L^1(0,\infty;W)}
\end{aligned}$$

$$= \sqrt{c} \int_0^\infty \|(Sz)(\tau)\|_W d\tau. \tag{3.152}$$

To this end, we proceed by duality. Let $\zeta \in L^2(0, \infty; Z')$, Z' is the dual of Z with H as a pivot space. Then, starting from (3.152) on N, we perform a change in the order of integration via Fubini's Theorem, we use Schwarz inequality and we invoke

$$\begin{aligned}
&\int_0^\infty ((Nz)(t), \zeta(t))dt \\
&= \int_0^\infty \left(\int_0^t e^{-A_F(t-\tau)}(Sz)(\tau)d\tau, \zeta(t) \right) dt \\
&\le \int_0^\infty \int_0^t \|e^{-A_F(t-\tau)}(Sz)(\tau)\|_Z \|\zeta(t)\|_{Z'} d\tau\, dt \\
&= \int_0^\infty \left\{ \int_\tau^\infty \|e^{-A_F(t-\tau)}(Sz)(\tau)\|_Z \|\zeta(t)\|_{Z'} dt \right\} d\tau \\
&\le \int_0^\infty \left\{ \left[\int_\tau^\infty \|e^{-A_F(t-\tau)}(Sz)(\tau)\|_Z^2 dt dt \right]^{\frac{1}{2}} \left[\int_0^\infty \|\zeta(t)\|_{Z'}^2 dt \right]^{\frac{1}{2}} \right\} d\tau \\
&= \|\zeta\|_{L^2(0,\infty;Z')} \int_0^\infty \left[\int_0^\infty \|e^{-A_F\sigma}(Sz)(\tau)\|_Z^2 d\sigma \right]^{\frac{1}{2}} d\tau \\
&\le \sqrt{c}\, \|\zeta\|_{L^2(0,\infty;Z')} \|Sz\|_{L^1(0,\infty;W)}.
\end{aligned}$$

(Here, we have also used Assumption (jj).) This yields (3.152), as claimed.

We recall that, by Lemma 3.3, we have

$$\|Sz\|_W \le K\|z\|_Z^2, \quad \forall z \in Z. \tag{3.153}$$

Using Estimate (3.153), the inequality yields

$$\begin{aligned}
\|Nz\|_{L^2(0,\infty;Z)} &\le \sqrt{c}\|Sz\|_{L^1(0,\infty;W)} \equiv \sqrt{c} \int_0^\infty \|(Sz)(t)\|_W dt \\
&\le \sqrt{c}\, K \int_0^\infty \|z(t)\|_Z^2 dt = \sqrt{c}\, K \|z\|_{L^2(0,\infty;Z)}^2
\end{aligned}$$

and so, (j) yields

$$\|\Lambda z\|_{L^2(0,\infty;Z)} \le \sqrt{c} \left[\|y_0\|_W + K \int_0^\infty \|z(t)\|_Z^2 dt \right].$$

We obtain, therefore, that, if

$$\int_0^\infty \|z(t)\|_Z^2 dt \le r^2, \qquad \|y_0\|_W \le \frac{r}{2\sqrt{c}},$$

where $r > 0$ is chosen to satisfy the constraints

$$r \le \frac{1}{2\sqrt{c}\,K},$$

we have

$$\|\Lambda z\|^2_{L^2(0,\infty;Z)} \equiv \int_0^\infty \|(\Lambda z)(t)\|^2_Z dt \le \frac{1}{2}r^2 + \frac{1}{2}r^2 = r^2,$$

that is, $\Lambda z \in \mathscr{K}_r$. Hence, the ball $\mathscr{K}_r$ is invariant under the action of the operator Λ. We have, also,

$$\begin{aligned} \|\Lambda z_1 - \Lambda z_2\|_{L^2(0,\infty;Z)} &= \|N z_1 - N z_2\|_{L^2(0,\infty;Z)} \\ &\le 2\sqrt{c}\,K r \|z_1 - z_2\|_{L^1(0,\infty;Z)}, \\ &\quad \forall z_1, z_2 \in S(0,r). \end{aligned} \tag{3.154}$$

Indeed, let $z_1, z_2 \in L^2(0,\infty;Z)$. Then, we have as above (see (3.152))

$$\begin{aligned} \|N z_1 - N z_2\|_{L^2(0,\infty;Z)} &\le \sqrt{c} \int_0^\infty \|(S z_1)(t) - (S z_2)(t)\|_W dt \\ &= \sqrt{c}\|S z_1 - S z_2\|_{L^1(0,\infty;W)}. \end{aligned}$$

On the other hand, for $z_1, z_2 \in \mathscr{K}_r$, we have by (3.153)

$$\begin{aligned} \|S z_1 - S z_2\|_W &= \|P[(z_1 \cdot \nabla) z_1 - (z_2 \cdot \nabla) z_2]\|_W \\ &= \|P[((z_1 - z_2) \cdot \nabla) z_1 + (z_2 \cdot \nabla)(z_1 - z_2)]\|_W \\ &\le K(\|z_1\|_Z + \|z_2\|_Z)\|z_1 - z_2\|_Z. \end{aligned}$$

This yields

$$\begin{aligned} &\|N z_1 - N z_2\|_{L^2(0,\infty;Z)} \\ &\quad \le \sqrt{c}\,K \int_0^\infty [\|z_1(\tau)\|_Z + \|z_2(\tau\|_Z]\|z_1(\tau) - z_2(\tau)\|_Z d\tau \\ &\quad \le 2\sqrt{c}\,K \left(\int_0^\infty [\|z_1(\tau)\|^2_Z + \|z_2(\tau)\|^2_Z] d\tau\right)^{\frac{1}{2}} \left(\int_0^\infty \|z_1(\tau) - z_2(\tau)\|^2_Z d\tau\right)^{\frac{1}{2}} \\ &\quad \le 4\sqrt{c}\,K \left(\int_0^\infty \|z_1(\tau) - z_2(\tau)\|^2_Z d\tau\right)^{\frac{1}{2}}. \end{aligned}$$

Then, by the Banach fixed-point theorem, we infer that, for $4Kr\sqrt{c} < 1$, the operator Λ has a unique fixed-point $z \in \mathscr{K}_r$. This implies that, for any $y_0 \in W$, $\|y_0\|_W \le \frac{r\sqrt{c}}{2}$, where $0 < r < (2\sqrt{c}k)^{-1}$ there is a unique solution $y \in \mathscr{K}_r \in L^2(0,\infty;Z)$

to (3.145). Clearly, such a solution, which can be seen as a "mild" solution to (3.145), satisfies also $y \in C([0,\infty); W) \cap L^2(0,\infty; Z)$ and

$$\|y(t)\|_W \le \sqrt{c}\, e^{-\delta t}\|y_0\|_W + \sqrt{c}\, K|y|^2_{L^2(0,\infty;Z)}. \tag{3.155}$$

Moreover, by (3.152) and (3.153), we have the estimate (recall that $y = \Lambda y$)

$$\|y\|^2_{L^2(0,\infty;Z)} \le 2c\|y_0\|^2_W + 2cK^2\|y\|^4_{L^2(0,\infty;Z)} \le 2c\|y_0\|^2_W + 2cK^2r^2\|y\|^2_{L^2(0,\infty;Z)}.$$

Hence,

$$\|y\|^2_{L^2(0,\infty;Z)} \le 2(1 - cK^2r^2)^{-1}c\|y_0\|^2_W,$$

and so, (3.155) yields

$$\|y(t)\|_W \le \sqrt{c}\, e^{-\delta t}\|y_0\|_W + \frac{2c\sqrt{c}\,K}{1 - 2cK^2r^2}\|y_0\|^2_W. \tag{3.156}$$

Now, if we take $t \ge T$ sufficiently large and $\|y_0\|_W \le \rho$ sufficiently small, we see by (3.156) that

$$\|y(t)\|_W \le \eta\|y_0\|_W, \quad \forall t \ge T, \tag{3.157}$$

where $0 < \eta < 1$. Taking into account that the flow $t \to y(t, y_0)$ is a semigroup, it follows by (3.157) that

$$\|y(nT)\|_W \le \eta\|y((n-1)T)\|_W \le \eta^n\|y_0\|_W, \quad n = 1, \dots.$$

This yields

$$\|y(t)\|_W \le Ce^{-\delta_0 t}\|y_0\|_W, \quad \forall t \ge 0,$$

where $\delta_0 > 0$. This completes the proof. □

We consider now a few special cases of feedback operators of the form (3.144) for which Assumptions (j) and (jj) hold and so, Theorem 3.9 is applicable.

1° Let $u = Fy$ be the feedback controller constructed in Theorem 3.8 and, more precisely, in Corollary 3.3, that is,

$$(Fy)(t) = \nu\sum_{i=1}^{M^*}\left(\frac{\partial}{\partial n}(R_0 y(t)), \frac{\partial \psi_i^*}{\partial n}\right)_{(L^2(\partial\mathcal{O}))^d}\frac{\partial \psi_i^*}{\partial n}, \quad t \ge 0, \tag{3.158}$$

where R_0 is the solution to the algebraic Riccati equation (3.131). As seen in Theorem 3.8 (see Remark 3.11), the operator A_F satisfies Conditions (i) and (ii) and so, by Theorem 3.9, we have that the boundary controller (3.158) stabilizes exponentially System (3.143) for all y_0 in a sufficiently small neighborhood of origin in the space $W = D(A^{\frac{1}{4}-\varepsilon})$, where $\varepsilon > 0$. Namely, we have (see (3.142)) the following theorem.

Theorem 3.10 *Assume that* $d = 2$, *Assumptions* (K1) *and* (K2) *hold and that* $\|y_0 - y_e\|_W \le \rho$. *If* ρ *is sufficiently small, then the closed-loop system*

$$\begin{aligned}
&\frac{\partial y}{\partial t} - \nu\Delta y + (y\cdot\nabla)y = f_e + \nabla p && \text{in } (0,\infty)\times\mathcal{O},\\
&y(0,x) = y_0(x) && \text{in } \mathcal{O},\\
&\nabla\cdot y = 0 && \text{in } (0,\infty)\times\mathcal{O},\\
&y = \nu\sum_{i=1}^{M^*}\left(\frac{\partial}{\partial n}(R_0(y-y_e)), \frac{\partial\psi_i^*}{\partial n}\right)_{(L^2(\partial\mathcal{O}))^d}\frac{\partial\psi_i^*}{\partial n} && \text{on } (0,\infty)\times\partial\mathcal{O},
\end{aligned} \tag{3.159}$$

has a unique solution $y \in C([0,\infty); W)$, *which satisfies*

$$\|y(t) - y_e\|_W \le Ce^{-\gamma t}\|y_0 - y_e\|_W, \quad \forall t > 0, \tag{3.160}$$

$$\int_0^\infty \|y(t) - y_e\|_Z^2 e^{2\gamma t}dt \le C\|y_0 - y_e\|_W^2, \tag{3.161}$$

where

$$\begin{aligned}
W &= D(A^{\frac{1}{4}-\varepsilon}) = (H^{\frac{1}{2}-2\varepsilon}(\mathcal{O}))^d \cap H \quad \textit{and}\\
Z &= D(A^{\frac{3}{4}-\varepsilon}) = (H^{\frac{3}{2}-2\varepsilon}(\mathcal{O}))^d \cap V.
\end{aligned}$$

Of course, the solution y to (3.146) is considered in the mild sense

$$y(t) - y_e = e^{-A_F t}(y_0 - y_e) - \int_0^t e^{-A_F(t-\tau)}S(y(\tau) - y_e)d\tau.$$

Theorem 3.10 is the main boundary stabilization result for the Navier–Stokes equation (3.142) and it amounts to saying that in 2-D, under quite reasonable assumptions on the spectrum of the corresponding linearized (Stokes–Navier) system, the equilibrium solutions are exponentially stabilizable by tangential feedback controllers having a finite-dimensional structure. As seen earlier in 3-D, such a result, at least in this form, is not possible due to the fact that the analysis of the nonlinear term $P(y\cdot\nabla)y$ (see the key estimate (3.153)) requires $W = (H^{\frac{1}{2}+\varepsilon}(\mathcal{O}))^d$, that is, $(H^{\frac{1}{2}+\varepsilon}(\mathcal{O}))^d \to (H^{\frac{3}{2}+\varepsilon}(\mathcal{O}))^d$ functional setting, which is in contradiction with the finite-dimensional structure of the boundary control. Of course, one might expect to have also in 3-D a similar result, but for feedback operators F which have a more general structure.

2° Consider now the boundary feedback controller (3.119), that is,

$$Fy = -\eta\sum_{j=1}^N (y,\varphi_j)\Phi_j, \quad \eta \ge \eta_0 > 0.$$

By Theorem 3.7, we already know that, in this case, the operator A_F satisfies Assumptions (j) and (jj). Then, Theorem 3.9 yields

Theorem 3.11 *Let $d=2$, $\eta \ge \eta_0 > 0$ sufficiently large and $W = D(A^{\frac{1}{4}-\varepsilon})$. Then, for all $y_0 \in W$, $\|y_0 - y_e\|_W \le \rho$, where ρ is sufficiently small, the closed-loop system*

$$\begin{aligned}
&\frac{\partial y}{\partial t} - \nu \Delta y + (y\cdot\nabla)y = f_e + \nabla p && \text{in } (0,\infty)\times\mathcal{O},\\
&y(0,x) = y_0(x) && \text{in } \mathcal{O},\\
&\nabla\cdot y = 0 && \text{in } (0,\infty)\times\mathcal{O},\\
&y = -\eta\sum_{j=1}^{N}(y-y_e,\varphi_j)\Phi_j && \text{on } (0,\infty)\times\partial\mathcal{O},
\end{aligned} \tag{3.162}$$

has a unique solution $y \in C([0,\infty); W)$, which satisfies

$$\|y(t)-y_e\|_W \le Ce^{-\gamma t}\|y_0-y_e\|_W, \quad \forall t>0, \tag{3.163}$$

$$\int_0^\infty e^{2\gamma t}\|y(t)-y_e\|_Z^2 dt \le C\|y_0-y_e\|_W^2, \tag{3.164}$$

where $Z = D(A^{\frac{3}{4}-\varepsilon})$.

Remark 3.12 As noticed earlier, the principal limitation of this stabilizable feedback controller is that it might be nonrobust to structural perturbations of the state-system.

3.5 Normal Stabilization of a Plane-periodic Channel Flow

Consider the model of a laminar flow in a two-dimensional channel with the walls located at $y=0,1$, already presented in Sect. 3.1. We assume that the velocity field $(u(t,x,y), v(t,x,y))$ and the pressure $p(t,x,y)$ are 2π periodic in $x \in (-\infty,\infty)$. The dynamics of flow is governed by the incompressible 2-D Navier–Stokes equation (see (3.12))

$$\begin{aligned}
&u_t - \nu\Delta u + uu_x + vu_y = p_x, \quad x\in R,\ y\in(0,1),\ t\ge 0,\\
&v_t - \nu\Delta v + uv_x + vv_y = p_y, \quad x\in R,\ y\in(0,1),\ t\ge 0,\\
&u_x + v_y = 0,\\
&u(t,x,0) = u(t,x,1) = 0, \quad x\in R,\ t\ge 0,\\
&v(t,x,0) = 0, \quad v(t,x,1) = v^*, \quad \forall x\in R,\ t\ge 0,\\
&u(t,x+2\pi,y) \equiv u(t,x,y), \quad v(t,x+2\pi,y) \equiv v(t,x,y),\\
&\quad y\in(0,1),\ t\ge 0.
\end{aligned} \tag{3.165}$$

Consider a steady-state flow governed by (3.165) with zero vertical velocity component, that is, $(U(x,y),0)$. (This is a stationary flow sustained by a pressure gradient in the x direction.)

Fig. 3.2

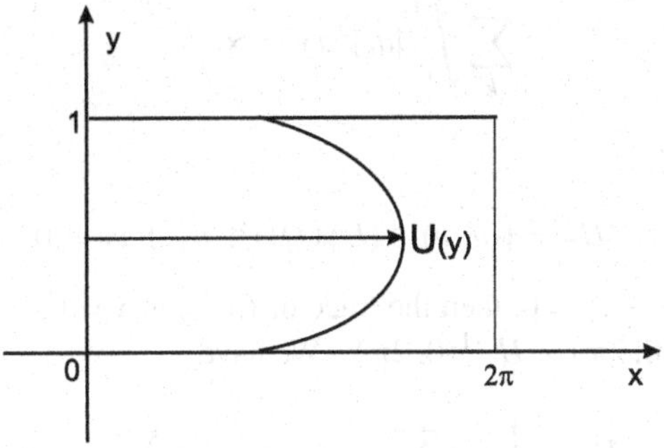

As seen earlier, we have $U(y) = C(y^2 - y)$, $\forall y \in (0,1)$, where $C < 0$. In the following, we take

$$C = -\frac{a}{2\nu},$$

where $a \in R^+$. (See Fig. 3.2.)

We recall that the stability property of the stationary flow $(U, 0)$ varies with the Reynolds number $\frac{1}{\nu}$; there is $\nu_0 > 0$ such that for $\nu > \nu_0$ the flow is stable while for $\nu < \nu_0$ it is unstable. Our aim here is the stabilization of this parabolic laminar flow profile by a boundary controller $v(t,x,1) = v^*(t,x)$, $t \geq 0$, $x \in R$, that is, only the normal velocity v is controlled on the wall $y = 1$.

The linearization of (3.165) around steady-state parabolic flow profile $(U(y), 0)$ leads to the following system

$$\begin{aligned}
&u_t - \nu\Delta u + u_x U + vU' = p_x, \quad y \in (0,1),\ x,t \in R,\\
&v_t - \nu\Delta v + v_x U = p_y,\\
&u_x + v_y = 0, \qquad u(t,x,0) = u(t,x,1) = 0,\\
&v(t,x,0) = 0, \qquad v(t,x,1) = v^*(t,x),\\
&u(t,x+2\pi,y) \equiv u(t,x,y), \qquad v(t,x+2\pi,y) \equiv v(t,x,y),
\end{aligned} \tag{3.166}$$

which governs the small perturbations to this equilibrium profile. Here the actuator v^* is a normal velocity boundary controller on the wall $y = 1$. However, there is no actuation in $x = 0$ or inside the channel.

The main advantage of the periodic control problem (3.166) is that it can be reduced, via Fourier analysis, to an infinite system of 1-D parabolic problems, which greatly reduces the complexity of the control system.

For this purpose, let us briefly describe the Fourier functional setting for Problem (3.166) (see, [74]).

Let $L^2_\pi(Q)$, $Q = (0, 2\pi) \times (0,1)$ be the space of all functions $u \in L^2_{\mathrm{loc}}(R \times (0,1))$ which are 2π-periodic in x. These functions are characterized by their Fourier series

$$u(x,y) = \sum_k a_k(y) e^{ikx}, \qquad a_k = \bar{a}_{-k}, \qquad a_0 = 0,$$

$$\sum_k \int_0^1 |a_k|^2 dy < \infty.$$

We set

$$H_\pi = \{(u,v) \in (L^2_\pi(Q))^2;\ u_x + v_y = 0,\ v(x,0) = v(x,1) = 0\}.$$

(If $u_x + v_y = 0$, then the trace of (u,v) at $y = 0, 1$ is well-defined as an element of $H^{-1}(0,2\pi) \times H^{-1}(0,2\pi)$.) We have

$$H_\pi = \Bigg\{ u = \sum_{k\neq 0} u_k(y)e^{ikx}, v = \sum_{k\neq 0} v_k(y)e^{ikx},\ v_k(0) = v_k(1) = 0,$$
$$\sum_{k\neq 0} \int_0^1 (|u_k|^2 + |v_k|^2)dy < \infty, iku_k(y) + v_k'(y) = 0,$$
$$\text{a.e., } y \in (0,1),\ k \in R \Bigg\}.$$

We now return to System (3.166) and rewrite it in terms of the Fourier coefficients $\{u_k\}_{k=-\infty}^{\infty}, \{v_k\}_{k=-\infty}^{\infty}$.

We have

$$\begin{aligned}
&(u_k)_t - \nu u_k'' + (\nu k^2 + ikU)u_k + U'v_k = ikp_k,\\
&\quad y \in (0,1),\ t \geq 0,\\
&(v_k)_t - \nu v_k'' + (\nu k^2 + ikU)v_k = p_k'\\
&iku_k + v_k' = 0, \quad y \in (0,1),\ k \neq 0,\ t > 0,\\
&u_k(t,0) = u_k(t,1) = 0,\\
&v_k(t,0) = 0, \quad v_k(t,1) = v_k^*(t), \quad t \geq 0,\\
&u_k(0,y) = u_k^0(y), \qquad v_k(0,y) = v_k^0(y),
\end{aligned} \tag{3.167}$$

where

$$p = \sum_{k\neq 0} p_k(t,y)e^{ikx}, \qquad u = \sum_{k\neq 0} u_k(t,y)e^{ikx},$$
$$v = \sum_{k\neq 0} v_k(t,y)e^{ikx}, \qquad u'' = \frac{\partial^2}{\partial y^2}u, \qquad u' = \frac{\partial}{\partial y}u.$$

Here, k is the wave number in streamwise direction and (3.167) is a completely decoupled system in state variables u_k, v_k and the boundary controller $v_k^*(t)$. This yields

$$\begin{aligned}
&ik(v_k)_t - ik\nu v_k'' + ik^2(\nu k + iU)v_k - (u_k')_t + \nu u_k'''\\
&\quad - k(\nu k + iU)u_k' - ikU'u_k - U'v_k' - U''v_k = 0.
\end{aligned}$$

Taking $u_k = -\frac{1}{ik}\, v'_k$, we obtain that

$$ik(v_k)_t - ikvv''_k + ik^2(vk + iU)v_k + \frac{1}{ik}\,(v''_k)_t$$
$$-\frac{v}{ik}\, v_k^{\text{iv}} + \frac{1}{i}\,(vk + iU)v''_k - U''v_k = 0, \quad t \geq 0,\ y \in (0, 1).$$

Finally,

$$\begin{aligned}
&(v''_k - k^2 v_k)_t - v v_k^{\text{iv}} + (2vk^2 + ikU)v''_k - k(vk^3 + ik^2U + iU'')v_k = 0,\\
&\quad t \geq 0,\ y \in (0, 1),\\
&v'_k(t, 0) = v'_k(t, 1) = 0,\\
&v_k(t, 0) = 0, \qquad v_k(t, 1) = v_k^*(t),\\
&v_k(0, y) = v_k^0(y), \quad y \in (0, 1).
\end{aligned} \tag{3.168}$$

System (3.168) is a linear parabolic control system in variable v_k on $(0, \infty) \times (0, 1)$ with the boundary controller v_k^* on $y = 1$.

In the following, we denote by H the complexified space $L^2(0, 1)$ with the norm $|\cdot|$ and product scalar denoted by $(\cdot, \cdot)$. We denote by $H^m(0, 1)$, $m = 1, 2, 3$, the standard Sobolev spaces on $(0, 1)$ and

$$H_0^1(0, 1) = \{v \in H^1(0, 1);\ v(0) = v(1) = 0\},$$
$$H_0^2(0, 1) = \{v \in H^2(0, 1) \cap H_0^1(0, 1);\ v'(0) = v'(1) = 0\}.$$

We set $\mathscr{H} = H^4(0, 1) \cap H_0^2(0, 1)$ and denote by $\mathscr{H}'$ the dual of $\mathscr{H}$ in the pairing with pivot space H, that is $\mathscr{H} \subset H \subset \mathscr{H}'$ algebraically and topologically. Denote by $(H^2(0, 1))'$ the dual of $H^2(0, 1)$ and by $H^{-1}(0, 1)$ the dual of $H_0^1(0, 1)$ with the norm denoted $\|\cdot\|_{-1}$. Denote also by $H_\pi^{-1}(Q)$ the space $L^2(0, 2\pi; H^{-1}(0, 1))$ with the norm $\|\cdot\|_{H_\pi^{-1}(Q)}$.

For each $k \in R$, we denote by $L_k : D(L_k) \subset H \to H$ and $F_k : D(F_k) \subset H \to H$ the second order differential operators

$$L_k v = -v'' + k^2 v, \quad v \in D(L_k) = H_0^1(0, 1) \cap H^2(0, 1), \tag{3.169}$$

$$\begin{aligned}
F_k v &= v v^{\text{iv}} - (2vk^2 + ikU)v'' + k(vk^3 + ik^2U + iU'')v,\\
&\quad \forall v \in D(F_k) = H^4(0, 1) \cap H_0^2(0, 1).
\end{aligned} \tag{3.170}$$

We set

$$\mathscr{F}_k v = v v^{\text{iv}} - (2vk^2 + ikU)v'' + k(vk^3 + ik^2U + iU'')v$$

and consider the solution V_k of the equation

$$\begin{aligned}
&\theta V_k + \mathscr{F}_k V_k = 0, \quad y \in (0, 1),\\
&V'_k(0) = V'_k(1) = 0, \qquad V_k(0) = 0, \qquad V_k(1) = v_k^*(t).
\end{aligned} \tag{3.171}$$

(As easily seen, for θ positive and sufficiently large, there is a unique solution V_k to (3.171).) Then, subtracting (3.168) and (3.171), we obtain that

$$(L_k v_k)_t + F_k(v_k - V_k) - \theta V_k = 0, \quad t \geq 0.$$

Equivalently,

$$\begin{aligned} &(L_k(v_k - V_k))_t + F_k(v_k - V_k) = \theta V_k - (L_k V_k)_t, \quad v_k - V_k \in D(F_k), \\ &v_k(0) = v_k^0. \end{aligned} \tag{3.172}$$

(The meaning of $L_k V_k$, which is a distribution on (0, 1), will be explained later on.)

In order to represent (3.172) as an abstract boundary control system, we consider the operator $\mathscr{A}_k : D(\mathscr{A}_k) \subset H \to H$ defined by

$$\mathscr{A}_k = F_k L_k^{-1}, \qquad D(\mathscr{A}_k) = \{u \in H;\ L_k^{-1} u \in D(F_k)\}. \tag{3.173}$$

We have the following lemma.

Lemma 3.5 *The operator $-\mathscr{A}_k$ generates a C_0-analytic semigroup on H and for each $\lambda \in \rho(-\mathscr{A}_k)$ (the resolvent set of $-\mathscr{A}_k$), $(\lambda I + \mathscr{A}_k)^{-1}$ is compact. Moreover, one has for each $\gamma > 0$*

$$\sigma(-\mathscr{A}_k) \subset \{\lambda \in \mathbb{C};\ \mathrm{Re}\,\lambda \leq -\gamma\}, \quad \forall |k| \geq M = \frac{1}{\sqrt{2\nu}}\left(\gamma + 1 + \frac{a}{\sqrt{2\nu}}\right)^{\frac{1}{2}}, \tag{3.174}$$

where $\sigma(-\mathscr{A}_k)$ is the spectrum of $-\mathscr{A}_k$.

Proof For $\lambda \in \mathbb{C}$ and $f \in H = L^2(0, 1)$, consider the equation

$$\lambda u + \mathscr{A}_k u = f$$

or, equivalently,

$$\lambda L_k v + F_k v = f. \tag{3.175}$$

Taking into account (3.169), (3.170) yields

$$\begin{aligned} &\mathrm{Re}\,\lambda \int_0^1 (|v'|^2 + k^2|v|^2)dy + \nu \int_0^1 |v''|^2 dy \\ &\quad + 2\nu k^2 \int_0^1 |v'|^2 dy + k^4 \nu \int_0^1 |v|^2 dy \\ &\quad + k \int_0^1 U'(\mathrm{Re}\, v' \,\mathrm{Im}\, v - \mathrm{Im}\, v' \,\mathrm{Re}\, v) dy = \mathrm{Re}\langle f, v\rangle, \end{aligned} \tag{3.176}$$

$$\mathrm{Im}\,\lambda \int_0^1 (|v'|^2 + k^2|v|^2)dy + k \int_0^1 |v'|^2 dy$$

$$+k\int_0^1\left(k^2U+\frac{1}{2}U''\right)|v|^2=\mathrm{Im}\langle f,v\rangle. \tag{3.177}$$

Taking into account that $\|u\|_{L^2(0,1)}\le C\|u\|_{H_0^1(0,1)}$, we see by (3.176), (3.177) that, for some $a>0$,

$$|(\lambda I+\mathscr{A}_k)^{-1}f|\le\frac{C}{|\lambda|-a}|f|\quad\text{for }|\lambda|>a,$$

which implies that $-\mathscr{A}_k$ is infinitesimal generator of C_0-analytic semigroup, $e^{-\mathscr{A}_kt}$ on H. Moreover, by (3.176), (3.177) we see that $(\lambda I+\mathscr{A}_k)^{-1}$ is compact in H and it follows also that all the eigenvalues λ of $-\mathscr{A}_k$ satisfy the estimate

$$\begin{aligned}
&\mathrm{Re}\,\lambda\int_0^1(|v_k'|^2+k^2|v_k|^2)dy+2\nu k^2\int_0^1|v_k'|^2dy\\
&\quad+\nu\int_0^1|v_k''|^2dy+\nu k^4\int_0^1|v_k|^2dy\\
&\le -k\int_0^1 U'(\mathrm{Re}\,v_k'\,\mathrm{Im}\,v_k-\mathrm{Im}\,v_k'\,\mathrm{Re}\,v_k)dy\\
&\le 2\nu k^2\int_0^1|v_k'|^2dy+\frac{1}{2\nu}\int_0^1|U'|^2|v_k|^2dy\\
&\le 2\nu k^2\int_0^1|v_k'|^2dy+\frac{a^2}{8\nu^3}\int_0^1|v_k|^2dy,\\
&\mathscr{A}_kv_k=-\lambda v_k.
\end{aligned}$$

Let $\gamma>0$ be arbitrary but fixed. Then, by the above estimate we see that

$$\mathrm{Re}\,\lambda\le-\gamma\quad\text{if }|k|\ge\frac{1}{\sqrt{2\nu}}\left(\gamma+1+\frac{a}{\sqrt{2}\,\nu}\right)^{\frac{1}{2}}.$$

This implies (3.174), as claimed. □

In particular, it follows by Lemma 3.5 that, for $|k|\ge M$, we have

$$\|e^{-\mathscr{A}_kt}\|_{L(H,H)}\le Ce^{-\gamma t},\quad\forall t\ge0.$$

More precisely, we have by (3.168) that

$$\begin{aligned}
&\frac{1}{2}\frac{d}{dt}(|v_k'(t)|^2_{L^2(0,1)}+k^2|v_k(t)|^2_{L^2(0,1)}+\nu(|v_k''(t)|^2_{L^2(0,1)}+k^2|v_k'(t)|^2_{L^2(0,1)})\\
&\quad+\nu k^4|v_k(t)|^2_{L^2(0,1)}=k\,\mathrm{Im}\int_0^1 Uv_k''(t)\bar{v}_k(t)dy
\end{aligned}$$

and this yields

$$\int_0^1 (|v_k'(t,y)|^2 + k^2|v_k(t,y)|^2)dy$$
$$\leq Ce^{-\nu k^2 t}\int_0^1 (|v_k'(0,y)|^2 + |v_k(0,y)|^2)dy, \quad t \geq 0, \tag{3.178}$$

for $|k| \geq M$. This implies that for the stabilization of (3.166) it suffices to stabilize System (3.167) (equivalently (3.168)) for $|k| \leq M$ only.

Now, coming back to System (3.172), we set

$$\widetilde{z}_k(t) = L_k(v_k(t) - V_k(t)) \tag{3.179}$$

and write it as

$$\begin{aligned}\widetilde{z}_k(t) &= e^{-\mathscr{A}_k t}\widetilde{z}_k(0) + \int_0^t e^{-\mathscr{A}_k(t-s)}(\theta V_k(s) - (L_k V_k(s))_s)ds \\ &= e^{-\mathscr{A}_k t}\widetilde{z}_k(0) - L_k V_k(t) + e^{-\mathscr{A}_k t} L_k V_k(0) \\ &\quad + \int_0^t e^{-\mathscr{A}_k(t-s)}(\theta V_k(s) + \widetilde{F}_k V_k(s))ds,\end{aligned} \tag{3.180}$$

where $\widetilde{F}_k : H \to \mathscr{H}'$ is the extension of F_k to all of H defined by

$$_{\mathscr{H}'}\langle \widetilde{F}_k v, \psi\rangle_{\mathscr{H}} = \int_0^1 v(y)F_k^*\psi(y)dy, \quad \forall \psi \in D(F_k^*),\ v \in H.$$

Here, F_k^* is the dual of F_k, that is,

$$\begin{aligned}F_k^* &= \nu\psi^{\text{iv}} - ((\nu k^2 - ik)U\psi)'' + (k - ik^2U - iU'')\psi, \\ D(F_k^*) &= H^4(0,1) \cap H_0^2(0,1).\end{aligned}$$

Define similarly $\widetilde{\mathscr{A}_k}$, the extension of $\mathscr{A}_k$ from H to $(D(\mathscr{A}_k^*))'$. Likewise $\mathscr{A}_k$, the operator $\widetilde{\mathscr{A}_k}$ generates a C_0-analytic semigroup on $(D(\mathscr{A}_k))' = (H^2(0,1))'$ (see (1.9)).

In the same way, the extension of L_k from (3.169) is given to an operator from H to $(H_0^1(0,1) \cap H^2(0,1))'$ again denoted L_k.

Then, (3.180) can be rewritten as

$$\frac{d}{dt} z_k(t) + \widetilde{\mathscr{A}_k} z_k(t) = (\theta + \widetilde{F}_k)V_k(t), \quad t \geq 0, \tag{3.181}$$

where $z_k = L_k v_k$.

For each $u \in R$, we denote by $V = D_k u \in H^4(0,1)$ the solution to the equation (see (3.171))

$$\begin{aligned}&\theta V + \mathscr{F}_k V = 0, \quad \forall y \in (0,1), \\ &V'(0) = V'(1) = 0, \qquad V(0) = 0, \qquad V(1) = u.\end{aligned} \tag{3.182}$$

The operator D_k is the Dirichlet map associated with $\theta + \mathscr{F}_k$ and it is easily seen that the dual $((\theta + \widetilde{F}_k)D_k)^*$ is given by

$$((\theta + \widetilde{F}_k)D_k)^*\varphi = \nu\varphi'''(1), \quad \forall \varphi \in D(F_k). \tag{3.183}$$

We note also that, by virtue of (3.171), we have $V_k = Dv_k^*(t)$ and so, (3.181) can be rewritten as

$$\frac{d}{dt} z_k(t) + \widetilde{\mathscr{A}}_k z_k(t) = (\theta + \widetilde{F}_k)D_k v_k^*(t), \quad \forall t \geq 0. \tag{3.184}$$

3.5.1 Feedback Stabilization

Let $\gamma > 0$ and let $k \in R$, $|k| \leq M$, be arbitrary but fixed. Here, M is given in Lemma 3.5. Then, the operator $-\mathscr{A}_k$ has a finite number $N = N_k$ of the eigenvalues $\lambda_j = \lambda_j^k$ with $\operatorname{Re}\lambda_j \geq -\gamma$. (In the following, since k is fixed, we omit the index k from $\mathscr{A}_k$ and λ_j^k.)

We denote by $\{\varphi_j\}_{j=1}^N$ the corresponding eigenfunctions and repeat each λ_j according to its algebraic multiplicity m_j. We have

$$\mathscr{A}_k \varphi_j = -\lambda_j \varphi_j, \quad j = 1, \ldots, N, \tag{3.185}$$

and recall that the geometric multiplicity of λ_j is the dimension of eigenfunction space corresponding to λ_j. (The eigenfunctions $\varphi_j = \varphi_j^k$ depend, of course, on k but, in agreement with the above convention, we omit k from φ_j^k.)

Here, we assume that the following assumption holds.

(A_1) *All the eigenvalues λ_j with $1 \leq j \leq N$ are semisimple.*

In each case, such a condition can be checked in part by taking into account that λ_j are eigenvalues λ of the boundary value problem

$$\begin{aligned}
&\lambda(-v'' + k^2 v) + \nu v^{\mathrm{iv}} - (2\nu k^2 + ikU)v'' + k(\nu k^3 + ik^2 U + iU'')v = 0, \\
&\quad y \in (0, 1), \\
&v(0) = v(1) = 0, \qquad v'(0) = v'(1) = 0.
\end{aligned}$$

We denote by φ_j^* the eigenfunctions to the dual operator $-\mathscr{A}_k^*$, that is,

$$\mathscr{A}_k^* \varphi_j^* = -\overline{\lambda}_j \varphi_j^*, \quad j = 1, \ldots, N. \tag{3.186}$$

As seen earlier, it follows by Assumption (A_1) that the system $\{\varphi_j^*\}$ can be chosen in such a way that

$$\langle \varphi_\ell, \varphi_j^* \rangle = \delta_{\ell j}, \quad \ell, j = 1, \ldots, N. \tag{3.187}$$

For the time being, we prove the following lemma.

Lemma 3.6 *Under Assumption* (A_1), *all the eigenvalues* λ_j, $j = 1, \dots, N$, *are simple and*

$$(\varphi_j^*)'''(1) \neq 0, \quad \forall j = 1, \dots, N. \tag{3.188}$$

Proof We first check (3.188). We have, by (3.186) and the expression of the dual operator $\mathscr{A}_k^*$, that

$$\begin{aligned} &\nu(\varphi_j^*)^{\text{iv}} - (2\nu k^2 - ikU + \overline{\lambda}_j)(\varphi_j^*)'' + 2ikU'(\varphi_j^*)' \\ &\quad + (k^2\overline{\lambda}_j + \nu k^4 - ik^3U)\varphi_j^* = 0 \quad \text{in } (0, 1), \\ &\varphi_j^*(0) = \varphi_j^*(1) = 0, \qquad (\varphi_j^*)'(0) = (\varphi_j^*)'(1) = 0. \end{aligned} \tag{3.189}$$

Let φ be any solution to the equation

$$\begin{aligned} &\mathscr{A}_k\varphi + \lambda\varphi = 0 \quad \text{in } (0, 1), \\ &\varphi(0) = \varphi'(0) = 0. \end{aligned} \tag{3.190}$$

Multiplying (3.189) by φ and integrating on (0, 1) yields

$$(\varphi_j^*)'''(1)\varphi(1) - (\varphi_j^*)''(1)\varphi'(1) = 0.$$

Hence, if $(\varphi_j^*)'''(1) = 0$, we must have $(\varphi_j^*)''(1)\varphi'(1) = 0$. However, it is easily seen that we can choose a solution φ to (3.190) such that $\varphi'(1) \neq 0$, which implies that necessarily $(\varphi_j^*)''(1) = 0$. Since φ_j^* is an analytic solution to the fourth-order differential equation (3.189) with the boundary conditions

$$\varphi_j^*(0) = \varphi_j^*(1) = (\varphi^*)'(0) = (\varphi_j^*)'(1) = (\varphi_j^*)''(1) = 0,$$

this clearly implies that $\varphi_j^* \equiv 0$ which is, of course, absurd.

Now, let us show that each eigenvalue λ_j (equivalently, $\overline{\lambda}_j$) is simple. Assume that the multiplicity m_j of λ_j is > 1 and argue from this to a contradiction. Let ψ_1^* and ψ_2^* be two linear independent eigenfunctions for λ_j, that is

$$\mathscr{A}_k^*\psi_i^* = -\overline{\lambda}_j\psi_i^*, \quad i = 1, 2.$$

By (3.188), we know that $(\psi_i^*)'''(1) \neq 0$, $i = 1, 2$, and so, $\mu = \frac{(\psi_1^*)'''(1)}{(\psi_2^*)'''(1)} \neq 0$ and is well-defined. Then, $\chi = \psi_1^* - \mu\psi_2^*$ is an eigenfunction for $\mathscr{A}_k^*$, that is,

$$\mathscr{A}_k^*\chi = -\overline{\lambda}_j\chi$$

and, in addition, $\chi'''(1) = 0$, which by (3.188) necessarily implies that $\chi \equiv 0$, that is, $\psi_1^* = \mu\psi_2^*$, which is absurd by assumption. □

Now, we prove, as an intermediate step of our approach, that System (3.172) (equivalently, (3.184)) is stabilizable for $|k| \leq M$.

Proposition 3.6 *For each $|k| \le M$, there is $v_k^* \in C^1([0,\infty))$ such that the corresponding solution v_k to* (3.172) *satisfies*

$$|v_k(t)| \le Ce^{-\gamma t}|v_k^0|, \quad \forall t \ge 0, \tag{3.191}$$

and

$$|v_k^*(t)| + \left|\frac{d}{dt}v_k^*(t)\right| \le Ce^{-\gamma t}\|v_k^0\|_{H_0^1(0,1)}. \tag{3.192}$$

Proof We proceed as in the previous cases. Namely, we set $y = L_k v_k$ and rewrite (3.184) (taking into account the biorthogonality relation (3.187)) as

$$\begin{aligned} &y = P_N y + (I - P_N)y, \quad y = \sum_{i=1}^{N} y_i \varphi_i, \quad y_s = (I - P_N)y, \\ &\frac{dy_j}{dt} + \lambda_j y_j = ((\widetilde{F}_k + \theta)D_k)^* \varphi_j^* v_k^*, \quad j = 1, \dots, N, \\ &y_j(0) = P_N(L_k v_k^0), \end{aligned} \tag{3.193}$$

$$\begin{aligned} &\frac{dy_s}{dt} + \widetilde{\mathscr{A}}_k^s y_s = (I - P_N)(\widetilde{F}_k + \theta)D_k v_k^*, \quad t \ge 0, \\ &y_s(0) = (I - P_N)(L_k v_k^0), \end{aligned} \tag{3.194}$$

where $\widetilde{\mathscr{A}_k} = \mathscr{A}_k|_{X_s}$, $X_s = (I - P_N)H$, and P_N is the projection on $X_u = \text{lin}\{\varphi_j\}_{j=1}^N = P_N(H)$. (In the following, we simply write $\mathscr{A}_k^s$ instead of $\widetilde{\mathscr{A}}_k^s$.)

By virtue of (3.183), we can rewrite System (3.193) as

$$\begin{aligned} &\frac{d}{dt} y_j + \lambda_j y_j = \mu_j v_k^*, \quad j = 1, \dots, N, \\ &y_j(0) = P_N(L_k v_k^0), \end{aligned} \tag{3.195}$$

where $\mu_j = \nu(\varphi_j^*)'''(1)$ and, by Lemma 3.6, $\mu_j \ne 0$ for all j. Taking into account that $\lambda_i \ne \lambda_j$ for $i \ne j$, we infer by (3.195) via Kalman's controllability criterion that there is a function v_k^* satisfying (3.192) and such that the solution $\{y_j\}_{j=1}^N$ to (3.195) satisfies

$$|y_j(t)| \le Ce^{-(\gamma+\delta)t}|y_j(0)|, \quad \forall j = 1, \dots, N, \ t \ge 0, \tag{3.196}$$

for some $\delta > 0$. As mentioned earlier, this choice of v_k^* is possible because it can be found in the feedback form

$$v_k^* = R_k \begin{pmatrix} y_1 \\ \vdots \\ y_N \end{pmatrix}.$$

Indeed, System (3.195) is of the form

$$\frac{d}{dt}z + \Lambda z = Bv_k^*, \quad t \in [0, T],$$

where

$$\Lambda = \operatorname{diag}\|\lambda_j\|, \qquad B = \begin{Vmatrix} \mu_1 \\ \vdots \\ \mu_N \end{Vmatrix} \quad \text{and}$$

$$\det \|B, \ldots, \Lambda^{N-1}B\| = \prod_{j=1}^{N} \mu_j \prod_{i \neq j} (\lambda_i - \lambda_j) \neq 0$$

and so, this implies that the pair (Λ, B) is exactly controllable.

Now, substituting this controller v_k^* into (3.194), we obtain (see (3.180))

$$\begin{aligned} y_s(t) &= e^{-\mathscr{A}_k^s t}(I - P_N)(L_k v_k^0) \\ &\quad + \mathscr{A}_k^s \int_0^t e^{-\mathscr{A}_k^s (t-s)}(I - P_N) L_k D_k v_k^*(s) ds \\ &\quad + (I - P_N)\theta \int_0^t e^{-\mathscr{A}_k^s (t-s)} L_k D_k v_k^*(s) ds. \end{aligned}$$

This yields

$$\begin{aligned} y_s(t) &= e^{-\mathscr{A}_k^s t}(I - P_N)(L_k(v_k^0 - D_k v_k^*(0))) \\ &\quad + (I - P_N) \int_0^t e^{-\mathscr{A}_k^s (t-s)} \left(L_k D_k \frac{d}{ds} v_k^*(s) \right) ds \\ &\quad + (I - P_N)\theta \int_0^t e^{-\mathscr{A}_k^s (t-s)} L_k D_k v_k^*(s) ds. \end{aligned} \tag{3.197}$$

□

We notice that (3.196) resembles (3.110) and so, to treat it we use a similar argument as in the proof of Theorem 3.6. In order to estimate the right-hand side of (3.197), we first prove the following lemma.

Lemma 3.7 *We have, for some $\delta > 0$,*

$$\left| L_k^{-1} e^{-\mathscr{A}_k^s t}(L_k h) \right| \leq C e^{-(\gamma+\delta)t} |h|, \quad \forall t \geq 0, \ h \in H. \tag{3.198}$$

Proof We set

$$z_k(t) = e^{-\mathscr{A}_k^s t}(L_k h), \quad t \geq 0.$$

We have, therefore,

$$z_k'(t) + \mathscr{A}_k z_k(t) = 0, \quad t \geq 0,$$
$$z_k(0) = L_k h.$$

We set $w_k(t) = (\lambda_0 + \mathscr{A}_k)^{-1} z_k$ and obtain that

$$w_k' + \mathscr{A}_k w_k = 0,$$
$$w_k(0) = \mathscr{A}_k^{-1} L_k h.$$

Hence, recalling that $\sigma(\mathscr{A}_k^s) \subset \{\lambda;\ \lambda > \gamma\}$, we get for some $\delta > 0$,

$$|w_k(t)| \leq Ce^{-(\gamma+\delta)t} |\mathscr{A}_k^{-1} L_k h|, \quad \forall t \geq 0. \tag{3.199}$$

On the other hand, it is easily seen that for $\lambda_0 > 0$ and sufficiently large, we have

$$|(\lambda_0 I + \mathscr{A}_k)^{-1} L_k h| \leq C|h|, \quad \forall h \in H. \tag{3.200}$$

Indeed, if set $(\lambda_0 I + \mathscr{A})^{-1} L_k h = \zeta$, we have

$$F_k L_k^{-1} \zeta + \lambda_0 \zeta = 0$$

and, taking into account (3.169), (3.170), we obtain that

$$\begin{aligned}
\int_0^1 h\bar{\zeta}\,dy &= \int_0^t F_k(L_k^{-1}\zeta)\overline{L_k^{-1}\zeta}\,dy + \lambda_0 \int_0^1 \zeta \cdot \overline{L_k^{-1}\zeta}\,dy \\
&= \nu \int_0^1 |(L_k^{-1}\zeta)''|^2 dy + \lambda_0 \int_0^1 \zeta \overline{L_k^{-1}\zeta}\,dy \\
&\quad - \int_0^1 (2\nu k^2 + ikU)(L_k^{-1}\zeta)''\overline{L_k^{-1}\zeta}\,dy \\
&\quad + \int_0^1 k(\nu k^3 + ik^2 U + iU'')|L_k^{-1}\zeta|^2 dy.
\end{aligned}$$

For λ_0 sufficiently large, the latter yields

$$\|L_k^{-1}\zeta\|^2_{H^2(0,1)} + \lambda_0 \|L_k^{-1}\zeta\|^2_{H_0^1(0,1)} \leq C|h|\,|\zeta|$$

and, taking into account that L_k is an isomorphism from $H^2(0,1) \cap H_0^1(0,1)$ to $L^2(0,1)$, we obtain that $|\zeta| \leq C|h|$, which implies (3.200), as claimed.

By (3.199) and (3.200), we obtain Estimate (3.198).

Now, coming back to (3.198), we get by (3.198) and (3.196) that

$$|L_k^{-1} y_s(t)| \leq Ce^{-(\gamma+\delta)t} |v_k^0 - D_k v_k^*(0)|$$

$$+C\int_0^t e^{-(\gamma+\delta)(t-s)}\left(|v_k^*(s)|+\left|\frac{d}{ds}\,v_k^*(s)\right|\right)ds$$
$$\le C_1e^{-\gamma t}|v_k^0-D_kv_k^*(0)|+C_1te^{-(\gamma+\delta)t}|v_k^0|.$$

Recalling that $v_k=L_k^{-1}y$, we obtain (3.191), as claimed. □

In particular, it follows by Estimate (3.178) the following corollary.

Corollary 3.5 *Let* $\{v_k^*\}_{k=1}^M$ *be as in Proposition* 3.6. *Then the boundary controller*

$$v^*(t,x)=\sum_{|k|\le M} e^{ikx}v_k^*(t),\quad x\in R,\ t\ge 0,$$

stabilizes exponentially System (3.166), *that is,*

$$\|u(t)\|_{L^2_\pi(Q)}+\|v(t)\|_{L^2(Q)}\le Ce^{-\gamma t}(\|u(0)\|_{L^2_\pi(Q)}+\|v(0)\|_{L^2(Q)}).$$

It turns out that the boundary controller $v^*(t)$ can be chosen in feedback form. To this aim, we note first that Proposition 3.2 implies also the following proposition.

Proposition 3.7 *For* $|k|\le M$, *there is* $v_k^*\in C^1[0,\infty)$ *satisfying* (3.192) *such that the solution* z_k *to System* (3.184) *satisfies*

$$|L_k^{-1}z_k(t)|\le Ce^{-\gamma t}|L_k^{-1}z_k(0)|,\quad \forall t\ge 0. \tag{3.201}$$

In other words, the controller v_k^* stabilizes (3.184) in the $X=(H^2(0,1)\cap H_0^1(0,1))'$ topology. This suggests us to consider System (3.185) in the space X and look for a feedback representation of the controller v_k^* via the linear quadratic control problem

$$\text{Min}\left\{\frac{1}{2}\int_0^\infty (e^{2\gamma t}|L_k^{-1}z(t)|^2+|v(t)|^2)dt\right\} \tag{3.202}$$

subject to $z\in L^2(0,\infty;X)$ and

$$\frac{dz}{dt}+\widetilde{\mathscr{A}}_kz=(\theta+\widetilde{F}_k)D_kv^*,$$
$$z(0)=z_0. \tag{3.203}$$

Without loss of generality we may assume that $\gamma=0$. Then, if we denote by $(R_kz_0,z_0)_X$, $R_k\in L(X,X)$, the infimum in (3.202), we have as in the standard theory of linear quadratic optimal control problems on $(0,\infty)$ that the optimal controller $v=v_k^*$ is given by (see Propositions 2.2 and 2.3)

$$v_k^*(t)=((\theta+\widetilde{F}_k)D_k)^*p=\nu p'''(1), \tag{3.204}$$

where

$$p_t - \mathscr{A}_k^* p = L_k^{-2} z, \quad t \ge 0, \tag{3.205}$$

and $z \in L^2(0,\infty; X)$ is optimal in (3.202). As seen in the previous cases, we also have

$$p(t) = -L_k^{-2} R_k z(t), \quad \forall t \ge 0, \tag{3.206}$$

and $R_k \in L(X, X)$ is the solution to the Riccati algebraic equation in X,

$$(L_k^{-1} R_k z_0, L_k^{-1} \mathscr{A}_k z_0) + \frac{1}{2} \nu^2 |(L_k^{-2} R_k z_0)'''(1)|^2 = \frac{1}{2} |L_k^{-1} z_0|^2, \quad \forall z_0 \in X. \tag{3.207}$$

By (3.204) and (3.206), the optimal control v_k^* is given in the feedback form

$$v_k^*(t) = -\nu (L_k^{-2} R_k z)'''(1) = -\nu (L_k^{-2} R_k L_k v_k(t))'''(1).$$

Then, clearly, the solution v_k to the closed-loop system with the above feedback-controller is exponentially asymptotically stable and, recalling (3.178), we obtain the main stabilizable result for System (3.166).

Theorem 3.12 *The feedback controller*

$$v^*(t,x) = -\nu \sum_{|k| \le M} (L_k^{-2} R_k L_k v_k(t))'''(1) e^{ikx}, \tag{3.208}$$

where $v_k(t,y) = \int_0^{2\pi} v(t,x,y) e^{-ikx} dx$, $|k| \le M$, *stabilizes exponentially System* (3.166), *that is,*

$$\|u(t)\|_{L^2_\pi(\mathscr{O})} + \|v(t)\|_{L^2_\pi(\mathscr{O})} \le C e^{-\gamma t} (\|u(0)\|_{L^2_\pi(\mathscr{O})} + \|v(0)\|_{L^2_\pi(\mathscr{O})}). \tag{3.209}$$

This theorem is a global stabilizable result for the linearization equation (3.166) with actuation on the wall $y = 1$ and the stabilizing effect is independent of how large is the Reynolds number.

For each wave number k, $|k| \le M$, there is one actuation variable on the wall $y = 1$ and one has a complete decoupling of system in separate modes.

It should also be emplasized that the Riccati equation (3.207), which provides the feedback controller (3.208), is easily manageable from the computational point of view since it is associated with a parabolic boundary control system on $(0, 1)$, whose structure is identical for different wave numbers k.

Remark 3.13 Having in mind the previous developments, one might suspect that the feedback controller (3.208) is exponentially stabilizable in (3.165). One might speculate that this follows by the fixed-point-type approach as in Theorem 3.9.

3.6 Internal Stabilization of Time-periodic Flows

Consider the control Navier–Stokes equations

$$\begin{aligned} &y_t(t,x) - \nu \Delta y(t,x) + (y\cdot\nabla)y(t,x) = m(x)u(t,x) + f_\pi(t,x) + \nabla p(t,x),\\ &\quad x\in\mathcal{O},\ t\in R,\\ &\nabla\cdot y = 0 \quad \text{in } R\times\mathcal{O},\\ &y=0 \quad \text{on } R\times\partial\mathcal{O}, \end{aligned} \tag{3.210}$$

on a bounded domain of R^d, $d=2,3$, with smooth boundary $\partial\mathcal{O}$.

Here, m is the characteristic function of an open subdomain $\mathcal{O}_0\subset\mathcal{O}$, and f_π is a given smooth function such that $f_\pi(t,x)\equiv f_\pi(t+T,x)$.

Now, we study the stabilization of time-periodic solutions $y_\pi = y_\pi(t,x)$ of the equation (3.210) by a feedback controller of the form

$$u(t,x) = \sum_{i=1}^N \Phi_i(y(t)-y_\pi(t))w_i(x),$$

where $\{w_i\}$ is a system of functions in $(L^2(\mathcal{O}))^d$ appropriately chosen.

We set $H=\{y\in(L^2(\mathcal{O}))^d;\ \nabla\cdot y=0,\ y\cdot n=0 \text{ on } \partial\mathcal{O}\}$, $V=H\cap(H_0^1(\mathcal{O}))^d$ and (see Sect. 3.1)

$$\begin{aligned} &Ay = -\nu P\Delta y, && y\in D(A)=V\cap(H^2(\mathcal{O}))^d,\\ &(Sy,w) = b(y,y,w), && \forall w\in V,\\ &b(y,z,w) = \int_{\mathcal{O}} y_i D_i y_j w_j, && \forall y,z,w\in V. \end{aligned}$$

(Here, $P:(L^2(\mathcal{O}))^d\to H$ is the Leray projector on H.) Then, we rewrite (3.210) as

$$\frac{dy}{dt}(t) + Ay(t) + Sy(t) = f(t) + Du(t), \quad t\ge 0, \tag{3.211}$$

where $f=Pf_\pi$, $Du=P(mu)$, $U=(L^2(\mathcal{O}))^d$. Let y_π be a time-periodic solution, that is,

$$\begin{aligned} &(y_\pi)_t - \nu\Delta y_\pi + (y_\pi\cdot\nabla)y_\pi = f_\pi + \nabla p && \text{in } R\times\mathcal{O},\\ &\nabla\cdot y_\pi = 0 && \text{in } R\times\mathcal{O},\\ &y_\pi = 0 && \text{on } R\times\partial\mathcal{O},\\ &y_\pi(t+T,x) = y_\pi(t,x), && \forall x\in\mathcal{O},\ t\in R. \end{aligned} \tag{3.212}$$

Equivalently,

$$\begin{aligned} &\frac{d}{dt}y_\pi + Ay_\pi + Sy_\pi = Pf_\pi, \quad \forall t\in R,\\ &y_\pi(t) = y_\pi(t+T). \end{aligned} \tag{3.213}$$

We assume here that

(ℓ) *$y_\pi \in C^\infty(R; D(A))$ and is analytic in time in a complex neighborhood of the real axis as a $D(A)$-valued function.*

It must be said that, if f_π is time-periodic and analytic in t as an H-valued function in a complex neighborhood of the real axis, then Assumption (ℓ) is automatically satisfied.

We are going to check Assumptions (kk) and (A1)$'$ of Theorem 2.12, where $B(t,y) = Sy - f(t)$ and the spaces H, U, and the operators D and A are as above. First, we note that

$$(S_y(y)\theta, w) = b(y,\theta,w) + b(\theta,y,w), \quad \forall w \in H,\ \theta \in V,$$

and so, by Proposition 1.7, we have that

$$|(S_y(y) - S_y(z))\theta| \le C(\|y-z\|_{m_1}\|\theta\|_{m_2+1} + \|\theta\|_{m_1^*}\|y-z\|_{m_2^*+1}),$$

where $m_1 + m_2, m_1^* + m_2^* \ge \frac{d}{2}$. We recall that, here, $\|y\|_\alpha = |A^{\frac{\alpha}{2}} y|$. Thus, Assumption (kk) holds with $\alpha = \frac{5}{8}$.

Let $\widetilde{S}(t,s)$ be the evolution generated on H by $\mathscr{A}(t) = A + S_y(y_\pi(t))$, that is,

$$\begin{aligned}
&\frac{\partial}{\partial t}\widetilde{S}(t,s)x + \mathscr{A}(t)\widetilde{S}(t,s)x = 0, \quad 0 \le s < t,\\
&\widetilde{S}(s,s)x = x,\\
&\mathscr{A}(t)y = P(-\nu\Delta y + (y\cdot\nabla)y_\pi(t) + (y_\pi(t)\cdot\nabla)y),\\
&D(\mathscr{A}(t)) = D(A), \quad \forall t \in R.
\end{aligned} \tag{3.214}$$

The dual operator $\mathscr{A}^*(t)$ is given by

$$\mathscr{A}^*(t)z = P(-\nu\Delta z + z\cdot\nabla y_\pi - (y_\pi\cdot\nabla)z).$$

Then, Assumption (A1)$'$ is implied by the following:

If z, w are two arbitrary solutions to the backward equation

$$\begin{aligned}
&z_t + \nu\Delta z - \nabla y_\pi\cdot z + (y_\pi\cdot\nabla)z = \nabla p && \text{in } (0,T)\times\mathscr{O},\\
&\nabla\cdot z = 0 && \text{in } (0,T)\times\mathscr{O},\\
&z = 0 && \text{on } (0,T)\times\partial\mathscr{O},\\
&z(x,0) = \lambda z(x,T) + w(x,0), && \forall x\in\mathscr{O},
\end{aligned} \tag{3.215}$$

where $\lambda \in \mathbb{C}$ is such that $z(x,T) = 0$ on ω, then $z(x,t) = 0$ on all of $(0,T)\times\mathscr{O}$.

The latter follows by Assumption (ℓ), and the unique continuation property of solutions to Stokes–Oseen equations (see Theorem 3.14) by the argument below. (See Example 2.1.)

We set $\eta = z - w$ and we extend, by periodicity, η as solution to (3.215) on $(-T, 0)$. Since, by smoothing effect on data, $z(x,0)$ and $w(x,0)$ are in $D(A)$, we have that $z(x,T) \in D(A)$. We may assume, therefore, that η is a strong solution to (3.215) on $(-T,T)$. Then, arguing as in Theorem 7.1 in [74], we infer that η is analytic in time on some interval $[-\delta, \delta]$. Hence,

$$\eta(x,t) = \sum_{k=0}^{\infty} \frac{1}{k!} \eta_x^{(k)}(x,0)t^k, \quad \forall s \in \mathcal{O},\ -\delta < t < \delta.$$

Since $\eta(x,0) \equiv 0$ on $\mathcal{O}_0$, the latter implying that $\eta(x,t) = 0$, for all $x \in \mathcal{O}_0$, $t \in (-\delta, \delta)$, and by the unique continuation property of solutions to (3.216), we conclude that $\eta(x,t) = 0$, $\forall x \in \mathcal{O}$, $t \in (-\delta,\delta)$. This implies that $z(x,T) \equiv 0$ and, since (3.215) is backward in time, we infer, therefore, that $z \equiv 0$, as claimed.

We may apply Theorem 2.12 in the present situation. Namely, we conclude that there is a system $\{w_i\}_{i=1}^N \subset H$ such that the feedback controller (see (2.202))

$$u(x,t) = -\sum_{i=1}^{N} w_i(x) \int_{\mathcal{O}_0} w_i(x) R(t)(y(\cdot,t) - y_\pi(\cdot,t))(x)dx \tag{3.216}$$

stabilizes exponentially System (3.210) in a neighborhood of $y_\pi(0)$.

Here, $R(t) \in L(D(A^{\frac{1}{4}}), (D(A^{\frac{1}{4}}))')$ is the solution to the Riccati equation

$$\begin{cases} \frac{d}{dt} R(t)y - \mathcal{A}^*(t)R(t)y - R(t)\mathcal{A}(t)y \\ \quad - \sum_{i=1}^N R(t)P(mw_i)\int_{\mathcal{O}_0} w_i R(t)y\,dx = -A^{\frac{3}{2}}y, \quad \forall y \in D(A),\ t \ge 0, \\ R(t) = R(t+T), \quad \forall t > 0, \end{cases}$$

where $\mathcal{A}(t)$ is Operator (3.214).

To summarize, we have the following stabilization result.

Theorem 3.13 *Let y_π be a periodic solution to* (3.210) *satisfying Assumption* (ℓ). *Then, there is a system $\{w_j\}_{i=1}^N \subset H$ and $\rho > 0$ such that the feedback controller* (3.216) *inserted in* (3.210) (*equivalently, in* (3.211)) *stabilizes exponentially y_π for $y(0) \in D(A^{\frac{5}{8}}) \cap \mathcal{U}$, where $\mathcal{U} = \{y_0 \in W;\ \|y_0 - y_\pi(0)\|_W < \rho\}$. More precisely, for such $y_0 = y(0)$, there is a unique strong solution $y \in C([0,\infty); H)$ to* (3.210) *such that*

$$\frac{dy}{dt}, Ay, Sy \in L^2_{\mathrm{loc}(0,\infty;H)},$$

$$\int_0^\infty |A^{\frac{3}{4}}(y(t) - y_\pi(t))|^2 dt \le C\|y(0) - y_\pi\|_W^2,$$

$$\|y(t) - y_\pi(t)\|_W \le Ce^{-\delta t}\|y(0) - y_\pi(0)\|_W, \quad \forall t > 0,$$

for some $\delta > 0$. Here, $W = D(A^{\frac{1}{4}})$.

Theorem 3.13 can be viewed also as a periodic forcing result for the periodic Navier–Stokes equation (3.210); the periodic solution y_π attracts exponentially every solution y which starts from a sufficiently small neighborhood of $y_\pi(0)$. The unstable eigenvalues $\lambda = \overline{\lambda}_j$, $j = 1, \dots, N$, are determined from the linearized system (3.215), via Floquet's transformation

$$\theta(x,t) = e^{\gamma t} z(x,t), \qquad q(x,t) = e^{\gamma t} p(x,t),$$

where θ satisfies the system

$$\begin{aligned} &\theta_t - \gamma\theta + \nu\Delta\theta - \theta \cdot \nabla y_\pi + (y_\pi \cdot \nabla)\theta = \nabla q \quad \text{in } R \times \mathcal{O}, \\ &\nabla \cdot \theta = 0, \qquad \theta = 0 \quad \text{on } R \times \partial\mathcal{O}, \qquad \theta(t+T) \equiv \theta(t). \end{aligned} \tag{3.217}$$

It is readily seen that $\varphi(x) = \theta(x,T)$ are the eigenfunctions of $U^*(T)$ with the corresponding eigenvalues $\lambda = e^{-\gamma T}$.

If $\{\theta_j\}_{j=1}^N$ is a system of solutions to (3.217) corresponding to $\gamma = \gamma_j$, $\operatorname{Re}\gamma_j \le 0$, we may choose $\{w_j\}$ from the system

$$\int_\omega w_j(x)\theta_i(x,T)dx = \delta_{ij}, \quad i,j = 1,\dots,M.$$

3.7 The Numerical Implementation of Stabilizing Feedback

We come back to the feedback controller (3.53) and look for a finite-dimensional approximation via finite element method. This can be achieved by approximating the infinite-dimensional Riccati equation (3.52) by a finite-dimensional approximation of Navier–Stokes equation. To this purpose, we repeat the analysis developed in the proof of Theorem 3.3 on a finite-dimensional approximation of the linear system (3.57).

The variational form of (3.57) is

$$\begin{aligned} &(y_t, \varphi) + \nu(\nabla y, \nabla\varphi) + ((y_e \cdot \nabla)y, \varphi) + ((y \cdot \nabla)y_e, \varphi) + (\nabla p, \varphi) \\ &\quad = \sum_{k=1}^{M^*} ((\psi_k, \varphi)_0 u_k, \quad \forall \varphi \in (H_0^1(\mathcal{O}))^d, \\ &(\nabla \cdot y, q) = 0, \quad \forall q \in (L_2(\mathcal{O}))^d, \qquad y(0) = y_0. \end{aligned} \tag{3.218}$$

We introduce the finite element spaces for the velocity y and pressure p, and we use the standard Taylor–Hood method. We divide $\mathcal{O}$ into small rectangles $\square_k$, that is, $\mathcal{O} = \sum_k \square_k$ such that no vertex of any rectangle lies on the interior of a side of another rectangle. Let h denote the maximal length of the sides of the rectangles and let $\mathcal{T}_h$ denote this partition. Let $S_h \subset H_0^1(\mathcal{O})$ denote the continuous piecewise quadratic polynomial defined on $\mathcal{T}_h$ such that $v_h = 0$ on $\partial\mathcal{O}$ for any $v_h \in S_h$. We denote the basis functions of S_h by $\{\varphi_i\}_{i=1}^{\mathbf{G}_h}$. Further, we let W_h denote the piecewise linear polynomial defined on $\mathcal{T}_h$. The basis functions of W_h are $\{\phi_j\}_{j=1}^{M_h}$. Then, (3.218) reduces to

$$\begin{aligned}
&(y_{h,t},\chi)+\nu(\nabla y_h,\nabla\chi)+((y_e\cdot\nabla)y_h,\chi)+((y_h\cdot\nabla)y_e,\chi)+(\nabla p_h,\chi)\\
&\quad=\sum_{k=1}^{M^*}((\psi_k,\chi)_\omega u_k,\quad \forall\chi\in(S_h)^d,\\
&(\nabla\cdot y_h,q_h)=0,\quad \forall q_h\in W_h,\\
&y_h(0)=y_{0,h},
\end{aligned}\tag{3.219}$$

where $y_{0,h}\in(S_h)^d$, $y_h\in(S_h)^d$ and $p_h\in W_h$.

For simplicity, we consider here the two-dimensional case. The velocity basis can be defined by $\{(\varphi_1,0)^T,\dots,(\varphi_{\mathbf{N}_h},0)^T,(0,\varphi_1)^T,\dots,(0,\varphi_{\mathbf{N}_h})^T\}$, and y can be approximated in the finite element space $S_h\times S_h$ by

$$y_h=\sum_{j=1}^{\mathbf{N}_h}z_j\begin{bmatrix}\varphi_j\\0\end{bmatrix}+\sum_{\mathbf{N}_h+1}^{2\mathbf{N}_h}z_j\begin{bmatrix}0\\\varphi_j\end{bmatrix},$$

and the pressure p can be approximated in the finite element space W_h by

$$p_h=\sum_{l=1}^{L}w_l\psi_l.$$

Now, we choose the test function $\chi=\left[\begin{smallmatrix}\varphi_j\\0\end{smallmatrix}\right]$, $j=1,\dots,\mathbf{N}_h$ and $\chi=\left[\begin{smallmatrix}0\\\varphi_j\end{smallmatrix}\right]$, $j=1,\dots,\mathbf{N}_h$ and $q_h=\psi_\ell$, $\ell=1,\dots,L$, in (3.219), we get the following matrix equation

$$\mathbf{M}z'(t)+\nu\widetilde{\mathbf{S}}z(t)+(\mathbf{N}+\mathbf{W})z(t)+\mathbf{P}w(t)=\mathbf{D}\mathbf{u}(t),\qquad \mathbf{P}^Tz(t)=0,$$

where

$$\widetilde{\mathbf{S}}=\begin{bmatrix}(\varphi_i,\varphi_j)&0\\0&(\varphi_i,\varphi_j)\end{bmatrix}_{2\mathbf{N}_h\times2\mathbf{N}_h},$$

$$\mathbf{M}=\begin{bmatrix}(\nabla\varphi_i,\nabla\varphi_j)&0\\0&(\nabla\varphi_i,\nabla\varphi_j)\end{bmatrix}_{2\mathbf{N}_h\times2\mathbf{N}_h},$$

$$\mathbf{N}=\begin{bmatrix}((y_e\cdot\nabla)\varphi_i,\varphi_j)&0\\0&((y_e\cdot\nabla)\varphi_i,\varphi_j)\end{bmatrix}_{2\mathbf{N}_h\times2\mathbf{N}_h},$$

$$\mathbf{W}=\begin{bmatrix}(\varphi_i\frac{\partial}{\partial x}(y_e)_x,\varphi_j)&(\varphi_i\frac{\partial}{\partial y}(y_e)_x,\varphi_j)\\(\varphi_i\frac{\partial}{\partial x}(y_e)_y,\varphi_j)&(\varphi_i\frac{\partial}{\partial y}(y_e)_y,\varphi_j)\end{bmatrix}_{2\mathbf{N}_h\times2\mathbf{N}_h},$$

$$\mathbf{P} = \begin{bmatrix} (\frac{\partial \psi_l}{\partial x}, \varphi_j) \\ (\frac{\partial \psi_l}{\partial x}, \varphi_j) \end{bmatrix}_{2\mathrm{N}_h \times L}, \qquad \mathbf{D} = \begin{bmatrix} (\psi_k, \begin{bmatrix} \varphi_j \\ 0 \end{bmatrix}) \\ (\phi_k, \begin{bmatrix} 0 \\ \varphi_j \end{bmatrix}) \end{bmatrix}_{2\mathrm{N}_h \times M^*},$$

$$z = [z_1, z_2, \dots, z_{2\mathrm{N}_h}]^T.$$

More precisely, we have

$$\begin{aligned} & z'(t) + \nu \tilde{\mathbf{A}}_h z(t) + \tilde{\mathbf{A}}_{0,h} z(t) + \mathbf{P}_h w(t) = \widetilde{\mathbf{D}}_h \mathbf{u}(t), \\ & \mathbf{P}^T z(t) = 0, \end{aligned} \tag{3.220}$$

where

$$\tilde{\mathbf{A}}_h = \mathbf{M}^{-1}(\mathbf{S}), \qquad \tilde{\mathbf{A}}_{0,h} = \mathbf{M}^{-1}(\mathbf{N} + \mathbf{W}), \qquad \mathbf{P}_h = \mathbf{M}^{-1}\mathbf{P}, \qquad \widetilde{\mathbf{D}}_h = \mathbf{M}^{-1}\mathbf{D}.$$

Multiplying $\mathbf{P}^T$ from the left in the both sides in (3.220), we get

$$\mathbf{P}(\nu \tilde{\mathbf{A}}_h + \tilde{\mathbf{A}}_{0,h}) z(t) + \mathbf{P}^T \mathbf{P}_h w(t) = \mathbf{P}^T \widetilde{\mathbf{D}}_h \mathbf{u}(t),$$

which implies that

$$w(t) = (\mathbf{P}^T \mathbf{P}_h)^{-1} \big(\mathbf{P}^T \widetilde{\mathbf{D}}_h \mathbf{u}(t) - \mathbf{P}(\nu \tilde{\mathbf{A}}_h + \tilde{\mathbf{A}}_{0,h}) z(t) \big).$$

We have, therefore,

$$z'(t) + \nu \mathbf{A}_h z(t) + \mathbf{A}_{0,h} z(t) = \mathbf{D}_h \mathbf{u}(t), \tag{3.221}$$

where

$$\begin{aligned} \mathbf{A}_h &= \tilde{\mathbf{A}}_h - \mathbf{P}_h (\mathbf{P}^T \mathbf{P}_h)^{-1} (\mathbf{P}^T \tilde{\mathbf{A}}_h), \\ \mathbf{A}_{0,h} &= \tilde{\mathbf{A}}_{0,h} - \mathbf{P}_h (\mathbf{P}^T \mathbf{P}_h)^{-1} (\mathbf{P}^T \tilde{\mathbf{A}}_{0,h}), \\ \mathbf{D}_h &= \widetilde{\mathbf{D}}_h - \mathbf{P}_h (\mathbf{P}^T \mathbf{P}_h)^{-1} (\mathbf{P}^T \widetilde{\mathbf{D}}_h). \end{aligned}$$

By the analysis above, we see that (3.221) is the finite element approximation of (3.57). Next, we compute the approximate gain operator R_N^h of R_N as solution to the following algebraic Riccati equation

$$\big(\nu \mathbf{A}_h + \mathbf{A}_{0,h}\big) R_N^h + R_N^h \big(\nu \mathbf{A}_h + \mathbf{A}_{0,h}\big) + R_N^h \mathbf{D}_h \mathbf{D}^* R_N^h = Q_h, \tag{3.222}$$

where

$$Q_h = (A^{3/2} \varphi_i, \varphi_j) = (A^{3/4} \varphi_i, A^{3/4} \varphi_j).$$

If A has the eigenpairs $\{(\mu_j, e_j)\}_{j=1}^{\infty}$, where $\mu_j > 0$, by the definition of the fractional operator, we can compute $A^{3/4} v$, $\forall v \in H$, in the following way

$$A^{3/4} v = \sum_{j=1}^{\infty} \mu_j^{3/4} (v, e_j) e_j, \quad \text{where} \sum_{j=1}^{\infty} \mu_j^{3/2} (v, e_j)^2 < \infty,$$

and we can approximate $A^{3/4}v_h$, $\forall v_h \in (S_h)^d$ by

$$A^{3/4}v_h \approx A_h^{3/4}v_h = \sum_{j=1}^{2\mathrm{N}_h}(v_h, e_{j,h})e_{j,h},$$

where $\{(\mu_{j,h}, e_{j,h})\}_{j=1}^{2\mathrm{N}_h}$ are eigenpairs of A_h.

The feedback controller $u_h(t) = -D_h^* y_h$ is, of course, stabilizable in the finite-dimensional system (3.219) and, by a similar analysis to that developed in Theorem 2.11, we see that, for N_h large enough, it still remains stabilizable in System (3.2).

We consider the following example

$$\begin{aligned}
&y_t(x,t) - \nu\Delta y(x,t) + (y\cdot\nabla)y(x,t) = m(x)u(x,t) + f_e(x) + \nabla p(x,t),\\
&\quad \text{in } \mathcal{O}\times(0,\infty),\\
&\nabla\cdot y = 0, \quad \text{in } Q,\\
&y = 0 \quad \text{on } \Sigma = \partial\mathcal{O}\times(0,\infty),\\
&y(x,0) = y_0(x) \quad \text{in } \mathcal{O},
\end{aligned}$$

where $\mathcal{O} = (0,1)\times(0,1)$, and

$$f_e(x_1,x_2) = \begin{bmatrix} f(x_1)g(x_2)\\ -f(x_2)g(x_1)\end{bmatrix},$$

$$f(r) = -256r^2(r-1)^2, \qquad g(r) = r(r-1)(4r-2).$$

We use the finite element method described above to approximate the Riccati equation corresponding to the linearized system. For $h = \frac{1}{2^4}$, we divide the $\mathcal{O} = (0,1)\times(0,1)$ into $\frac{1}{h}\times\frac{1}{h}$ squares. We use the piecewise quadratic polynomial to approximate the velocity and the piecewise linear polynomial to approximate the pressure. We choose $K = 2$, and

$$\psi_1 = -0.01\varphi_{20}, \qquad \psi_2 = -0.01\varphi_{21}, \qquad \psi_3 = -0.01\varphi_{22}, \qquad \psi_4 = -0.01\varphi_{23},$$

where φ_k is the kth velocity basis functions.

Now, we present the numerical results.

We choose the viscosity $\nu = 0.02$ and the initial value is $y_0 = 1.1y_e$, where y_e is the steady-state solution which we can solve numerically. We plot the L^2 norm of $y(t) - y_e$ without and with the controller. We see that the uncontrolled solution blows up near $t = 3$. But the controlled solution is stable for very large t. In our case, we plot till $T = 10$ and the results are presented in Figs. 3.3, 3.4, 3.5.

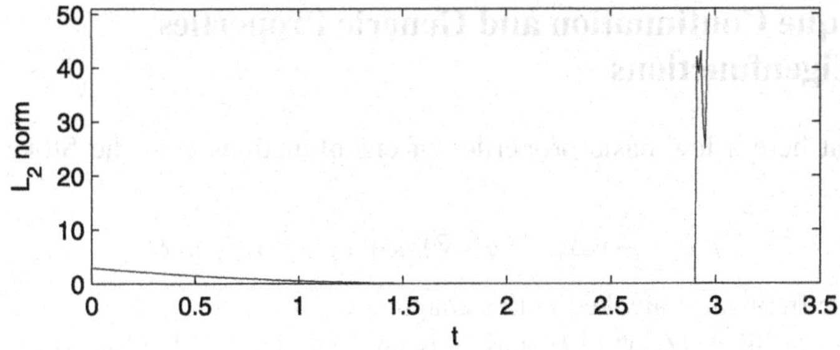

Fig. 3.3 L_2 norm of $y_h(t,x) - y_e$ in uncontrolled case

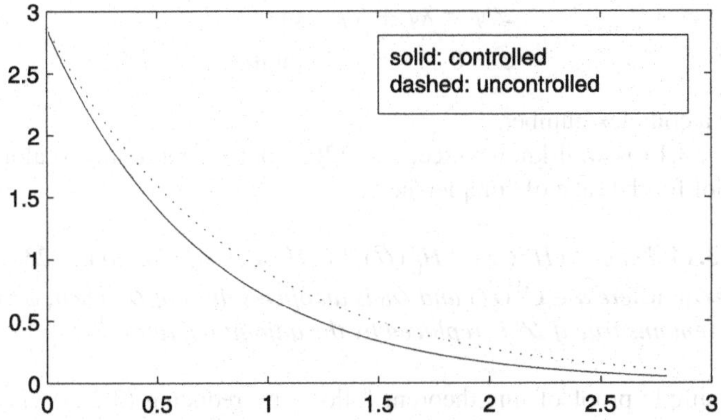

Fig. 3.4 L_2 norm of $y_h(t,x) - y_e$

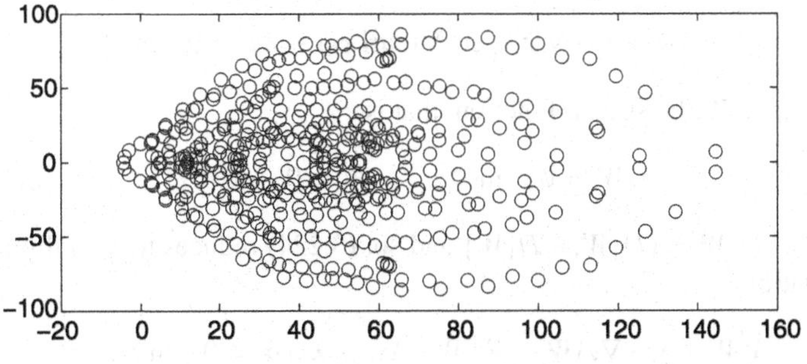

Fig. 3.5 The spectrum of the approximation matrix $\nu \mathbf{A}_h + \mathbf{A}_{0,h}$ of the linearized operator $\mathscr{A}$

3.8 Unique Continuation and Generic Properties of Eigenfunctions

We present here a few basic properties of eigenfunctions φ to the Stokes–Oseen operator

$$\mathcal{L}\varphi = -\nu\Delta\varphi + (\varphi\cdot\nabla)y_e + (y_e\cdot\nabla)\varphi \quad \text{in } \mathcal{O},$$

which were frequently invoked in this chapter.

Here, $y_e \in (W^{2,\infty}(\mathcal{O}))^d \cap H$ and $\mathcal{O}$ is an open, bounded subset of R^d. In the examples treated so far, y_e arose as a stationary solution to Navier–Stokes equations, but this fact is not essential for the results to be presented below.

The first result refers to the unique continuation property of the eigenfunctions $\varphi \in (H^2(\mathcal{O}))^d \cap V$ to the operator $\mathcal{L}$, that is,

$$\begin{aligned} \mathcal{L}\varphi &= \lambda\varphi + \nabla p \quad && \text{in } \mathcal{O}, \\ \varphi &= 0 && \text{on } \partial\mathcal{O}, \end{aligned} \tag{3.223}$$

where λ is a complex number.

Theorem 3.14 is well-known (see, e.g., [26]), but we reproduce it along with a simple proof for the sake of completeness.

Theorem 3.14 *Let $\varphi \in (H^2(\mathcal{O}) \cap H_0^1(\mathcal{O}))^d \cap H$ be a solution to* (3.223) *such that $\varphi \equiv \nabla q$ in $\mathcal{O}_0$, where $q \in C^1(\overline{\mathcal{O}})$ and $\mathcal{O}_0$ is an open subset of $\mathcal{O}$. Then, $\varphi \equiv 0$ on $\mathcal{O}$. The result remains true if $\mathcal{L}$ is replaced by the adjoint operator $\mathcal{L}^*$.*

Proof A simple proof of this theorem follows by reducing (3.223) via vorticity transformation to a fourth-order elliptic equation and applying after the classical Carleman's inequality.

For simplicity, we consider first the case $d = 2$ and take $\nu = 1$. Taking the curl operator in (3.223) and setting $\psi = \operatorname{curl}\varphi = D_2\varphi_1 - D_1\varphi_2$, $\varphi = \{\varphi_1, \varphi_2\}$, we obtain the vorticity equation

$$-\Delta\psi + y_e\cdot\nabla\psi + \varphi\cdot\nabla(\operatorname{curl} y_e) - \lambda\psi = 0 \quad \text{in } \mathcal{O}, \tag{3.224}$$

and, if we take W the stream function, that is,

$$\Delta W = \psi \quad \text{in } \mathcal{O}; \qquad W = 0 \quad \text{on } \partial\mathcal{O}, \tag{3.225}$$

we get $\varphi = \nabla^{\perp}W = \{D_2W, -D_1W\}$ and so, (3.224) reduces to the elliptic fourth-order equation

$$-\Delta^2 W + y_e\cdot\nabla\Delta W - \nabla^{\perp}W\cdot\Delta y_e - \lambda\Delta W = 0 \quad \text{in } \mathcal{O}. \tag{3.226}$$

(Without loss of generality, we may assume that the domain $\mathcal{O}$ is simple-connected and so, the stream function W is well-defined.) We have $W = 0$ on $\partial\mathcal{O}$ and

$$\Delta W = 0 \quad \text{on } \mathcal{O}_0. \tag{3.227}$$

Lemma 3.8 *By* (3.226) *and* (3.227) *it follows that* $\Delta W \equiv 0$ *in* $\mathcal{O}$.

By Lemma 3.8 and by (3.225) we see that $W \equiv 0$ in $\mathcal{O}$ and so $\varphi \equiv 0$ in $\mathcal{O}$, as claimed.

Proof of Lemma 3.8 We rewrite (3.226) as

$$P(x, D)V = \nabla^{\perp} W \cdot \Delta y_e, \quad V = \Delta W \text{ in } \mathcal{O}, \tag{3.228}$$

where $P(x, D)V = -\Delta V + y_e \cdot \nabla V - \lambda V$. We proceed as in the proof of Theorem 8.2.1 in [55].

Let $\psi \in C^{\infty}(\overline{\mathcal{O}})$ be such that $|\nabla \psi(x)| \neq 0$, $\forall x \in \mathcal{O}$ and $\sum_{i,j=1}^{2} D_{ij}^2 \psi(x) > 0$ in $\overline{\mathcal{O}}$. Let $x_0 \in \partial \mathcal{O}_0$ be arbitrary and let $\widetilde{\mathcal{O}}_0$ be a neighborhood of x_0 such that

$$\{x \in \widetilde{\mathcal{O}}_0;\ \psi(x) \geq \psi(x_0),\ x \neq x_0\} \subset \omega, \tag{3.229}$$

where ω is an open subset of $\mathcal{O}_0$. This implies the existence of $\varepsilon > 0$ such that

$$\psi(x) \leq \psi(x_0) - \varepsilon, \quad \forall x \in (\widetilde{\mathcal{O}}_0 \setminus \widetilde{\mathcal{O}}_0') \cap \omega^c, \tag{3.230}$$

where $\widetilde{\mathcal{O}}_0' \subset \widetilde{\mathcal{O}}_0$ is another neighborhood of x_0. Now, we choose $\chi \in C_0^{\infty}(\widetilde{\mathcal{O}}_0)$ such that $\chi = 1$ on $\widetilde{\mathcal{O}}_0'$ and set $u = \chi V$.

We have

$$\begin{aligned} P(x, D)u &= \chi P(x, D)V + \sum_{0<|\alpha|\leq 2} C_\alpha D^\alpha \chi P^{(\alpha)}(x, D)V \\ &= \chi \nabla^{\perp} W \cdot \Delta y_e + H \quad \text{in } \mathcal{O}, \end{aligned}$$

where $H \equiv 0$ in $\widetilde{\mathcal{O}}_0' \cup \widetilde{\mathcal{O}}_0^c$. We set $\widetilde{\mathcal{O}}_0^c = \mathcal{O} \setminus \widetilde{\mathcal{O}}_0$, $\mathcal{O}_0^c = \mathcal{O} \setminus \mathcal{O}_0$. We apply the Carleman inequality to the elliptic operator $P(x, D)$ (see [55, Theorem 8.3.1]). We have

$$\begin{aligned} &\sum_{|\alpha|\leq 2} \tau^{2(2-|\alpha|)} \int |D^\alpha u|^2 e^{2\tau\psi} dx \\ &\quad \leq K\tau \int |P(x, D)u|^2 e^{2\tau\psi} dx \\ &\quad \leq 2K\tau \int (|H|^2 + |\chi \nabla^{\perp} W \cdot \Delta y_e|^2) e^{2\tau\psi} dx \\ &\quad \leq K_1 \tau \left(\int_{(\widetilde{\mathcal{O}}_0 \setminus \widetilde{\mathcal{O}}_0') \cap \mathcal{O}_0^c} e^{2\tau\psi} dx + \int |\nabla^{\perp}(\chi W)|^2 e^{2\tau\psi} dx \right) \end{aligned} \tag{3.231}$$

because, by (3.226) and (3.227),

$$\chi \nabla^{\perp} W \cdot \Delta y_e = 0 \quad \text{in } \mathcal{O}_0, \qquad \nabla^{\perp} \chi = 0 \quad \text{in } \widetilde{\mathcal{O}}_0'.$$

Now, we can estimate the first order term in (3.231) as follows

$$\int |\nabla^\perp(\chi W)|^2 e^{2\tau\psi}dx \le C\int (|\Delta(\chi W)|^2+\tau|\chi W|^2)e^{2\tau\psi}dx$$
$$\le C\left(\int (\tau|u|^2+|\Delta u|^2)e^{2\tau\psi}dx+\int_{(\widetilde{\mathcal{O}}_0\setminus\widetilde{\mathcal{O}}'_0)\cap\mathcal{O}_0^c} e^{2\tau\psi}dx\right)$$

and, inserting into (3.231), yields

$$\tau^3\int |u|^2 e^{2\tau\psi}dx \le K_2\int_{(\widetilde{\mathcal{O}}_0\setminus\widetilde{\mathcal{O}}'_0)\cap\mathcal{O}_0^c} e^{2\tau\psi}dx$$

for τ large enough. Recalling that, by (3.229),

$$\psi(x)\le\psi(x_0)-\varepsilon,\quad \forall x\in(\widetilde{\mathcal{O}}_0\setminus\widetilde{\mathcal{O}}'_0)\cap\mathcal{O}_0^c,$$

we obtain that

$$\int |u|^2 e^{2\tau(\psi(x)-\psi(x_0)+\varepsilon)}dx \le \frac{K_1}{\tau^2}. \tag{3.232}$$

We set $\mathcal{O}_\varepsilon=\{x\in\mathcal{O};\ \psi(x)>\psi(x_0)-\varepsilon\}$. Then, by (3.232) we see that

$$\int_{\mathcal{O}_\varepsilon}|u|^2 d\chi \le \frac{K_2}{\tau^2},\quad \forall\tau>0$$

and, letting $\tau\to\infty$, we obtain that $u\equiv 0$ on $\mathcal{O}_\varepsilon$ and we mention that $\mathcal{O}_\varepsilon\cap\widetilde{\mathcal{O}}'_0$ has nonempty intersection with $\mathcal{O}_0^c$. Since $u=\Delta W$ in $\widetilde{\mathcal{O}}'_0$, it follows that $\Delta W\equiv 0$ in $\mathcal{O}_\varepsilon\cap\widetilde{\mathcal{O}}'_0$. Hence, $\Delta W\equiv 0$ in an open neighborhood of x_0 outside $\mathcal{O}_0$ and this, clearly, implies that $\Delta W\equiv 0$ in $\mathcal{O}$, as claimed. (It suffices to repeat the above argument on a larger domain $\mathcal{O}_0^*\supset\mathcal{O}_0$.)

The 3-*D* case. We set $\psi=\nabla\times\varphi=\operatorname{curl}\varphi$ and recall that $\varphi=\nabla\times(\widetilde{S}\varphi)$, where $\widetilde{S}$ is the Biot–Savart operator

$$(\widetilde{S}\varphi)(x)=\frac{1}{4\pi}\int_\Omega \varphi(y)\cdot\frac{x-y}{|x-y|^3}dy,\quad \forall x\in\Omega.$$

Taking into account that

$$\nabla\times(\nabla\times z)=-\Delta z\quad \text{if } \nabla\cdot z=0,$$

we have that

$$\nabla\times\varphi=-\Delta(\widetilde{S}\varphi)\quad \text{in } \mathcal{O}.$$

We set $W=\widetilde{S}\varphi$ and get, therefore,

$$-\Delta W=\nabla\times\varphi\quad \text{in } \mathcal{O},$$
$$\varphi=\nabla\times W\qquad \text{in } \mathcal{O}.$$

Applying the curl operator in (3.223), we obtain that

$$\begin{aligned} \Delta^2 W + \nabla \times (y_e \cdot \nabla)(\nabla \times W) + \nabla \times ((\nabla \times \varphi) \cdot \nabla) y_e + \lambda \Delta W = 0 \quad &\text{in } \mathscr{O}, \\ \nabla \times \Delta W = \Delta \varphi = 0 \quad &\text{in } \mathscr{O}_0, \end{aligned}$$

because $\nabla \cdot \varphi = 0$ and $\varphi = \nabla q$ in $\mathscr{O}_0$ and, therefore, $\Delta q = 0$ in $\mathscr{O}_0$.

We set $V = \nabla \times \Delta W$ and obtain that

$$-\Delta V + \Delta((y_e \cdot \nabla)(\nabla \times W)) + \Delta((\Delta W \cdot \nabla) y_e) - \lambda V = 0 \quad \text{in } \mathscr{O},\ V = 0 \text{ in } \mathscr{O}_0.$$

If we put, as in the previous case, $u = \chi V$, we obtain via the Carleman inequality (3.231) that $V = 0$ in $\mathscr{O}$ and so $\nabla \times \Delta W = 0$ in $\mathscr{O}$. Hence, $\Delta \varphi = 0$ in $\mathscr{O}$, as claimed. □

Theorem 3.15 *Let $\{\varphi_j\}_{j=1}^m$, $m > 1$, be a system of eigenfunctions for the Stokes–Oseen operator $\mathscr{L}$ and let $\mathscr{O}_0$ be an open subset of $\mathscr{O}$. Then, the system $\{\varphi_j\}_{j=1}^m$ is linearly independent in $(L^2(\mathscr{O}_0))^d$. The same result holds for the dual operator $\mathscr{L}^*$.*

Let $\{\varphi_j\}_{j=1}^N$ be eigenfunctions corresponding to eigenvalues λ_j, that is,

$$\begin{cases} \mathscr{L}\varphi_j = \lambda_j \varphi_j + \nabla p_j & \text{in } \mathscr{O}, \\ \nabla \cdot \varphi_j = 0 & \text{in } \mathscr{O}, \\ \varphi_j = 0 & \text{on } \partial\mathscr{O}. \end{cases}$$

One must prove that each system $\{\varphi_1, \dots, \varphi_m\}$, $1 \le m \le N$, is linearly independent in $\mathscr{O}_0$. As mentioned earlier, this is immediate if all φ_j are eigenfunctions corresponding to the same eigenvalue λ_j and so, it suffices to prove this for distinct eigenvalues λ_j. For $m = 1$, this follows by Theorem 3.14. Let $m = 2$ and let φ_1 and φ_2 be two eigenfunctions with corresponding eigenvalues λ_1, λ_2. Then, we have

$$\mathscr{L}(\lambda_2\varphi_1 - \lambda_1\varphi_2) = \lambda_1\lambda_2(\varphi_1 - \varphi_2) + \lambda_2 \nabla p_1 - \lambda_1 \nabla p_2 = \nabla p \quad \text{in } \mathscr{O}. \tag{3.233}$$

Assume that $\alpha_1\varphi_1 + \alpha_2\varphi_2 \equiv 0$ on $\mathscr{O}_0$ for $\alpha_1, \alpha_2 \neq 0$ and argue from this to a contradiction. Indeed, in this case, replacing φ_1 by $\frac{\alpha_1}{\lambda_2}\varphi_1$ and φ_2 by $-\frac{\alpha_2}{\lambda_1}\varphi_2$, we see that $\lambda_2\varphi_1 - \lambda_1\varphi_2 \equiv 0$ on $\mathscr{O}_0$. Hence, by (3.233), $\varphi_1 = \alpha \nabla q$ in $\mathscr{O}_0$ for some α and so, by Theorem 3.14, $\varphi_1 \equiv 0$ in $\mathscr{O}$, which is absurd. We treat, now, the case $m = 3$. We have as above, besides (3.233), that

$$\mathscr{L}(\lambda_3\varphi_1 - \lambda_1\varphi_3) = \lambda_1\lambda_3(\varphi_1 - \varphi_3) + \nabla p \quad \text{in } \mathscr{O},$$

and, therefore,

$$\begin{aligned} &\mathscr{L}((\lambda_2 - \lambda_3)\varphi_1 - \lambda_1\varphi_2 + \lambda_1\varphi_3) \\ &\quad = \lambda_1\lambda_2(\varphi_1 - \varphi_2) - \lambda_1\lambda_3(\varphi_1 - \varphi_3) + \nabla q \quad \text{in } \mathscr{O}. \end{aligned} \tag{3.234}$$

If $\alpha_1\varphi_1+\alpha_2\varphi_2+\alpha_3\varphi_3\equiv 0$ on $\mathscr{O}_0$, then, replacing $\varphi_1,\varphi_2,\varphi_3$ by $\frac{\alpha_1}{\lambda_2-\lambda_3}\varphi_1, -\frac{\alpha_2}{\lambda_1}\varphi_2$, $\frac{\alpha_3}{\lambda_1}\varphi_3$, respectively, we obtain that

$$(\lambda_2-\lambda_3)\varphi_1-\lambda_1\varphi_2+\lambda_1\varphi_3\equiv 0 \quad \text{in } \mathscr{O}_0,$$

which, by virtue of (3.234), implies that

$$(\lambda_2-\lambda_3)\varphi_1-\lambda_2\varphi_2+\lambda_3\varphi_3\equiv \nabla q \quad \text{in } \mathscr{O}_0.$$

This yields $\widetilde{\alpha}_1\varphi_1+\widetilde{\alpha}_2\varphi_2=\nabla q$ in $\mathscr{O}_0$, which, by the previous step, is absurd. The argument extends *mutatis mutandis* to all m and this completes the proof.

Now, we prove that, for the Stokes–Oseen operator $\mathscr{L}$, the property of eigenvalues λ_j to be simple is generic. We have seen earlier in this chapter that this property is essential to determine the dimension of the stabilizing feedback, but it has, however, an intrinsic interest.

Let $\mathscr{A}=\mathscr{A}_{y_e}$ be the Stokes–Oseen operator

$$\mathscr{A}_{y_e}\varphi=P(-\Delta\varphi+(y_e\cdot\nabla)\varphi+(\varphi\cdot\nabla)y_e)=P\mathscr{L}\varphi,$$
$$\nabla\cdot\varphi=0 \quad \text{in } \mathscr{O}, \qquad \varphi=0 \quad \text{on } \partial\mathscr{O},$$

where $P:(L^2(\mathscr{O}))^d\to H$ is, as usual, the Leray projection.

Consider the eigenvalue problem for $\mathscr{A}_{y_e}$, that is,

$$\mathscr{A}_{y_e}(\varphi)=\lambda\varphi \quad \text{in } \mathscr{O}.$$

As seen earlier, for each $y_e\in\mathscr{W}=(H^2(\mathscr{O}))^d\cap(H_0^1(\mathscr{O}))^d\cap H$, $H=\{y\in(L^2(\mathscr{O}))^d;\ \nabla\cdot y=0,\ y\cdot n=0 \text{ on } \partial\mathscr{O}\}$ there is a countable set of eigenvalues $\{\lambda=\lambda_j(y_e)\}_{j=1}^\infty$ and each eigenvalue $\lambda_j(y_e)$ has a finite algebraic multiplicity $m_j=m_j(y_e)$.

Theorem 3.16 amounts to saying that the property $m_j(y_e)=1$ for all j is generic with respect to y_e. In other words, the set $\{y_e\in\mathscr{W};\ m_j(y_e)=1,\ \forall j\}$ is a *residual*.

Recall that a set $\mathscr{X}$ of a topological space Y is called *residual* or *of second category* if it is a countable intersection of open dense sets. According to the Baire category theorem, a residual subset of a complete metric space Y is dense in Y.

Theorem 3.16 *The set*

$$\mathscr{M}=\{y_e\in\mathscr{W};\ \textit{all the eigenvalues } \lambda \textit{ of } \mathscr{A}_{y_e} \textit{ are simple}\}$$

is residual in the space $\mathscr{W}=(H^2(\mathscr{O}))^d\cap(H_0^1(\mathscr{O}))^d\cap H$.

In other words, for "almost all" y_e, the eigenvalues of the Stokes–Oseen operator $\mathscr{A}_{y_e}$ are simple. Of course, the fact that y_e is an equilibrium solution to the Navier–Stokes equation is not relevant because each function $y_e\in\mathscr{W}$ can be viewed as such an equilibrium solution to (3.6) for a suitable f_e.

Before proceeding with the proof, let us briefly recall a few definitions related to the Sard–Smale theorem.

Let X, Y be two Banach manifolds and let $T : X \to Y$ be a C^1-mapping with differential $T'(x) : X \to L(X, Y)$. The mapping T is called *Fredholm* if $T'(x)$ has for each $x \in X$ finite-dimensional kernel and its range has finite codimension, that is, it is a Fredholm map. A point $f \in Y$ is called *regular value* for T if $T'(x)$ is surjective for all $x \in T^{-1}(f)$. The index of T is the difference (independent of x)

$$\text{index}(T) = \dim \operatorname{Ker} T'(x) - \operatorname{codim} R(T'(x)).$$

The Smale infinite-dimensional extension of the classical Sard's theorem is formulated below (see [69]).

Theorem 3.17 *Let X, Y be separable Banach manifolds and let $T : X \to Y$ be a C^k-Fredholm mapping such that* $\text{index}(T) < k < \infty$. *Then, the set*

$$\{f \in Y;\ f \text{ is a regular value of } T\}$$

is residual in Y.

Theorem 3.18 is a variant of the Sard–Smale theorem known in literature as the *transversality theorem.*

Theorem 3.18 *Let W, X, Y be separable Banach manifolds and let $F : W \times X \to Y$ be a C^k mapping such that, for each $w \in W$, $F_w = F(w, \cdot, \cdot)$ is a Fredholm map from X to Y of index less than k. Then, if $f \in Y$ is a regular value of F, the set*

$$\{w \in W;\ F_w \text{ has } f \text{ as regular value}\}$$

is residual in W.

Proof of Theorem 3.16 The idea of the proof was already used by Uhlenbeck [76] to prove that the eigenvalues of a Laplace operator on a compact manifold with metric g and without boundary are simple for g in a residual set.

In our case, we apply the transversality Theorem 3.18 to the spaces

$$\begin{aligned} X &= ((H^2(\mathcal{O}))^d \cap (H_0^1(\mathcal{O}))^d \cap H) \times \mathbb{C}, \\ Y &= H, \qquad W = W_m^M, \end{aligned} \tag{3.235}$$

where

$$W_m^M = \left\{ \sum_{j=1}^m w_j e_j;\ \sum_{j=1}^m |w_j|^2 < M \right\}$$

and $\{e_j\}$ is an orthonormal basis in H which is made precise later on. Here, the spaces X, Y are endowed with their natural topology and the operator F :

$W \times X \to H$ is defined by

$$F(y_e, \varphi, \lambda) = P(-\Delta\varphi + (y_e \cdot \nabla)\varphi + (\varphi \cdot \nabla)y_e - \lambda\varphi) = \mathscr{A}_{y_e}\varphi - \lambda\varphi,$$

where $\mathbb{C}$ is the complex field.

We denote by $W_m = \text{lin span}\{e_j\}_{j=1}^m$, that is, the tangent space to the manifold W_m^M. We note that the differential F' of F is given by

$$\begin{aligned} F'(y_e, \varphi, \lambda)(z, \widetilde{\varphi}, \lambda) = P[&-\Delta\widetilde{\varphi} + (y_e \cdot \nabla)\widetilde{\varphi} + (\widetilde{\varphi} \cdot \nabla)y_e \\ &- \lambda\widetilde{\varphi} + (z \cdot \nabla)\varphi + (\varphi \cdot \nabla)z - \widetilde{\lambda}\varphi] \end{aligned}$$

and that F is a Fredholm operator.

Indeed, $\text{Ker}[F'(y_e, \varphi, \lambda)] = \{z, \widetilde{\varphi}, \widetilde{\lambda}\}$, where

$$P[-\Delta\widetilde{\varphi} + (y_e \cdot \nabla)\widetilde{\varphi} + (\widetilde{\varphi} \cdot \nabla)y_e - \lambda\widetilde{\varphi} + (z \cdot \nabla)\varphi + (\varphi \cdot \nabla)z - \widetilde{\lambda}\varphi] = 0$$

and, since $z \in W_m$ (a finite-dimensional space) and the operator

$$\widetilde{\varphi} \to P[-\Delta\widetilde{\varphi} + (y_e \cdot \nabla)wt\varphi + (\widetilde{\varphi} \cdot \nabla)y_e - \lambda\widetilde{\varphi}]$$

is Fredholm, we infer that

$$\dim \text{Ker}[F'(y_e, \varphi, \lambda)] < \infty, \quad \forall (y_e, \varphi, \lambda) \in W_m^M \times X \times C.$$

For the same reason, it follows that $\text{codim}\, F'(y_e, \varphi, \lambda) < \infty$.

By a similar argument, it follows that the operator $F_{y_e} : X \to H$,

$$F_{y_e}(\varphi, \lambda) = P[-\Delta\varphi + (y_e \cdot \nabla)\varphi + (\varphi \cdot \nabla)y_e - \lambda\varphi], \quad \forall y_e \in W_m^M$$

is a Fredholm operator.

Indeed, its differential $F'_{y_e}(\varphi, \lambda)(\widetilde{\varphi}, \widetilde{\lambda})$ in direction $(\widetilde{\varphi}, \widetilde{\lambda}) \in X$ is given by

$$F'_{y_e}(\varphi, \lambda)(\widetilde{\varphi}, \widetilde{\lambda}) = P(-\Delta\widetilde{\varphi} + (y_e \cdot \nabla)\widetilde{\varphi} + (\widetilde{\varphi} \cdot \nabla)y_e - \lambda\widetilde{\varphi} - \widetilde{\lambda}\varphi), \quad \forall (\widetilde{\varphi}, \widetilde{\lambda}) \in X.$$

Then, $\text{Ker}\, F'_{y_e}(\varphi, \lambda)$ is the set of all $(\widetilde{\varphi}, \widetilde{\lambda}) \in X$ such that

$$\mathscr{A}_{y_e}\widetilde{\varphi} - \lambda\widetilde{\varphi} = P(-\Delta\widetilde{\varphi} + (y_e \cdot \nabla)\widetilde{\varphi} + (\widetilde{\varphi} \cdot \nabla)y_e - \lambda\widetilde{\varphi}) = \widetilde{\lambda}\varphi \tag{3.236}$$

and since, by the Riesz–Schauder theory, $R(\mathscr{A}_{y_e} - \lambda I)$ is closed and $\dim\{\widetilde{\lambda}\varphi;\ \widetilde{\lambda} \in \mathbb{C}\} = 1$, it follows by the closed-range theorem that

$$\begin{aligned} &\dim\{\widetilde{\varphi}\} = 1 \quad \text{if } \lambda \text{ is not eigenvalue for } \mathscr{A}_{y_e}, \\ &\dim\{\widetilde{\varphi}\} = 1 \quad \text{if } \lambda \text{ is not eigenvalue and } \varphi \in (N(\overline{\lambda}))^\perp, \\ &\dim\{\widetilde{\varphi}\} = 0 \quad \text{if } \lambda \text{ is not eigenvalue and } \varphi \overline{\in} (N(\overline{\lambda}))^\perp. \end{aligned}$$

(Here, $(\widetilde{\varphi}, \widetilde{\lambda})$ is solution to (3.236) and $N(\overline{\lambda}) = N(\overline{\lambda} I - \mathscr{A}_{y_e}^*)$ is the eigenspace of the eigenvalue $\overline{\lambda}$ for the adjoint operator $\mathscr{A}_{y_e}^*$.) Hence,

$$\dim \operatorname{Ker} F'_{y_e}(\varphi, \lambda) \leq 1, \quad \forall (\varphi, \lambda) \in X.$$

As regards the codimension of $R(F'_{y_e}(\varphi, \lambda))$, it follows by the same argument that

$$R(F'_{y_e}(\varphi, \lambda)) = H$$

if λ is not eigenvalue for $\mathscr{A}_{y_e}$ and

$$\operatorname{codim} R(R'_{y_e}(\varphi, \lambda)) = M(\lambda) + 1$$

if λ is eigenvalue for $\mathscr{A}_{y_e}$, where $M(\lambda) = \dim N(\overline{\lambda} I - \mathscr{A}_{y_e}^*)$. Hence,

$$\operatorname{index} F_{y_e}(\varphi, \lambda) \leq 1, \quad \forall (\varphi, \lambda) \in X, \ \forall y_e \in W.$$

To complete the proof, the following observation is crucial.

Lemma 3.9 *$\{0\}$ is regular value for F_{y_e} if and only if λ is a simple eigenvalue for $\mathscr{A}_{y_e}$.*

Proof We consider $(\varphi, \lambda) \in X$, which satisfy the equation

$$F_{y_e}(\varphi, \lambda) = 0 \tag{3.237}$$

and consider, for any $f \in Y = H$, the equation

$$F'_{y_e}(\varphi, \lambda)(\widetilde{\varphi}, \widetilde{\lambda}) = f, \tag{3.238}$$

that is,

$$\begin{aligned} &P(-\Delta\widetilde{\varphi} + (y_e \cdot \nabla)\widetilde{\varphi} + (\widetilde{\varphi} \cdot \nabla) y_e - \lambda\widetilde{\varphi}) = \widetilde{\lambda}\varphi + f, \\ &\widetilde{\varphi} \in (H^2(\mathscr{O}))^d \cap H_0^1(\mathscr{O}); \ \widetilde{\varphi} \in H. \end{aligned} \tag{3.239}$$

We note that, by (3.237), it follows that φ is eigenfunction for the operator $\mathscr{A}_{y_e}$ with the eigenvalue λ. Then, by the closed-range theorem, (3.238) has solution if and only if

$$(\widetilde{\lambda}\varphi + f, \varphi^*) = 0, \quad \forall \varphi^* \in N(\mathscr{A}_{y_e}^* - \overline{\lambda} I), \tag{3.240}$$

where $(\cdot, \cdot)$ is the scalar product in H. If λ is a simple eigenvalue for $\mathscr{A}_{y_e}$ with eigenfunctions $\{\alpha\varphi;\ \alpha \in \mathbb{C}\}$, then $N(\mathscr{A}_{y_e}^* - \overline{\lambda} I) = \{\alpha\varphi^*;\ \alpha \in \mathbb{C}\}$ and so, there is $\widetilde{\lambda} \in \mathbb{C}$ which satisfies (3.239). More precisely, $\lambda = -(f, \varphi^*)(\varphi, \varphi^*)^{-1}$. Then, coming back in (3.239), we find $(\widetilde{\varphi}, \widetilde{\lambda})$ which satisfies (3.238), that is, $F'_{y_e}(\varphi, \lambda)$ is surjective. Hence, $\{0\}$ is a regular value.

Conversely, if (3.238) has solution $(\widetilde{\varphi}, \widetilde{\lambda}) \in Z$ for all $f \in H$, then (3.240) holds. Assume that λ is not simple and argue from this to a contradiction.

Let $\{\varphi_i^*\}_{i=1}^m$ be an independent system in the space of all the generalized eigenfunctions of $\mathscr{A}_{y_e}^*$ corresponding to the eigenvalue $\overline{\lambda}$. (Here, m is the algebraic multiplicity of λ and $\overline{\lambda}$.) We have, therefore,

$$(\mathscr{A}_{y_e}^* - \overline{\lambda} I)^i \varphi_i^* = 0, \quad \forall i = 1, \dots, m.$$

Then, by (3.240) we obtain that

$$(f + \widetilde{\lambda}\varphi, (\mathscr{A}_{y_e}^* - \overline{\lambda} I)^{i-1} \varphi_i^*) = 0, \quad \forall i = 1, \dots, m.$$

Since, by (3.238), φ is eigenfunction for $\mathscr{A}_{y_e}$, this yields

$$(f, (\mathscr{A}_{y_e}^* - \overline{\lambda} I)^{i-1} \varphi_i^*) = 0, \quad \forall i = 1, \dots, m,$$

and, because f is arbitrary in H, we obtain that

$$(\mathscr{A}_{y_e}^* - \overline{\lambda} I)^{i-1} \varphi_i^* = 0, \quad \forall i = 1, \dots, m.$$

Hence, φ_2^* is eigenfunction for $\mathscr{A}_{y_e}^*$.

Then, step by step, it follows that

$$(\mathscr{A}_{y_e}^* - \overline{\lambda} I) \varphi_i^* = 0, \quad \forall i = 1, \dots, m, \tag{3.241}$$

that is, the eigenvalue $\overline{\lambda}$ is semisimple. Moreover, by (3.240) and (3.241), we see that there is $\widetilde{\lambda} \in \mathbb{C}$ (independent of φ_i^*) such that

$$(\widetilde{\lambda}\varphi + f, \varphi_i^*) = 0, \quad \forall i = 1, \dots, m,$$

that is, $\widetilde{\lambda} = -(f, \varphi_i^*)$. This yields

$$(f, \varphi_i^*)(\varphi, \varphi_k^*) - (f, \varphi_k^*)(\varphi, \varphi_i^*) = 0, \quad \forall i, k = 1, \dots, m,$$

and, since f is arbitrary, we obtain that

$$(\varphi, \varphi_i^*)\varphi_k^* - (\varphi, \varphi_k^*)\varphi_i^* = 0, \quad \forall i, k = 1, \dots, m.$$

Clearly, this implies that system $\{\varphi_i^*\}_{i=1}^m$ is linearly dependent. The contradiction we arrived at concludes the proof of Lemma 3.9. □

Proof of Theorem 3.16 (continuation). First, let us check that 0 is a regular value for F, that is, the equation

$$P[-\Delta\widetilde{\varphi} + (y_e \cdot \nabla)\widetilde{\varphi} + (\widetilde{\varphi} \cdot \nabla) y_e - \lambda\widetilde{\varphi} + (z \cdot \nabla)\varphi + (\varphi \cdot \nabla) z - \widetilde{\lambda}\varphi] = f$$

has, for each $f \in H$, a solution $(z, \widetilde{\varphi}, \widetilde{\lambda}) \in W_m \times X \times C$. To this end, we write $f = f_1 + f_2$, where $f_2 \in N(\mathscr{A}_{y_e}^* - \overline{\lambda} I)$ and $f_1 \in R(\mathscr{A}_{y_e} - \lambda I)$. Then, the above equation reduces to

$$(z \cdot \nabla)\varphi + (\varphi \cdot \nabla) z - \widetilde{\lambda}\varphi = f_2. \tag{3.242}$$

Let $p_{y_e} = \dim N(\mathscr{A}^*_{y_e} - \widetilde{\lambda} I)$. For y_e in the ball of radius M (in the space $H^2(\mathcal{O}) \cap H$), clearly, we have $p_{y_e} \leq m$ if m is large enough. Then, we look for a solution z of (3.242) of the form

$$z = \sum_{i=1}^{m} \beta_i e_i, \qquad \widetilde{\lambda} = 0.$$

If $f_2 = \sum_{i=1}^{m} \alpha_i e_i$, we rewrite (3.242) as

$$\sum_{i=1}^{m} (b(e_i, \varphi, e_j) + b(\varphi, e_i, e_j))\beta_i = \alpha_j, \quad j = 1, \dots, m. \tag{3.243}$$

(Here b is the trilinear form (1.53).)

Clearly, (3.243) has a solution $\{\beta_i\}_{i=1}^{m}$ if

$$\det \|b(e_i, \varphi, e_j) + b(\varphi, e_i, e_j)\|_{i,j=1}^{m} \neq 0.$$

We recall that $F(y_e, \varphi, \lambda) = 0$, that is, (φ, λ) is an eigenvalue pair corresponding to y_e and $\{e_i\}_{i=1}^{\infty}$ is an arbitrary orthonormal basis in H. Of course, it can be taken independent of y_e and φ and so, it can be arranged in such a way that

$$\det \|b(e_i, \varphi, e_j) + b(\varphi, e_i, e_j)\|_{i,j=1}^{m} \neq 0.$$

Hence, there is $z \in \text{lin span}\{e_i\}_{i=1}^{m} = W_m$, which satisfies System (3.242).

Applying the transversality Theorem 3.18 to the operator F and to the spaces X, Y, W, we conclude that the set

$$\{y_e \in W_m^M;\ 0 \text{ is regular for } F_w\}$$

is residual in W_m^M. Then, by Lemma 3.9, it follows that the set

$$\{y_e \in W_m^M;\ \text{each eigenvalue } \lambda = \lambda(y_e) \text{ of } \mathscr{A}_{y_e} \text{ is simple}\}$$

is residual in W_m^M, for all m, M.

Taking into account that

$$W = \bigcup_{m,M} W_m^m,$$

we conclude that

$$\{y_e \in \mathscr{W}, \lambda(y_e) \text{ is simple}\}$$

is a residual in W, as claimed. □

3.9 Comments on Chap. 3

The results from Sect. 3.3 were established first in Barbu and Triggiani [26] (for some earlier result in this direction, see also Barbu [12]) but here they are given in a slightly different form in order to simplify their presentation and, more exactly, to make precise the dimension of the stabilizable controller.

In [61], Lefter has obtained a similar result for the internal stabilization of the 2-D Navier–Stokes equations (3.1) with Navier slip boundary conditions

$$\operatorname{curl} y + \gamma y \cdot \tau = 0 \quad \text{on } \partial\mathcal{O}.$$

(Here, τ is the tangent vector to $\partial\mathcal{O}$.) In the work [24] of Barbu, Rodriguez and Shirikyan, the stabilization of a nonstationary trajectory to a Navier–Stokes equation is studied by similar arguments.

Section 3.4, which is devoted to the tangential boundary stabilization of Navier–Stokes equations, is based on the works [21, 22] by Barbu, Lasiecka and Triggiani. However, also in this case, the presentation is somewhat different from that in [21, 22], and is confined to the essential features of the tangential stabilization theory and to the main results which can be obtained via spectral decomposition. A related approach to the boundary stabilization is due to Fursikov [51, 52]. The Fursikov results are, in a certain sense, more general since no condition of the nature of Assumption (K2) is assumed, but are confined to open-loop boundary stabilization controllers and the nature of boundary controller (tangential or not) is not made precise. On these lines, we must mention also the boundary stabilization results of Raymond [70, 71] who was the first to consider a low-gain Riccati-based approach to the construction of stabilizable feedback controllers and has also designed stabilizable boundary feedback controllers under general assumptions. In this context, we cite also the works of Bedra [30, 31]. Though their results are of the same nature and the methods quite similar, we did not present them in details, our option for boundary stabilization being oriented to a more direct approach which avoids tedious calculation and arguments. However, these results deserved special attention for their generality. Theorem 3.12 on normal stabilization of a periodic flow in a 2-D channel is one of the few results on normal boundary stabilization of Navier–Stokes equations for large value of the Reynolds number

$$Re = \frac{1}{\nu}.$$

In this case, other results, somewhat different from Theorem 3.12, were previously given by Barbu [13] and Triggiani [75]. The main difficulty with the normal boundary stabilization problem is to get rid of the pressure p in the system. In [13], this follows by a Fourier approach similar to that used above while in [75] this is achieved via the vorticity function $v_x - u_y$, but the final result is quite different from that in [13]. The boundary stabilization of parabolic equilibrium profile periodic fluids in 2-D channels was extensively studied in the last decade and notable advances have been made by Krstic and coworkers (see [1, 2, 7, 80, 81] for a few results in this direction). Their results refer to the design of normal or tangential stabilizable controllers as well as to their numerical implementation. The only severe limitation of these results is that, with few exceptions, these are obtained for low-value Reynolds numbers $\frac{1}{\nu}$. Theorem 3.12 is essentially due to Munteanu [65].

The stabilization of time-periodic flows (Theorem 3.13) was established first in Barbu and Wang [29] and seems to be new in this context. There is, however, a large literature on stability of time-periodic flows. (See, e.g., Joseph's book [57].)

In stabilization analysis of Navier–Stokes equations, proportional linear feedback controllers of the form $u = k(y - y_e)$ are quite popular though they are efficient for low Reynolds numbers only. In Barbu and Lefter [23], it is shown that an internal feedback controller of this form, namely

$$u = -k(y - y_e)\mathbf{1}_{\mathcal{O}},$$

where $\mathcal{O}_0 \subset \mathcal{O}$, is exponentially stabilizable in Navier–Stokes System (3.1) if $k \geq k_0$ is sufficiently large and

$$\lambda_1^*(\mathcal{O}_0) > \nu^{-1}\gamma^*(y_e),$$

where $\lambda_1^*(\mathcal{O}_0)$ is the first eigenvalue of the Stokes operator A on $\mathcal{O}_0$ with Dirichlet boundary conditions and

$$\gamma^*(y_e) = \sup\{|b(y, y, y_e)|;\ |y| = 1\}.$$

For a survey on stabilization control techniques for Navier–Stokes systems, we refer the reader to [37]. From the computational point of view, the present approach is close to the technique of reduced-order controller design for fluid flows (see, e.g., [68]). The numerical computation given in Sect. 3.7 is from the work [19].

We did not discuss, here, other important topics related to the control of Navier–Stokes equations, for instance, exact and approximate controllability, and optimal control of Navier–Stokes equations which though close remain, however, beyond the purposes of this presentation. In fact, there is a close connection between internal or boundary stabilizability of Navier–Stokes equation (3.1) and its local exact controllability (Fursikov, Imanuvilov [53], Imanuvilov [56]) since the exact controllability implies stabilization by open-loop controllers. However, it does not imply the internal stabilization by finite-dimensional controllers, which is the key result established in this chapter. We refer to Coron's book [42] for significant recent results in this direction, which also cover the case of Euler equations.

As it is apparent from the analysis developed in this chapter, for internal or boundary stabilizations via spectral linearization method, the unique continuation property of eigenfunctions to the Stokes–Oseen operator is the key instrument for a rigorous approach. This is the reason we presented in Sect. 3.8 a few basic results in this context. Theorem 3.16 seems to be new and has an intrinsic interest which exceeds the needs of this chapter.

Chapter 4
Stabilization by Noise of Navier–Stokes Equations

The stochastic stabilization of Navier–Stokes equations is an alternative approach to stabilization techniques described in Chap. 3, which have two important advantages: the simplicity of the stabilizable feedback law and its robustness to (deterministic and stochastic) perturbations. A long time ago, it was observed that the noise might stabilize the finite and infinite-dimensional dynamical systems and several empirical observations in fluid dynamics suggested that noise might have a dissipation effect comparable with increasing the viscosity of fluid. This is exactly what will be rigorously proven here by designing stabilizing noise feedback controller with internal or boundary support.

4.1 Internal Stabilization by Noise

We prove here that the equilibrium solution y_e to the Navier–Stokes controlled equation (3.1), that is,

$$\begin{aligned} &\frac{\partial X}{\partial t} - \nu\Delta X + (X\cdot\nabla)X = \nabla p + f_e + mu && \text{in } (0,\infty)\times\mathcal{O},\\ &X(0) = x && \text{in } \mathcal{O},\\ &\nabla\cdot X = 0 && \text{in } (0,\infty)\times\mathcal{O}, \end{aligned} \tag{4.1}$$

is exponentially stabilizable in probability by a stochastic feedback controller of the form

$$u(t) = \eta\sum_{j=1}^{N}(X(t)-y_e,\varphi_j^*)P(m\phi_j)\dot\beta_j(t),$$

or, equivalently, the stochastic controller

$$u(t) = \eta\sum_{j=1}^{N}(X(t),\varphi_j^*)P(m\phi_j)\dot\beta_j(t) \tag{4.2}$$

V. Barbu, *Stabilization of Navier–Stokes Flows*,
Communications and Control Engineering,
DOI 10.1007/978-0-85729-043-4_4,

stabilizes in probability the control system

$$\begin{aligned}&\frac{d}{dt}X(t)+\mathscr{A}X(t)+S(X(t))=u(t),\quad t\ge 0,\\&X(0)=x,\end{aligned}\tag{4.3}$$

for a suitable chosen system $\{\phi_j\}\subset H$. Here and everywhere in the following, $\beta_j,\ j=1,\dots,N$, are independent real Brownian motions in a filtered probability space $(\Omega,\mathbb{P},\mathscr{F},\{\mathscr{F}_t\}_{t>0})$ and we refer to Sect. 4.5 for definition and basic results on stochastic analysis of differential systems and spaces of stochastic processes adapted to filtration $\{\mathscr{F}_t\}_{t>0}$. The scalar product of H is denoted $(\cdot,\cdot)$ and the norm $|\cdot|$. We denote by $\widetilde{H}$ the complexified space $H+iH$ with scalar product, again denoted by $(\cdot,\cdot)$, and norm by $|\cdot|_{\widetilde{H}}$. $C_W([0,T];L^2(\Omega,\widetilde{H}))$ is the space of all adapted square-mean $\widetilde{H}$-valued continuous processes on $[0,T]$. By $\dot{\beta}_j$ we have denoted the white noise associated with the Brownian motion β_j. We adopt the notation of Chap. 3. In particular, $\mathscr{A}$ is the Stokes–Oseen operator associated with the Navier–Stokes equation (4.1), N is the number of eigenvalues λ_j of $\mathscr{A}$ with $\operatorname{Re}\lambda_j\le\gamma$ and φ_j are the eigenfunctions of $\mathscr{A}$, while φ_j^* are the eigenfunctions of the dual operator $\widetilde{\mathscr{A}}^*$ in the space $\widetilde{H}$.

Equation (4.3) with the stochastic controller (4.2) should be viewed as the stochastic differential equation

$$\begin{aligned}&dX+(\mathscr{A}X+S(X))dt=\eta\sum_{j=1}^{N}(X,\varphi_j^*)P(m\phi_j)d\beta_j,\\&X(0)=x.\end{aligned}\tag{4.4}$$

Here η is a real number, $|\eta|>0$ and $m=\mathbf{1}_{\mathscr{O}_0}$ is the characteristic function of the open subset $\mathscr{O}_0\subset\mathscr{O}$, and $\{\phi_j\}_{j=1}^N\subset\widetilde{H}$ is a system of functions to be made precise below.

In 2-D, the stochastic differential equation (4.4) has a global mild solution $X\in C_W([0,T];L^2(\Omega,\widetilde{H}))$ for all $T>0$ (see Theorem 4.9 in Sect. 4.5).

The closed-loop system (4.4) can be written, equivalently, as

$$\begin{aligned}&dX(t)-\nu\Delta X(t)dt+(X(t)\cdot\nabla)y_e\,dt+(y_e\cdot\nabla)X(t)dt+(X(t)\cdot\nabla)X(t)dt\\&\quad=\eta m\sum_{j=1}^{N}(X(t),\varphi_j^*)\phi_j d\beta_j(t)+\nabla p(t)dt\quad\mathbb{P}\text{-a.s. in }(0,\infty)\times\mathscr{O},\\&\nabla\cdot X(t)=0\quad\text{in }\mathscr{O},\qquad X(t)\big|_{\partial\mathscr{O}}=0,\quad\mathbb{P}\text{-a.s., }\forall t\ge 0,\\&X(0)=x\quad\text{in }\mathscr{O}.\end{aligned}\tag{4.5}$$

Hence, in the space $(L^2(\mathscr{O}))^d$, the feedback controller $\{u_j=\eta m(X,\varphi_j^*)\phi_j\}_{j=1}^N$ has the support in $\mathscr{O}_0$ and (4.4) can be viewed as a stochastic perturbation of the (deterministic) Navier–Stokes equation (4.3).

Let us, briefly, recall other notation related to the operator $\mathscr{A}$ (see Sect. 3.1).

We set

$$\mathscr{X}_u = \text{lin span}\{\varphi_j\}_{j=1}^N$$

and denote by P_N the corresponding projector.

We have $\mathscr{X}_u = P_N \widetilde{H}$ and

$$P_N = \frac{1}{2\pi i} \int_\Gamma (\lambda I - \mathscr{A})^{-1} d\lambda,$$

where Γ is a closed smooth curve in $\mathbb{C}$, which is the boundary of a domain containing in interior the eigenvalues $\{\lambda_j\}_{j=1}^N$.

Let $\mathscr{A}_u = P_N \mathscr{A}$, $\mathscr{A}_s = (I - P_N)\mathscr{A}$. Then, as seen earlier, $\mathscr{A}_u$, $\mathscr{A}_s$ leave invariant the spaces $\mathscr{X}_u$ and $\mathscr{X}_s = (I - P_N)\widetilde{H}$ and the spectra $\sigma(\mathscr{A}_u)$, $\sigma(\mathscr{A}_s)$ are given by

$$\sigma(\mathscr{A}_u) = \{\lambda_j\}_{j=1}^N, \qquad \sigma(\mathscr{A}_s) = \{\lambda_j\}_{j=N+1}^\infty.$$

Since $\sigma(\mathscr{A}_s) \subset \{\lambda \in \mathbb{C};\ \operatorname{Re}\lambda > \gamma\}$ and $\mathscr{A}_s$ generates an analytic C_0-semigroup on $\widetilde{H}$ and, as seen earlier, we have

$$|e^{-\mathscr{A}_s t} x|_{\widetilde{H}} \le C e^{-\gamma t} |x|_{\widetilde{H}}, \quad \forall x \in \widetilde{H},\ t \ge 0. \tag{4.6}$$

Herein, we assume that the following hypothesis holds.

(J1) *All the eigenvalues* λ_j, $j = 1, \dots, N$, *are semisimple.*

As seen earlier, (J1) allows to take the systems $\{\varphi_j\}_{j=1}^N$ and $\{\varphi_j^*\}_{j=1}^N$ as biorthogonal systems, that is,

$$(\varphi_i, \varphi_j^*) = \delta_{ij}, \quad i, j = 1, \dots, N. \tag{4.7}$$

We denote by $(\cdot,\cdot)_0$ the scalar product in $(L^2(\mathscr{O}_0))^d$, that is,

$$(u, v)_0 = \int_{\mathscr{O}_0} u \cdot \bar{v}\, d\xi, \quad \forall u, v \in (L^2(\mathscr{O}_0))^d.$$

Now, we define ϕ_j, $j = 1, \dots, N$, as follows.

$$\phi_j(\xi) = \sum_{\ell=1}^N \alpha_{\ell j} \varphi_\ell^*(\xi), \quad \xi \in \mathscr{O}, \tag{4.8}$$

where $\alpha_{\ell j}$ are chosen in such a way that

$$\sum_{\ell=1}^N \alpha_{\ell j} (\varphi_\ell^*, \varphi_k^*)_0 = \delta_{jk}, \quad j, k = 1, \dots, N. \tag{4.9}$$

The latter is possible because, by unique continuation property of eigenfunctions φ_j^*, the system $\{\varphi_j^*\}_{j=1}^m$ is linearly independent on $\mathscr{O}_0$ (see Theorem 3.15) and so, $\det \|(\varphi_j^*, \varphi_k^*)_0\|_{j,k=1}^m \ne 0$. This yields

$$(\phi_j, \varphi_k^*)_0 = \delta_{jk} \quad \text{for } j, k = 1, \dots, N. \tag{4.10}$$

4.1.1 Stabilization by Noise of the Linearized Navier–Stokes System

Consider the linear system

$$\begin{aligned} &dX + \mathscr{A}X dt = \eta \sum_{i=1}^{N} (X, \varphi_i^*) P(m\phi_i) d\beta_i, \\ &X(0) = x, \end{aligned} \tag{4.11}$$

where $\eta \in R$.

Here $\{\phi_i\}_{i=1}^N \subset \widetilde{H}$ is the system of functions made precise by (4.8), (4.9).

We may rewrite (4.11) as

$$X(t) = e^{-\mathscr{A}t}x + \eta \sum_{i=1}^{N} \int_0^t (X(s), \phi_i) e^{-\mathscr{A}(t-s)} P(m\phi_i) d\beta_i(s),$$
$$\mathbb{P}\text{-a.s.}, t \geq 0,$$

which, by the standard existence theory (see Theorem 4.9), has a unique solution $X \in C_W([0, T]; L^2(\Omega, \widetilde{H}))$, $\forall T > 0$.

The closed-loop system (4.11) can be written, equivalently, as

$$\begin{aligned} &dX(t) - \nu_0 \Delta X(t) dt + (X(t) \cdot \nabla) y_e dt + (y_e \cdot \nabla) X(t) dt \\ &\quad = \eta m \sum_{i=1}^{N} (X(t), \varphi_i^*) \phi_i d\beta_i(t) + \nabla p(t) dt, \quad \mathbb{P}\text{-a.s. in } (0, \infty) \times \mathscr{O}, \\ &\nabla \cdot X(t) = 0 \quad \text{in } \mathscr{O}, \qquad X(t)\big|_{\partial\mathscr{O}} = 0, \quad \mathbb{P}\text{-a.s.}, \ \forall t \geq 0, \\ &X(0) = x \quad \text{in } \mathscr{O}. \end{aligned} \tag{4.12}$$

Hence, in the space $(L^2(\mathscr{O}))^d$, the feedback controller $\{u_i = \eta m(X, \varphi_i^*)\phi_i\}_{i=1}^N$ has the support in $\mathscr{O}_0$.

Theorem 4.1 is the main result of this section.

Theorem 4.1 *Under Hypothesis* (J1), *the solution* X *to* (4.11), *for* $|\eta|$ *sufficiently large, satisfies*

$$\mathbb{P}\left[\lim_{t \to \infty} e^{\gamma t} |X(t, x)|_{\widetilde{H}} = 0\right] = 1, \quad \forall x \in H. \tag{4.13}$$

Remark 4.1 If we set $X_1(t) = \operatorname{Re} X(t)$, $X_2(t) = \operatorname{Im} X(t)$, System (4.12) can be rewritten as a real system in (X_1, X_2). In this case, Controller (4.2) is an implicit stabilizable feedback controller with support in $\mathscr{O}_0$ for the real Stokes–Oseen equation corresponding to y_e. Of course, if λ_j, $j = 1, \dots, N$, are real, then we may view $X(t)$ as a real-valued function and so, in (4.13), $|X|_{\widetilde{H}} = |X|$.

4.1.2 Proof of Theorem 4.1

The idea, already used several times so far, is to decompose (4.11) in a finite-dimensional system and an infinite-dimensional exponentially stable system. To this end, we set $X_u = P_N X$, $X_s = (I - P_N)X$ and we rewrite (4.11) as

$$dX_u(t)+\mathscr{A}_u X_u(t)dt = \eta P_N \sum_{i=1}^{N} (X_u(t), \varphi_i^*) P(m\phi_i) d\beta_i(t), \quad \mathbb{P}\text{-a.s.}, \ t \geq 0, \tag{4.14}$$

$$X_u(0) = P_N x,$$

$$dX_s(t)+\mathscr{A}_s X_s(t)dt = \eta (I-P_N) \sum_{i=1}^{N} (X_u(t), \varphi_i^*) P(m\phi_i) d\beta_i(t), \quad \mathbb{P}\text{-a.s.}, \ t \geq 0, \tag{4.15}$$

$$X_s(0) = (I - P_N)x.$$

Then, we may represent $X_u(t) = \sum_{i=1}^{N} y_i(t)\varphi_i$ and reduce (4.11) via biorthogonal relations (4.7) and (4.10) to the finite-dimensional complex system

$$\begin{aligned} &dy_j + \lambda_j y_j dt = \eta y_j d\beta_j, \quad \mathbb{P}\text{-a.s.}, \ t \geq 0, \ j = 1, \dots, N, \\ &y_j(0) = y_j^0, \end{aligned} \tag{4.16}$$

where $y_j^0 = (P_N x, \varphi_j^*)$.

As seen in Sect. 2.4, the stochastic differential equation (4.16) can be explicitly solved by the formula (see (2.88))

$$y_j(t) = e^{-\lambda_j t - \frac{\eta^2}{2} t + \eta \beta_j(t)} y_j^0, \quad j = 1, \dots, N,$$

and the proof concludes as in Theorem 2.7. However, we use below a different approach based on Ito's formula which, though longer, provides sharper informations on the stabilizing performance of the stochastic controller and is also applicable in a variety of situations when the solution to System (4.16) cannot be found explicitly. Applying Ito's formula in (4.16) to $\varphi(y) = (e^{\gamma t}|y|)^2$, we obtain that (see Theorem 4.8)

$$\begin{aligned} &d\left(e^{2\gamma t}|y_j(t)|^2\right) + 2e^{2\gamma t}\left(\operatorname{Re}\lambda_j - \gamma\right)|y_j(t)|^2 dt \\ &= \eta^2 e^{2\gamma t}|y_j(t)|^2 dt + 2\eta e^{2\gamma t}|y_j(t)|^2 d\beta_j(t), \quad \text{for } j=1,\dots,N. \end{aligned} \tag{4.17}$$

Now, in (4.17) we take $z(t) = e^{2\gamma t}|y_j(t)|^2$ and obtain that

$$\begin{aligned} &dz + 2e^{2\gamma t}\left(\operatorname{Re}\lambda_j - \gamma\right)|y_j|^2 dt = \eta^2 e^{2\gamma t}|y_j|^2 dt + 2\eta e^{2\gamma t}|y_j|^2 d\beta_j, \\ &j = 1, \dots, N. \end{aligned}$$

In the latter equation, we apply Ito's formula to the function

$$\Phi(r)=(\varepsilon+r)^{\delta}, \quad \text{where } 0<\delta<\frac{1}{2} \text{ and } \varepsilon>0.$$

We have

$$\Phi'(r)=\delta(\varepsilon+r)^{\delta-1}, \quad \Phi''(r)=\delta(\delta-1)(\varepsilon+r)^{\delta-2}, \quad r>0$$

and, therefore, we obtain that

$$d\Phi(z)=\Phi'(z)dz+2\eta^2 e^{4\gamma t}\Phi''(z)|y_j|^4 dt.$$

This yields

$$\begin{aligned}d\Phi(z)=&-\delta e^{2\gamma t}(\varepsilon+z)^{\delta-1}[2(\operatorname{Re}\lambda_j-\gamma)|y_j(t)|^2dt-\eta^2|y_j|^2dt\\&-2\eta|y_j|^2d\beta_j]+2\eta^2\delta(\delta-1)e^{4\gamma t}(\varepsilon+z)^{\delta-2}|y_j|^4dt.\end{aligned}$$

Now, if we replace z by $e^{2\gamma t}|y_j|^2$, we obtain that

$$\begin{aligned}d((\varepsilon+e^{2\gamma t}|y_j|^2)^{\delta})&+2\delta(\varepsilon+e^{2\gamma t}|y_j|^2)^{\delta-1}e^{2\gamma t}(\operatorname{Re}\lambda_j-\gamma)|y_j(t)|^2dt\\=&\ 2\eta^2(\delta-1)\delta e^{4\gamma t}(\varepsilon+e^{2\gamma t}|y_j|^2)^{\delta-2}|y_j|^4dt\\&+\eta^2\delta e^{2\gamma t}(\varepsilon+e^{2\gamma t}|y_j|^2)^{\delta-1}|y_j|^2dt\\&+2\eta\delta e^{2\gamma t}(\varepsilon+e^{2\gamma t}|y_j|^2)^{\delta-1}|y_j|^2d\beta_j, \quad j=1,\dots,N.\end{aligned} \tag{4.18}$$

We set

$$\begin{aligned}K_\varepsilon^j(t)=&\ 2\delta\, e^{2\gamma t}(\varepsilon+e^{2\gamma t}|y_j|^2)^{\delta-1}(\operatorname{Re}\lambda_j-\gamma)|y_j(t)|^2\\&-\delta\eta^2e^{2\gamma t}(\varepsilon+e^{2\gamma t}|y_j(t)|^2)^{\delta-1}|y_j(t)|^2\\&-2\delta(\delta-1)\eta^2e^{4\gamma t}(\varepsilon+e^{2\gamma t}|y_j(t)|^2)^{\delta-2}|y_j(t)|^4, \quad j=1,\dots,N.\end{aligned} \tag{4.19}$$

Taking into account (4.19), we may rewrite (4.18) as

$$\begin{aligned}&(\varepsilon+e^{2\gamma t}|y_j|^2)^{\delta}+\int_0^t K_\varepsilon^j(s)ds=(\varepsilon+|y_j^0|^2)^{\delta}+M_\varepsilon^j(t),\\&\mathbb{P}\text{-a.s.}, \ t\ge 0, \ j=1,\dots,N,\end{aligned} \tag{4.20}$$

where M_ε^j is the stochastic process

$$M_\varepsilon^j(t)=2\delta\eta\int_0^t e^{2\gamma s}|y_j(s)|^2(\varepsilon+e^{2\gamma s}|y_j(s)|^2)^{\delta-1}d\beta_j(s), \quad j=1,\dots,N.$$

Taking into account that

$$\lim_{\varepsilon\to 0}|y_j(s)|^2(\varepsilon+e^{2\gamma s}|y_j(s)|^2)^{\delta-1}e^{2\gamma s}=e^{2\gamma\delta s}|y_j(s)|^{2\delta}, \quad \mathbb{P}\text{-a.s.}$$

uniformly on $[0, T]$, we may pass to limit into (4.20) to get that

$$e^{2\gamma\delta t}|y_j(t)|^{2\delta} + \int_0^t K_j(s)ds = |y_j^0|^{2\delta} + M_j(t), \quad \mathbb{P}\text{-a.s., } t > 0, \tag{4.21}$$

where

$$K_j(t) = \lim_{\varepsilon\to 0} K_\varepsilon^j(t) = 2\delta(\operatorname{Re}\lambda_j - \gamma)e^{2\gamma\delta t}|y_j(t)|^{2\delta} + 2\delta(1-2\delta)\eta^2 e^{2\gamma\delta t}|y_j(t)|^{2\delta},$$

$$M_j(t) = 2\delta\eta \int_0^t e^{2\gamma\delta s}|y_j(s)|^{2\delta}d\beta_j(s), \quad \mathbb{P}\text{-a.s.}$$

If in (4.21) we take the expectation E, we obtain that

$$e^{2\gamma\delta t}E|y_j(t)|^{2\delta} + E\int_0^t K_j(s)ds = |y_j^0|^{2\delta}, \quad \forall t \ge 0.$$

This yields

$$2\delta(\eta^2(1-2\delta) + \operatorname{Re}\lambda_j - \gamma)E\int_0^t e^{2\gamma\delta s}|y_j(s)|^{2\delta}ds \le |y_j^0|^{2\delta}, \quad j = 1, \dots, N,$$

and, since $0 < \delta < \frac{1}{2}$, for all $j = 1, \dots, N$, we get therefore, for η sufficiently large,

$$E\int_0^t e^{2\gamma\delta s}|y_j(s)|^{2\delta}ds \le C, \quad \forall t \ge 0, \ j = 1, \dots, N.$$

This yields

$$E\int_0^\infty e^{2\gamma\delta s}|y_j(s)|^{2\delta}ds < \infty, \quad \forall j = 1, \dots, N,$$

and, in particular, it follows that

$$\int_0^\infty e^{2\gamma\delta s}|y_j(s)|^{2\delta}ds < \infty, \quad \mathbb{P}\text{-a.s., } j = 1, \dots, N. \tag{4.22}$$

It should be said, however, that the latter does not imply automatically that $e^{2\gamma\delta t}|y_j(t)|^{2\delta}$ is $\mathbb{P}$-a.s. convergent to zero as $t \to \infty$ and for this we need to invoke some sharp stochastic arguments.

We write

$$\int_0^t K_j(s)ds = I_j(t) - (I_j)_1(t), \quad \mathbb{P}\text{-a.s., } \forall t \ge 0, \ j = 1, \dots, N,$$

where

$$I_j(t) = 2(1-2\delta)\delta\eta^2 \int_0^t e^{2\gamma\delta s}|y_j(s)|^{2\delta}ds,$$

$$(I_j)_1(t) = 2\delta(\gamma - \mathrm{Re}\,\lambda_j)\int_0^t e^{2\gamma\delta s}|y_j(s)|^{2\delta}ds.$$

Then, we may rewrite (4.21) as

$$e^{2\gamma\delta t}|y_j(t)|^{2\delta} + I_j(t) = |y_j^0|^{2\delta} + (I_j)_1(t) + M_j(t), \quad \mathbb{P}\text{-a.s., } t \ge 0. \tag{4.23}$$

Taking into account that, for $j = 1, \ldots, N$, $M_j(t)$ is a local martingale and $t \to I_j(t)$, $t \to (I_j)_1(t)$ are nondecreasing processes, we see by (4.23) that $t \to e^{2\gamma\delta t}|y_j(t)|^{2\delta}$ is a semimartingale, as the sum of a local martingale and of an adapted finite variation process.

We are going to apply Lemma 4.5 to the processes

$$Z(t) = e^{2\gamma\delta t}|y_j(t)|^{2\delta}, \qquad I = I_j, \qquad I_1 = (I_j)_1, \qquad M = M_j,$$

defined above.

By virtue of (4.22), $(I_j)_1(\infty) < \infty$. This implies, by Lemma 4.5, that there exists the limit

$$\lim_{t\to\infty}(e^{2\gamma\delta t}|y_j(t)|^{2\delta}) < \infty, \quad \mathbb{P}\text{-a.s., } j = 1, \ldots, N, \tag{4.24}$$

and $I_j(\infty) < \infty$. It follows, therefore, that there exists

$$\lim_{t\to\infty} e^{\gamma t}|y(t)| = 0, \quad \mathbb{P}\text{-a.s.,} \tag{4.25}$$

where $|y|^2 = \sum_{j=1}^N |y_j|^2$. We have shown, therefore, that

$$\lim_{t\to\infty} e^{2\gamma t}|X_u(t)|^2_{\widetilde{H}} = 0, \quad \mathbb{P}\text{-a.s.} \tag{4.26}$$

By (4.22) and (4.25), it follows also that

$$\int_0^\infty e^{2\gamma t}|y(t)|^2 dt < \infty, \quad \mathbb{P}\text{-a.s.,}$$

because, by (4.24), it follows that $e^{2\gamma\delta t}|y|^{2\delta} \in L^\infty(0,\infty)$, $\mathbb{P}$-a.s. This yields

$$\int_0^\infty e^{2\gamma t}|X_u(t)|^2_{\widetilde{H}} dt < \infty, \quad \mathbb{P}\text{-a.s.} \tag{4.27}$$

Next, we come back to the infinite-dimensional system (4.15). Since, as seen earlier, the operator $-\mathscr{A}_s$ generates a γ-exponentially-stable C_0-semigroup on $\widetilde{H}$, by the Lyapunov theorem there is $Q \in L(\widetilde{H}, \widetilde{H})$, $Q = Q^* \ge 0$ such that

$$\mathrm{Re}(Qx, \mathscr{A}_s x - \gamma x) = \frac{1}{2}\,|x|^2_{\widetilde{H}}, \quad \forall x \in D(\mathscr{A}_s).$$

(We note that we have $(Qx, x) > 0$ for all $x \ne 0$.)

Applying Ito's formula in (4.15) to the function $\varphi(x) = \frac{1}{2}(Qx, x)$, we obtain that

$$\frac{1}{2}d(QX_s(t), X_s(t)) + \frac{1}{2}|X_s(t)|^2_{\widetilde{H}}dt + \gamma(QX_s(t), X_s(t))dt$$
$$= \frac{1}{2}\eta^2\sum_{i=1}^{N}(QY_i(t), Y_i(t))dt + \eta\sum_{i=1}^{N}((\mathrm{Re}(QX_s(t)), \mathrm{Re}\, Y_i(t))$$
$$+ (\mathrm{Im}(QX_s(t)), \mathrm{Im}\, Y_i(t)))d\beta_i(t),$$

where Y_i are processes defined by

$$Y_i(t) = (X_u(t), \varphi_i^*)(I - P_N)P(m\phi_i) = y_i(I - P_N)P(m\phi_i), \quad i = 1, \dots, N.$$

This yields

$$e^{2\gamma t}(QX_s(t), X_s(t)) + \int_0^t e^{2\gamma s}|X_s(s)|^2_{\widetilde{H}}ds$$
$$= (Q(I - P_N)x, (I - P_N)x)$$
$$+ \eta^2\sum_{i=1}^{N}\int_0^t e^{2\gamma s}(QY_i(s), Y_i(s))ds$$
$$+ 2\eta\sum_{i=1}^{N}\int_0^t e^{2\gamma s}((\mathrm{Re}(QX_s(s)), \mathrm{Re}\, Y_i(s))$$
$$+ (\mathrm{Im}(QX_s(s)), \mathrm{Im}\, Y_i(s))_H)d\beta_i(s), \quad \mathbb{P}\text{-a.s., } t \geq 0. \tag{4.28}$$

Once again we apply Lemma 4.5 to the processes Z, I, M defined below

$$Z(t) = e^{2\gamma t}(QX_s(t), X_s(t)),$$
$$I(t) = \int_0^t e^{2\gamma s}|X_s(s)|^2_{\widetilde{H}}ds, \qquad I_1(t) = \eta^2\sum_{i=1}^{N}\int_0^t e^{2\gamma s}(QY_i, Y_i)ds,$$
$$M(t) = 2\eta\sum_{i=1}^{N}\int_0^t e^{2\gamma s}((\mathrm{Re}(QX_s(s)), \mathrm{Re}\, Y_i(s))$$
$$+ (\mathrm{Im}(QX_s(s)), \mathrm{Im}\, Y_i(s)))d\beta_j(s), \quad \mathbb{P}\text{-a.s., } t \geq 0.$$

Since, by the first step of the proof (see (4.27)), $I_1(\infty) < \infty$, we conclude that

$$\lim_{t\to\infty} e^{2\gamma t}(QX_s(t), X_s(t)) = 0, \quad \mathbb{P}\text{-a.s.},$$

and this implies that

$$\lim_{t\to\infty} e^{\gamma t}|X_s(t)|_{\widetilde{H}} = 0, \quad \mathbb{P}\text{-a.s.}$$

Recalling that $X = X_u + X_s$ and invoking (4.26), the latter implies (4.14), thereby completing the proof of Theorem 4.1.

4.1.3 Stabilization by Noise of Navier–Stokes Equations

One might expect that, when plug the stochastic feedback controller (4.2) into the Navier–Stokes controlled system (4.3), it stabilizes in probability the system. We see below that this is indeed the case but in a certain precise sense.

Theorem 4.2 *Let* $d = 2, 3,\ y_e \in C^2(\overline{\mathcal{O}})$ *and*

$$|\eta| \geq \max_{1 \leq j \leq N} \sqrt{6\gamma - 2\,\mathrm{Re}\,\lambda_j}. \tag{4.29}$$

Then, there is $C^* > 0$ *independent of* $\omega \in \Omega$ *such that, for each* $x \in W$, $\|x\|_W \leq (C^*)^2$ *there is* $\Omega_x^* \subset \Omega$ *with*

$$\mathbb{P}(\Omega_x^*) \geq 1 - 2\left(\frac{C^*}{\sqrt{\|x\|_W}} - 1\right)^{-\frac{\gamma}{2(\eta N)^2}}, \tag{4.30}$$

the solution $X(t, x)$ *to* (4.4) *satisfies*

$$\lim_{t \to \infty} \left(e^{\frac{\gamma t}{4}} |X(t, x)|_{\widetilde{H}}\right) = 0, \quad \mathbb{P}\text{-a.s. in } \Omega_x^*. \tag{4.31}$$

Here, $W = D(A^{\frac{1}{4}})$ if $d = 2$ and $W = D(A^{\frac{1}{4}+\varepsilon})$ if $d = 3$, and $\varepsilon > 0$ is arbitrarily small.

In particular, Theorem 4.2 implies that if $\|x\|_W \leq \rho_0 < (C^*)^2$, then $X = X(t, x)$ is exponentially decaying to 0 on a set Ω_x^* of probability greater than

$$1 - 2\left(\frac{C^*}{\sqrt{\|x\|_W}} - 1\right)^{-\frac{\gamma}{2(\eta N)^2}}$$

and we see that $\mathbb{P}(\Omega_x^*) \to 1$ as $\|x\|_W \leq \rho_0 \to 0$. The constant C^* might depend, however, on η. One might say, therefore, that the stochastic feedback (4.2) exponentially stabilizes with high probability System (4.5) if x is taken in a sufficiently small neighborhood of origin.

Remark 4.2 As mentioned earlier, System (4.4) is written here in the complex space $\widetilde{H}$. If we set $X_1(t) = \mathrm{Re}\,X(t)$, $X_2(t) = \mathrm{Im}\,X(t)$, it can be rewritten as a real system in (X_1, X_2). In this case, the feedback controller is an implicit stabilizable feedback controller with support in $\mathcal{O}_0$ for the real Navier–Stokes equation.

More precisely, we have

$$
\begin{aligned}
&dX_1 - \nu \Delta X_1 dt + (X_1 \cdot \nabla) y_e dt + (y_e \cdot \nabla) X_1 dt + (X_1 \cdot \nabla) X_1 dt - (X_2 \cdot \nabla) X_2 dt \\
&\quad = \eta m \sum_{j=1}^{N} ((X_1, \operatorname{Re} \varphi_j^*) + (X_2, \operatorname{Im} \varphi_j^*)) d\beta_j + \nabla p_j^1 dt, \\
&dX_2 - \nu \Delta X_2 dt + (X_1 \cdot \nabla) X_2 dt + (X_2 \cdot \nabla) X_1 dt \\
&\quad = \eta m \sum_{j=1}^{N} ((X_1, \operatorname{Im} \varphi_j^*) - (X_2, \operatorname{Re} \varphi_j^*)) d\beta_j + \nabla p_j^2 dt, \\
&\nabla \cdot X_i = 0, \quad i = 1, 2, \\
&X_i = 0 \quad \text{on } (0, \infty) \times \mathcal{O}, \quad i = 1, 2, \\
&X_1(0) = x, \qquad X_2(0) = 0.
\end{aligned}
\tag{4.32}
$$

System (4.32) can be viewed as the Navier–Stokes equation with two sets of feedback inputs U_1 and U_2

$$
U_1 = \eta m \sum_{j=1}^{N} ((X_1, \operatorname{Re} \varphi_j^*) + (X_2, \operatorname{Im} \varphi_j^*)) d\beta_j,
$$

$$
U_2 = \eta m \sum_{j=1}^{N} ((X_1, \operatorname{Im} \varphi_j^*) - (X_2, \operatorname{Re} \varphi_j^*)) d\beta_j.
$$

Of course, if λ_j, $j = 1, \dots, N$, are real, then we may view $X(t)$ as a real-valued function and so, in (4.32), $|X|_{\tilde{H}} = |X|$.

In particular, by Theorem 4.2 we have the corollary below.

Corollary 4.1 *Under the assumptions of Theorem* 4.2, *the feedback controller*

$$
\eta m \sum_{j=1}^{N} (X - y_e, \varphi_j^*) \phi_j
$$

stabilizes in the sense of (4.31) *the stationary solution* y_e *to the Navier–Stokes system* (4.1), $\mathbb{P}$-*a.s. in* Ω_x^*.

Remark 4.3 An interesting feature of Theorem 4.2, we have already encountered in Theorem 3.4, is that the stabilization space W is different in function of dimension d. A biproduct of this theorem is also the fact that the stochastic Navier–Stokes equation (4.4) in 3-D is with high probability globally well-posed in $(H^{\frac{1}{2}+\varepsilon}(\mathcal{O}))^3$ in a neighborhood of the equilibrium solution.

4.1.4 Proof of Theorem 4.2

The idea of the proof is to reduce (4.4) to a deterministic equation with random coefficients, via substitution

$$y(t)=\prod_{j=1}^{N} e^{-\beta_j(t)\Gamma_j}X(t), \quad t\ge 0,$$

where $\Gamma_j:\widetilde{H}\to\widetilde{H}$ is the linear operator

$$\Gamma_j x:=\eta(x,\varphi_j^*)P(m\phi_j), \quad x\in\widetilde{H},\ j=1,\dots,N$$

and $e^{s\Gamma_j}\in L(\widetilde{H},\widetilde{H})$ is the C_0-group generated by Γ_j that is,

$$\frac{d}{ds}e^{s\Gamma_j}x-\Gamma_j e^{s\Gamma_j}x=0, \quad \forall s\in R,\ x\in\widetilde{H}.$$

We have

$$\Gamma_j\Gamma_k x=\eta^2(x,\varphi_j^*)P(m\phi_j)\delta_{jk}, \quad \forall j,k=1,\dots,N,$$

and, therefore, the operators $\Gamma_1,\dots,\Gamma_N$ commute.

Then, by Theorem 4.9, (4.4) reduces to the random differential equation

$$\begin{cases}\frac{dy(t)}{dt}+\mathscr{A}y(t)+\frac{1}{2}\sum_{j=1}^{N}\Gamma_j^2 y(t)+e^{-\sum_{j=1}^{N}\beta_j(t)\Gamma_j}S(e^{\sum_{j=1}^{N}\beta_j(t)\Gamma_j}y(t))\\ \quad +F(t)y(t)=0, \quad \mathbb{P}\text{-a.s.},\ \forall t\ge 0,\\ y(0)=x,\end{cases} \tag{4.33}$$

where

$$F(t)=e^{-\sum_{j=1}^{N}\beta_j(t)\Gamma_j}\mathscr{A}e^{\sum_{j=1}^{N}\beta_j(t)\Gamma_j}-\mathscr{A}.$$

By a solution of (4.33) we mean a function $y\in C(([0,\infty);D(A^{\frac{1}{4}}))\cap L^2(0,\infty;D(A)))$ which satisfies it $\mathbb{P}$-a.s. in the mild sense (see Lemma 4.3).

Conversely, if y is an adapted C^1 solution to (4.33), then the process

$$X(t)=\prod_{j=1}^{N} e^{\beta_j(t)\Gamma_j}y(t), \quad t\ge 0, \tag{4.34}$$

belongs to $C_W([0,T];L^2(\Omega,\mathbb{P};D(A^{\frac{1}{4}}))\cap L^2(\Omega;C[0,T];D(A^{\frac{3}{4}}))$ and satisfies (4.4). (See Theorem 4.10.)

In the following, we study the existence and exponential convergence in probability to solutions y to the deterministic random equation (4.33).

By a little calculation, we see that

$$e^{s\Gamma_j}y=\eta^{-1}\Gamma_j y(e^{\eta s}-1)+y=(e^{\eta s}-1)(y,\varphi_j^*)P(m\phi_j)+y,$$

$$\forall s > 0,\ j = 1, \dots, N,\ y \in H, \tag{4.35}$$

and

$$e^{-s\Gamma_j} y = \eta^{-1}\Gamma_j y(e^{-\eta s} - 1) + y = (e^{-\eta s} - 1)(y, \varphi_j^*)P(m\phi_j) + y,$$
$$\forall s > 0,\ j = 1, \dots, N,\ y \in H. \tag{4.36}$$

This yields

$$F(t)y = \sum_{j=1}^{N} (e^{\beta_j(t)} - 1)(y, \varphi_j^*)(\mathscr{A} P(m\phi_j) - \lambda_j P(m\phi_j)). \tag{4.37}$$

Now, we consider the operator

$$\mathscr{A}_\Gamma y := \mathscr{A} y + \frac{1}{2}\sum_{j=1}^{N} \Gamma_j^2 y, \quad \forall y \in D(\mathscr{A}),$$

which generates an analytic C_0-semigroup $e^{-\mathscr{A}_\Gamma t}$ on $\widetilde{H}$. We note also that the operator $\mathscr{A}_\Gamma + F(t)$ generates an evolution $U(t,\tau)$ on $\widetilde{H}$, that is (see Sect. 2.6.1)

$$\frac{d}{dt}U(t,\tau) + (\mathscr{A}_\Gamma + F(t))U(t,\tau) = 0, \quad 0 \le \tau \le t,$$
$$U(\tau,\tau) = I.$$

In Lemmas 4.1 and 4.2, we collect a few asymptotic properties of $U(t,\tau)$.

Lemma 4.1 *We have, for* $\eta \ge \max_{1\le j\le N} \sqrt{6\gamma - 2\,\mathrm{Re}\,\lambda_j}$,

$$\|U(t,\tau)\|_{L(\widetilde{H},\widetilde{H})} \le C e^{-\gamma(t-\tau)}(1+\eta^2)|x|\left(1 + \int_\tau^t e^{-\gamma(\tau+2s)}\zeta(s)ds\right),$$
$$\mathbb{P}\text{-a.s.},\ \forall t \ge \tau, \tag{4.38}$$

where C is independent of ω and $\zeta(t) = \sum_{j=1}^N e^{\beta_j(t)}$, respectively.

Proof We use, as in Sect. 3.3, the spectral decomposition of the system

$$\begin{cases} \frac{dy}{dt} + \mathscr{A}_\Gamma y + F(t)y = 0, & t \ge \tau, \\ y(\tau) = x, \end{cases} \tag{4.39}$$

in the direct sum $\mathscr{X}_u \oplus \mathscr{X}_s$ of γ-unstable and γ-stable spaces of the operator $\mathscr{A}$. Namely, we set

$$y_u = P_N y, \qquad y_s = (I - P_N)y$$

and so, we rewrite System (4.39) as

$$\begin{cases} \frac{dy_u}{dt} + \mathscr{A}_u y_u + \frac{1}{2} P_N \sum_{j=1}^N \Gamma_j^2 y_u = 0, & t \geq \tau, \\ y_u(\tau) = P_N x, \end{cases} \tag{4.40}$$

and, respectively,

$$\begin{cases} \frac{dy_s}{dt} + \mathscr{A}_s y_s + \frac{1}{2}(I - P_N)\sum_{j=1}^N \Gamma_j^2 y_u + (I - P_N)F(t)y_u = 0, & t \geq \tau, \\ y_s(\tau) = (I - P_N)x, \end{cases} \tag{4.41}$$

because, as seen by (4.37), $P_N F(t) = 0$.

We have

$$y = y_u + y_s, \quad y_u = \sum_{j=1}^N y_j \varphi_j$$

and, since $\sigma(\mathscr{A}_u) = \{\lambda_j\}_{j=1}^N$, we have

$$\mathscr{A}_u \varphi_j = \lambda_j \varphi_j, \quad j = 1, \ldots, N.$$

Recalling that

$$\Gamma_j^2 y = \eta \Gamma_j y = \eta^2 (y, \varphi_j^*) P(m\phi_j),$$

we may rewrite (4.40) as

$$\begin{cases} \frac{dy_j}{dt} + \lambda_j y_j + \frac{1}{2}\eta^2 y_j (P(m\phi_j), \varphi_j^*) = 0, & t \geq 0,\ j = 1, \ldots, N, \\ y_j(0) = (x, \varphi_j^*). \end{cases}$$

By (4.7), it follows that

$$\begin{cases} \frac{dy_j}{dt} + \lambda_j y_j + \frac{1}{2}\eta^2 y_j = 0, & t \geq \tau,\ j = 1, \ldots, N, \\ y_j(\tau) = (x, \varphi_j^*). \end{cases}$$

This yields,

$$y_j(t) = e^{-(\lambda_j + \frac{1}{2}\eta^2)t}(x, \varphi_j^*), \quad j = 1, \ldots, N,\ t \geq 0,$$

and, therefore, for $\eta^2 \geq 6\gamma - 2\,\mathrm{Re}\,\lambda_j$, $j = 1, \ldots, N$, we have

$$|y_u(t)|_{\widetilde{H}} \leq Ce^{-3\gamma t}|x|_{\widetilde{H}}, \quad \forall t \geq 0.$$

Now, coming back to System (4.41), we rewrite it as

$$\begin{cases} \frac{dy_s}{dt} + \mathscr{A}_s y_s + \frac{1}{2}\eta^2 \sum_{j=1}^N y_j (I - P_N)(m\phi_j) \\ \quad + \sum_{j=1}^N (e^{\beta_j(t)} - 1) y_j (I - P_N)(\mathscr{A} P(m\phi_j) - \lambda_j P(m\phi_j)) = 0, & t \geq \tau, \\ y_s(\tau) = (I - P_N)x, \end{cases} \tag{4.42}$$

and, since $e^{-\mathscr{A}_s t}$ is exponentially stable, we have that

$$\begin{aligned}|y_s(t)|_{\widetilde{H}} &\le |e^{-\mathscr{A}_s(t-\tau)}(I-P_N)x|_{\widetilde{H}}\\&\quad+\frac{1}{2}\eta^2\int_\tau^t\sum_{j=1}^N(e^{\beta_j(s)}-1)|e^{-\mathscr{A}_s(t-s)}y_j(s)(I-P_N)[P(m\phi_j)\\&\quad+\mathscr{A}P(m\phi_j)-\lambda_jP(m\phi_j)]|_{\widetilde{H}}ds\\&\le Ce^{-\gamma(t-\tau)}|x|+C\frac{\eta^2}{2}|x|\int_0^t\sum_{j=1}^N|e^{-3\gamma s}e^{-\gamma(t-s)}\zeta(s)|_{\widetilde{H}}ds\\&\le Ce^{-\gamma(t-\tau)}(1+\eta^2)|x|\int_\tau^t e^{-\gamma(2s+\tau)}\zeta(s)ds,\quad \mathbb{P}\text{-a.s., }\forall t\ge\tau,\end{aligned}$$

for some constant C independent of x and $\omega\in\Omega$. This completes the proof of (4.38). □

In the following, we fix η such that (4.29) holds.

Lemma 4.2 *We have, for* $0\le\varepsilon<\frac{1}{4}$,

$$\int_\tau^\infty e^{\gamma(t-\tau)}\|U(t,\tau)x\|_W^2dt\le C\|x\|_W^2\left(1+\int_\tau^\infty e^{-\gamma(\tau+2s)}\zeta(s)ds\right)^2,\quad \forall x\in W, \tag{4.43}$$

where C *is independent of* $\omega\in\Omega$.

Proof We set $z(t):=e^{\frac{\gamma}{2}(t-\tau)}U(t,\tau)x$, $0<\tau<t$. Then, by Lemma 4.1 we have

$$\int_\tau^\infty|z(t)|_{\widetilde{H}}^2dt\le C|x|^2\left(1+\int_\tau^\infty e^{-\gamma(\tau+2s)}\zeta(s)ds\right)^2,\quad\forall x\in H,$$

while

$$\frac{dz}{dt}+\nu Az+A_0z+\frac{1}{2}\sum_{j=1}^N\Gamma_j^2z+F(t)z=\frac{\gamma}{2}z,\quad t\ge\tau.$$

Multiplying the latter by z and $A^{\frac{1}{2}+2\varepsilon}z$ (scalarly in $\widetilde{H}$), we have the standard estimates in 3-D (see Proposition 1.7)

$$|(A_0z,z)|=|b(z,y_e,z)|\le C|z|_{\frac{1}{4}+\varepsilon}|y_e|_1|z|\le C|z|_{\frac{1}{4}+\varepsilon}|z|$$

and

$$|(A_0z,A^{\frac{1}{2}+2\varepsilon}z)|=|b(z,y_e,A^{\frac{1}{2}+2\varepsilon}z)|+|b(y_e,z,A^{\frac{1}{2}+2\varepsilon}z)|$$

$$\le C(|z|_{\frac14+\varepsilon}|y_e|_1|A^{\frac12+2\varepsilon}z|+|y_e|_1|z|_{\frac12}|A^{\frac12+2\varepsilon}z|)\le C|z|^2_{\frac12+2\varepsilon}.$$

We get that

$$\frac12\frac{d}{dt}|z(t)|^2+\nu|z(t)|^2_{\frac12}\le C(|z(t)|_{\frac12}|z(t)|+|z(t)|^2)+|(F(t)z,z)|,$$

$$\begin{aligned}\frac12\frac{d}{dt}|z(t)|^2_{\frac14+\varepsilon}+\nu|z(t)|^2_{\frac34+\varepsilon}&\le C(|z(t)|_{\frac14+\varepsilon}|z(t)|+|z(t)|(|z(t)|_{\frac12+2\varepsilon})\\&\quad+|(F(t)z,A^{\frac12+2\varepsilon}z)|.\end{aligned}$$

(Here, $|\cdot|=|\cdot|_{\widetilde H}$).)

We recall that $|(F(t)z,A^{\frac12+2\varepsilon}z)|\le C|z||z|_{\frac12+2\varepsilon}$. This yields (see (4.38))

$$\int_\tau^\infty|z(t)|^2_{\frac34+\varepsilon}dt\le C|x|_{\frac14+\varepsilon}\left(1+\int_\tau^\infty e^{-\gamma(\tau+2s)}\zeta(s)ds\right)^2,$$

as claimed. The case $d=2$ follows completely similarly by taking $\varepsilon=0$. □

We come back to the nonlinear part S of system (4.33) and set

$$G(t,y):=e^{-\sum_{j=1}^N\beta_j(t)\Gamma_j}S(e^{\sum_{j=1}^N\beta_j(t)\Gamma_j}y),\quad\forall y\in\widetilde H,\ t\ge 0.$$

Recalling the definition and properties of Γ_j, we see that

$$\begin{aligned}S(e^{\beta_j(t)\Gamma_j}y)&=S(y)+(y,\varphi_j^*)^2(e^{\eta\beta_j}-1)^2S(P(m\phi_j))\\&\quad+(e^{\eta\beta_j(t)}-1)(y,\varphi_j^*)[S_1(y,P(m\phi_j))+S_2(y,P(m\phi_j))],\end{aligned}$$

where $S(y)=P((y\cdot\nabla)y)$, and

$$S_1(y,z)=P((y\cdot\nabla)z),\quad S_2(y,z)=P((z\cdot\nabla)y),\quad\forall y,z\in D(\mathscr A).$$

Therefore, we have for all $j,k=1,\dots,N$

$$\begin{aligned}e^{-\beta_k(t)\Gamma_k}S(e^{\beta_j(t)\Gamma_j}y)&=e^{-\beta_k(t)\Gamma_k}[S(y)+(y,\varphi_j^*)^2(e^{\eta\beta_j}-1)^2S(P(m\phi_j))\\&\quad+(e^{\eta\beta_j(t)}-1)(y,\varphi_j^*)[S_1(y,P(m\phi_j))+S_2(y,P(m\phi_j))]].\end{aligned}$$

But, by virtue of (4.37), we have

$$e^{-\beta_k(t)\Gamma_k}y=\eta^{-1}\Gamma_ky(e^{-\eta\beta_k(t)}-1)+y=(e^{-\eta\beta_k(t)}-1)(y,\varphi_k^*)P(m\phi_k)+y.$$

Therefore,

$$\begin{aligned}e^{-\beta_k(t)\Gamma_k}S(e^{\beta_j(t)\Gamma_j}y)&=S(e^{\beta_j(t)\Gamma_j}y)+(e^{-\eta\beta_j(t)}-1)(S(e^{\beta_j(t)\Gamma_j}y),\varphi_k^*)P(m\phi_k),\\e^{-\beta_k(t)\Gamma_k}S(e^{\beta_j(t)\Gamma_j}y)&=S(y)+(y,\varphi_j^*)^2(e^{\eta\beta_j(t)}-1)^2S(P(m\phi_j))\end{aligned}$$

$$+(e^{\eta\beta_j(t)}-1)(y,\varphi_j^*)[S_1(y,P(m\phi_j))+S_2(y,P(m\phi_j))]$$
$$+(e^{-\eta\beta_j(t)}-1)(S(e^{\beta_j(t)\Gamma_j}y),\varphi_j^*)P(m\phi_j).$$

Since φ_j^*, φ_k^* are smooth, we may write

$$e^{-\beta_k(t)\Gamma_k}S(e^{\beta_j(t)\Gamma_j}y)=S(y)+\Theta_{j,k}(t,y),\quad j,k=1,\dots,N, \tag{4.44}$$

where

$$\begin{aligned}|\Theta_{j,k}(t,y)|_\alpha &\le C(1+\delta(t))(|(y,\varphi_j^*)|^2+|S_1(y,P(m\phi_j))|_\alpha^2\\ &\quad+|S_2(P(m\phi_j),y)|_\alpha^2+|(S(y),\varphi_j^*)|^2),\\ &\quad \forall t\ge 0,\ y\in D(A),\ j,k=1,\dots,N,\end{aligned}$$

where $0<\alpha<1$ (recall that $|x|_\alpha=|A^\alpha x|_{\widetilde{H}}$) and

$$\delta(t)=\sup_{1\le j\le N}\max\{e^{-4\eta\beta_j(t)},e^{4\eta\beta_j(t)}\}. \tag{4.45}$$

To summarize, we have that

$$G(t,y)=S(y)+\Theta(t,y),\quad \forall t\ge 0,\ y\in D(A),$$

where

$$\begin{aligned}&|\Theta(t,y)|_\alpha\\ &\le C(1+\delta^N(t))\left(\max_{1\le j\le N}\{|S_1(y,P(m\phi_j))|_\alpha^2+|S_2(P(m\phi_j),y)|_\alpha^2\}+|S(y)|_{\widetilde{H}}\right)\end{aligned} \tag{4.46}$$

where δ is given by (4.45) and C is independent of t, y and ω.

We write (4.33) as

$$\frac{dy(t)}{dt}+\mathscr{A}_\Gamma y(t)+G(t,y(t))+F(t)y(t)=0,\quad \mathbb{P}\text{-a.s., } \forall t\ge 0.$$

We set $z(t)=e^{\frac{1}{2}\gamma t}y(t)$ and rewrite it as

$$\begin{cases}\frac{dz(t)}{dt}+(\mathscr{A}_\Gamma-\frac{1}{2}\gamma)z(t)+e^{-\frac{1}{2}\gamma t}G(t,z(t))+F(t)z(t)=0,\\ z(0)=x.\end{cases} \tag{4.47}$$

Equivalently, z is the solution to the integral equation

$$z(t)=S(t,0)x-\int_0^t\widetilde{S}(t,s)e^{-\frac{1}{2}\gamma s}G(s,z(s))ds,\quad \forall t\ge 0, \tag{4.48}$$

where

$$\widetilde{S}(t,s)=U(t,s)e^{-\frac{\gamma}{2}(t-s)},\quad 0\le s\le t.$$

Lemma 4.3 *There is $\Omega_x \subset \Omega$, with*

$$\mathbb{P}(\Omega_x) \geq 1 - \left(\frac{C^*}{\sqrt{\|x\|_W}} - 1\right)^{-\frac{\gamma}{8(\eta N)^2}},$$

where $C^* > 0$ is independent of ω such that, for each $x \in X$, $\|x\|_W \leq (C^*)^2$, (4.48) has a unique solution

$$z \in C([0,\infty); W) \cap L^2(0,\infty; W).$$

Here, $W = D(A^{\frac{1}{4}})$, $Z = D(A^{\frac{3}{4}})$ if $d = 2$ and $W = D(A^{\frac{1}{4}+\varepsilon})$, $Z = D(A^{\frac{3}{4}+\varepsilon})$ if $d = 3$.

Proof We proceed as in the proof of Theorem 3.4. Namely, we rewrite (4.48) as

$$z(t) = \widetilde{S}(t,0)x + \mathcal{N} z(t) := \Lambda z(t), \quad t \geq 0,$$

where $\mathcal{N} : L^2(0,\infty; Z)$ is the integral operator

$$\mathcal{N} z(t) = -\int_0^t \widetilde{S}(t,s) e^{-\frac{1}{2}\gamma s} G(s, z(s)) ds.$$

First, we prove the following estimate

$$\|\mathcal{N} z\|_{L^2(0,\infty;Z)} \leq C \int_0^\infty e^{-\frac{1}{2}\gamma t} \|G(t, z(t))\|_W dt. \tag{4.49}$$

Indeed, for any $\zeta \in L^2(0,\infty; Z')$ (Z' is the dual of Z), we have via Fubini's theorem

$$\begin{aligned}
&\int_0^\infty (\mathcal{N} z(t), \zeta(t)) dt \\
&= \int_0^\infty dt \left(\int_0^t \widetilde{S}(t,s) e^{-\frac{1}{2}\gamma s} G(s, z(s)) ds, \zeta(t)\right) \\
&\leq \int_0^\infty dt \int_0^t \|\widetilde{S}(t,s) e^{-\frac{1}{2}\gamma s} G(s, z(s))\|_Z ds\, |\zeta(t)|_{Z'} \\
&= \int_0^\infty d\tau \int_\tau^\infty \|\widetilde{S}(t,\tau) e^{-\frac{1}{2}\gamma\tau} G(\tau, z(\tau))\|_Z \|\zeta(t)\|_{Z'} dt \\
&\leq \int_0^\infty d\tau \left(\int_\tau^\infty \|S(t,\tau) e^{-\frac{1}{2}\gamma\tau} G(\tau, z(\tau))\|_Z^2 dt\right)^{\frac{1}{2}} \|\zeta\|_{L^2(0,\infty;Z')}.
\end{aligned}$$

Now, we set

$$I := \int_0^\infty d\tau \left(\int_\tau^\infty \|\widetilde{S}(t,\tau) e^{-\frac{1}{2}\gamma\tau} G(\tau, z(\tau))\|_Z^2 dt\right)^{\frac{1}{2}}.$$

By Lemma 4.2 we have

$$\int_\tau^\infty \|\widetilde{S}(t,\tau)x\|_Z^2 dt \le C\|x\|_W^2 \left(1+\int_\tau^\infty e^{-\gamma(t+2s)}\zeta(s)ds\right)^2, \quad \forall x \in W.$$

Next, we apply this for $x = e^{-\frac{1}{2}\gamma\tau}G(\tau, z(\tau))$ and get

$$\begin{aligned}\int_\tau^\infty & \|\widetilde{S}(t,\tau)e^{-\frac{1}{2}\gamma\tau}G(\tau,z(\tau))\|_Z^2 dt \\ &\le C\|G(\tau,z(\tau))\|_W^2 e^{-\gamma\tau}\left(1+\int_\tau^\infty e^{-\gamma(\tau+2s)}\zeta(s)ds\right)^2, \\ &\forall x \in W,\end{aligned}$$

and, therefore,

$$I \le C\int_0^\infty \|G(\tau,z(\tau))\|_W e^{-\frac{1}{2}\gamma\tau}d\tau \left(1+\int_0^\infty e^{-2\gamma s}\zeta(s)ds\right)^2,$$

as claimed.

Next, by (4.49) and Lemma 4.2, we have

$$\begin{aligned}\|Az\|&_{L^2(0,\infty;Z)} \\ &\le C\left(\|x\|_W + \left(1+\int_0^\infty e^{-2\gamma s}\zeta(s)ds\right)^2 \int_0^\infty e^{-\frac{1}{2}\gamma\tau}\|G(\tau,z(\tau))\|_W d\tau\right).\end{aligned} \tag{4.50}$$

On the other hand, we have

$$\|G(t,y)\|_W \le \|Sy\|_W + \|\Theta(t,y)\|_W.$$

By Lemma 3.3, we have

$$\|Sy\|_W \le C\|y\|_Z^2, \quad \forall y \in Z,$$

and, similarly, by (4.46),

$$\|\Theta(t,y)\|_W \le C(1+\delta^N(t))\|y\|_Z^2, \quad \forall y \in Z.$$

(As in the previous cases, the key estimate (3.89) in Lemma 3.3 determines the choice $w = D(A^{\frac{1}{4}+\varepsilon})$ in 3-D.) Then, (4.50) yields

$$\|Az\|_{L^2(0,\infty;Z)} \le C_1^*\left(\|x\|_W + \int_0^\infty (1+\delta^N(t))e^{-\frac{1}{2}\gamma t}\|z(t)\|_Z^2 dt\right), \quad \mathbb{P}\text{-a.s.}, \tag{4.51}$$

where C_1^* is a positive constant independent of ω. By (4.45) we have

$$\sup_{t\geq 0}(1+\delta^N(t)(\omega))e^{-\frac{1}{2}\gamma t} = 1+\sup_{t\geq 0}\max_{0\leq j\leq N}\{e^{4\eta N\beta_j(t)-\frac{1}{2}\gamma t}\} = 1+\mu(\omega),$$
$$\omega\in\Omega. \tag{4.52}$$

Similarly,

$$\int_0^\infty e^{-2\gamma s}\zeta(s) \leq \frac{1}{\gamma}\sup_{1\leq j\leq N}\sup_{t\geq 0} e^{\beta_j(t)-\gamma t} \leq \frac{1}{\gamma}\mu(\omega).$$

So, (4.51) yields

$$\|\Lambda z\|_{L^2(0,\infty;Z)} \leq C_1^*\left(\|x\|_W + (1+\mu(\omega))^2\|z\|^2_{L^2(0,\infty;Z)}\right),\quad \mathbb{P}\text{-a.s.} \tag{4.53}$$

By Lemma 4.6, for each $\lambda>0$, we have

$$\mathbb{P}\left(\sup_{t\geq 0} e^{\beta(t)-\lambda t}\geq r\right) = \mathbb{P}\left(e^{\sup_{t>0}(\beta(t)-\lambda t)}\geq r\right)$$
$$= \mathbb{P}\left(\sup_{s>0}(\beta(s)-\lambda s)\geq \log r\right) = r^{-2\lambda}. \tag{4.54}$$

By (4.52), it follows that

$$\mathbb{P}\left(\sup_{t\geq 0} e^{4\eta N\beta_j(t)-\frac{1}{2}\gamma t}\leq r\right) = 1-r^{-\frac{\gamma}{8(N\eta)^2}},\quad j=1,\dots,N, \tag{4.55}$$

and, therefore, again by (4.52),

$$\mathbb{P}(1+\mu\leq r)\geq 1-(r-1)^{-\frac{\gamma}{8(N\eta)^2}},\quad \forall r\geq 1. \tag{4.56}$$

We set

$$\mathscr{U}(\omega) := \{z\in L^2(0,\infty;Z) : \|z\|_{L^2(0,\infty;Z)}\leq R(\omega)\},$$

where $R:\Omega\to R^+$ is a random variable such that

$$\frac{2C_1^*\|x\|_W}{1+\sqrt{1+4(C_1^*)^2\|x\|_W(1+\mu)^2}}$$
$$\leq R(\omega) \leq \frac{2C_1^*\|x\|_W}{1-\sqrt{1-4(C_1^*)^2\|x\|_W(1+\mu)}},\quad \omega\in\Omega. \tag{4.57}$$

Then, as easily follows from (4.53) and (4.57), if

$$\|x\|_W\leq \rho_1(\omega) := [8(1+\mu)^2(C_1^*)^2]^{-1}, \tag{4.58}$$

we have $\Lambda\mathscr{U}(\omega)\subset\mathscr{U}(\omega)$.

Now, we apply the Banach fixed-point theorem for Λ on the set $\mathscr{U}(\omega)$.
Let $z_1, z_2 \in \mathscr{U}(\omega)$. Arguing as in the proof of (4.53), we find

$$\begin{aligned}
&\|\Lambda z_1 - \Lambda z_2\|_{L^2(0,\infty;Z)} \\
&\le C_1^* \int_0^\infty e^{-\frac{1}{2}\gamma t}\|G(t,z_1) - G(t,z_2)\|_W\, dt \left(1 + \int_0^\infty e^{-2\gamma t}\zeta(t)dt\right) \\
&\le C_1^* C_2^* \int_0^\infty (1+\delta^N(t))e^{-\frac{1}{2}\gamma t}\|z_1(t) - z_2(t)\|_Z(\|z_1(t)\|_Z + \|z_2(t)\|_Z)dt \\
&\le C_1^* C_2^* \left(\int_0^\infty \|z_1(t) - z_2(t)\|_Z^2 dt\right)^{\frac{1}{2}} \left(\int_0^\infty (\|z_1(t)\|_Z^2 + \|z_2(t)\|_Z^2)dt\right)^{\frac{1}{2}} \\
&\quad \times \left(1 + \int_0^\infty e^{-2\gamma t}\zeta(t)dt\right) \\
&\le 2C_1^* C_2^* (1+\mu(\omega))^2 R(\omega)\|z_1 - z_2\|_{L^2(0,\infty;Z)},
\end{aligned}$$

where C_1^*, C_2^* are independent of ω.

Now, if we choose x such that, besides (4.58), it also has

$$\|x\|_W \le \frac{\sqrt{2}+1}{2\sqrt{2}(C_1^*)^2 C_2^*(1+\mu)^2} =: \rho_2(\omega),$$

we see that there is $R = R(\omega)$ satisfying (4.57) and such that $2C_1^* C_2^*(1+\mu)^2 < 1$.

Now, we take

$$\|x\|_W \le \rho(\omega) := \min\{\rho_1(\omega), \rho_2(\omega)\} = ((C^*)^2(1+\mu)^2)^{-1}, \tag{4.59}$$

where C^* is a suitable constant independent of ω. Then, for x satisfying (4.59), Λ is a contraction on $\mathscr{U}(\omega)$.

We set

$$\Omega_x = \{\omega \in \Omega : \|x\|_W \le \rho(\omega)\}. \tag{4.60}$$

Hence, for each $\omega \in \Omega_x$ (4.48) has a unique solution satisfying the conditions in Lemma 4.3. On the other hand, by (4.56) and (4.60), we see that

$$\mathbb{P}(\Omega_x) \ge 1 - \left(\frac{C^*}{\sqrt{\|x\|_W}} - 1\right)^{-\frac{\gamma}{8(\eta N)^2}},$$

as claimed.

We have also

$$\lim_{t\to\infty} |z(t)|_{\widetilde{H}} = 0, \quad \mathbb{P}\text{-a.s. in } \Omega_x. \tag{4.61}$$

Indeed, by (4.47) it follows as in the proof of Lemma 4.2 that

$$\begin{aligned}&\frac{1}{2}\frac{d}{dt}\|z(t)\|_{\widetilde{H}}^2+\frac{\nu}{2}|z(t)|_{\frac{1}{2}}^2\\&\quad\le C_1\|z(t)\|_{\widetilde{H}}^2+e^{-\frac{\gamma t}{2}}|(G(t,z(t)),z(t))|+|(F(t,z(t)),z(t))|.\end{aligned}$$

Taking into account that

$$|e^{-\gamma t}(G(t,z(t)),z(t))|=e^{-\gamma t}|(\Theta(t,z(t)),z(t))|\le C_2\|z(t)\|_Z^2$$

and that $z\in L^2(0,\infty;D(A^{\frac{3}{4}}))$, we infer that

$$\frac{d}{dt}|z(t)|_{\widetilde{H}}^2\in L^\infty(0,\infty),$$

and, together with $z\in L^2(0,\infty;\widetilde{H}))$, this implies (4.61), as claimed. □

Proof of Theorem 4.2 (continued). By (4.61) we have that

$$\lim_{t\to\infty}|y(t)|_{\widetilde{H}}e^{\frac{1}{2}\gamma t}=0,\quad \forall\omega\in\Omega_x. \tag{4.62}$$

Then, as seen earlier,

$$X(t)=\prod_{j=1}^N e^{\beta_j(t)\Gamma_j}\,y(t),\quad \mathbb{P}\text{-a.s.}$$

is the solution to (4.4). On the other hand, by (4.36) and (4.50), we see that

$$|X(t)|_{\widetilde{H}}e^{\frac{\gamma t}{4}}\le C_1^*\left(1+\max_{1\le j\le N}\left\{e^{N\eta\beta_j(t)-\frac{\gamma t}{4}},e^{-N\eta\beta_j(t)-\frac{\gamma t}{4}}\right\}\right)|y(t)|_{\widetilde{H}}e^{\frac{\gamma t}{2}}. \tag{4.63}$$

We set

$$\Omega_x^r=\left\{\omega\in\Omega:\sup_{t\ge0}\max_{1\le j\le N}\left\{e^{N\eta\beta_j(t)-\frac{\gamma t}{4}},e^{-N\eta\beta_j(t)-\frac{\gamma t}{4}}\right\}\le r\right\},$$

where $r>0$. By (4.55), we have that

$$\mathbb{P}(\Omega_x^r)\ge 1-r^{-\frac{\gamma}{2(\eta N)^2}}, \tag{4.64}$$

and this yields

$$\mathbb{P}(\Omega_x\cap\Omega_x^r)\ge 1-\left(\frac{C^*}{\sqrt{\|x\|_W}}-1\right)^{-\frac{\gamma}{2(\eta N)^2}}-r^{-\frac{\gamma}{2(\eta N)^2}}, \tag{4.65}$$

for any $r > 0$. We set $\Omega_x^* = \Omega_x \cap \Omega_x^r$, where r is given by

$$r = \left(\frac{C^*}{\sqrt{\|x\|_W}} - 1 \right)^{\frac{1}{4}}.$$

Then, by (4.62) and (4.63), we obtain (4.30) and

$$\lim_{t\to\infty} |X(t)|_{\widetilde{H}} e^{\frac{\gamma t}{4}} = 0 \quad \mathbb{P}\text{-a.s. in } \Omega_x^*.$$

This completes the proof of Theorem 4.2. □

Analyzing the proof of Theorem 4.2, we see that the stochastic perturbation (4.2) in System (4.1) has a dissipation effect due to the presence in (4.33) of the positive linear operator $\frac{1}{2}\sum_{j=1}^N \Gamma_j^2 y$, which is enforcing the dissipation. This is the source of the stabilizing effect of the Gaussian noise (4.2) added into system as well as the source of the robustness of the closed-loop system. More will be said about below.

Remark 4.4 One might design a feedback controller of the above form in the absence of Assumption (J1).

Indeed, if we replace $\{\varphi_j\}_1^N$ by its Schmidt's orthogonalization $\{\widetilde{\varphi}_j\}_1^N$, we still have $\mathscr{X}_u = \text{lin span}\{\widetilde{\varphi}_j\}_1^N$ and $\mathscr{X}_s = \text{lin span}\{\widetilde{\varphi}_j\}_{M+1}^\infty$. Then, we consider the feedback controller

$$u = \eta \sum_{j=1}^N (X, \phi_j^*)_{\widetilde{H}} P(m\widetilde{\Phi}_j)\dot{\beta}_j, \tag{4.66}$$

where $\{\widetilde{\Phi}_j\}$ are determined by the condition

$$(\widetilde{\Phi}_j, \widetilde{\varphi}_k)_0 = \delta_{jk}, \quad j, k = 1, \dots, N.$$

By Proposition 3.2, it follows that system $\{\widetilde{\varphi}_j\}_1^N$ is linearly independent on $\mathscr{O}_0$ and so, such a system $\{\widetilde{\Phi}_j\}_1^N$ exists. Then, the proof of Theorem 4.2 applies with minor modifications to show that the controller u defined by (4.66) is exponentially stabilizable in the sense of Theorem 4.2. The details are omitted.

4.1.5 Stochastic Stabilization Versus Deterministic Stabilization

As seen in Sect. 3.3, the feedback controller

$$u = -\eta \sum_{j=1}^N (X, \varphi_j^*) P(m\phi_j), \tag{4.67}$$

where η is sufficiently large, stabilizes exponentially the linearization of System (4.3) in a neighborhood $\{x \in W : \|x\|_W < \rho\}$, where ϕ_j are chosen as in (4.8).

Apparently, the feedback controller (4.67) is simpler than its stochastic counterpart (4.2) above, while the stabilization performances are comparable.

It should be said, however, that Controller (4.67), though stabilizable, is not robust, while the stochastic one is. In fact, it is easily seen that (4.67) is very sensitive to structural perturbations in System (4.1), because small variations of the spectral system $\{\phi_j\}$, $\{\varphi_j^*\}$ might break the orthogonality condition (4.10) from which ϕ_j are determined.

In this way, the deterministic linear closed-loop equation

$$dX + \mathscr{A}Xdt = -\eta \sum_{j=1}^{N} (X, \varphi_j^*) P(m\phi_j) dt$$

might become unstable even for $\eta > 0$ and very large. By contrary, this does not happen for the stochastic system

$$dX + AXdt = -\eta \sum_{j=1}^{N} (X, \varphi_j^*) P(m\phi_j) d\beta_j,$$

because its unstable part that is

$$X = \sum_{j=1}^{N} X_j \phi_j,$$

where

$$dX_j + \lambda_j X_j dt = -\eta \sum_{j=1}^{N} X_j (\phi_j, \varphi_j^*)_0 P(m\phi_j) d\beta_j,$$

$$\operatorname{Re} \lambda_j \le \gamma, \; j = 1, \dots, N, \tag{4.68}$$

still remains exponentially stable with probability one to small perturbations of $\{\varphi_j^*\}$.

Indeed, in this case, instead of (4.10), we have

$$|(\phi_j, \varphi_k^*)_0 - \delta_{jk}| \le \varepsilon, \quad \forall j, k = 1, \dots, N,$$

and, therefore,

$$\sum_{j=1}^{N} \sum_{i=1}^{N} |(\phi_j, \varphi_j^*)_0|^2 |X_j|^2 \ge \mu \sum_{j=1}^{N} |X_j|^2,$$

which, as seen from the analysis developed in the proof of Theorem 4.1, implies the stabilization of (4.68) for sufficiently large $|\eta|$.

As mentioned in Chap. 3, starting from (4.67) one might design a robust stabilizable controller via infinite-dimensional Riccati equations associated with the linear system, but this involves hard numerical computations.

4.2 Stabilization of the Stokes–Oseen Equation by Impulse Feedback Noise Controllers

The analysis developed in Sect. 3.3 was confined to the design of a stabilizable feedback controller with support in a subset $\mathcal{O}_0 \subset \mathcal{O}$ with nonempty interior. One might ask if such a stabilizable controller can be designed for subsets $\mathcal{O}_0$ with empty interior and, in particular, for $\mathcal{O}_0 = \{\xi_k\}_{k=1}^M$, that is for impulse stabilizing controller. This problem is discussed in this section.

To be more specific, we address here the problem of linear stabilization of steady-state solutions X_e to Navier–Stokes equations by mean of a noise internal controller with support in discrete set of points of the domains. More precisely, the stabilizable feedback controller proposed here is of the form

$$u = \left(\sum_{k=1}^{M} \mu_k \delta(\xi_k)\right) \sum_{j=1}^{N} (X - y_e, \varphi_j^*)\dot{\beta}_j, \tag{4.69}$$

where $\mu_k \subset R^d$, $\delta(\xi_k)$ is the Dirac measure concentrated in the point $\xi_k \in \mathcal{O} \subset R^d$, $d = 2, 3$, $\{\beta_j\}$ is a system of independent Brownian motions and $\{\varphi_j^*\}_{j=1}^N$ are eigenfunctions to the dual Stokes–Oseen operator corresponding to the eigenvalues $\{\lambda_j;\ \mathrm{Re}\,\lambda_j \le \gamma\}$, $j = 1, \dots, N$. The main result amounts to saying that the feedback controller (4.69) exponentially stabilizes in probability the linearized Navier–Stokes system in a certain weak sense to be discussed below.

As mentioned earlier, this is quite surprising if one takes into account that, for the previous internal stabilization results, the condition int $\mathcal{O}_0 \neq \emptyset$ was essential and was required by the unique continuation theorem.

Here, we come back to the Oseen–Stokes equation associated with the equilibrium solution y_e to Navier–Stokes equation (4.1), that is,

$$\begin{aligned}
&y_t - \nu \Delta y + (y_e \cdot \nabla) y + (y \cdot \nabla) y_e = \nabla p && \text{in } (0,\infty) \times \mathcal{O}, \\
&\nabla \cdot y = 0 && \text{in } (0,\infty) \times \mathcal{O}, \\
&y = 0 && \text{on } (0,\infty) \times \partial\mathcal{O}, \\
&y(0,\xi) = x(\xi), && \xi \in \mathcal{O}.
\end{aligned} \tag{4.70}$$

Our objective here is to design an internal feedback controller u with support in a finite number of points $\{\xi_k\}_{k=1}^M \subset \mathcal{O}$, which exponentially stabilizes System (4.70). As mentioned earlier, this is a fundamental problem in the linear theory of fluid dynamics and can be viewed as a first step to the stabilization of the stationary solution y_e to Navier–Stokes equation (4.1).

Consider, as above, the Stokes–Oseen operator

$$\mathscr{A} = \nu A + A_0, \qquad D(\mathscr{A}) = D(A),$$

where $A_0 y = P((y_e \cdot \nabla) y + (y \cdot \nabla) y_e)$.

We fix $\gamma > 0$ and $N \in \mathbb{N}$ such that $\mathrm{Re}\,\lambda_j \le \gamma$, $j = 1, \dots, N$, where λ_j are the eigenvalues of $\mathscr{A}$.

For each λ_j, consider the corresponding eigenfunction φ_j, each λ_j being repeated according to its (algebraic) multiplicity m_j. Recall that the adjoint operator $\mathscr{A}^*$ with $D(\mathscr{A}^*) = D(\mathscr{A})$ has the eigenvalues $\overline{\lambda}_j$ with corresponding eigenfunction φ_j^*.

For simplicity, in this case we assume also Hypothesis (J1), that is,

(J1) *Each* λ_j, $j = 1, \ldots, N$, *is semisimple.*

Choose, as above, φ_j and φ_j^* such that

$$(\varphi_i, \varphi_j^*) = \delta_{ij}, \quad i, j = 1, \ldots, N. \tag{4.71}$$

(For simplicity, denote again by $(\cdot, \cdot)$ the scalar product in $\widetilde{H}$.)

Consider now a probability space $\{\Omega, \mathscr{F}, \mathscr{F}_t, \mathbb{P}\}_{t>0}$ and a system of independent complex Brownian motion $\{\beta_j = \beta_j^1 + i\beta_j^2\}_{j=1}^N$ in this probability space. We use the standard notation for the spaces of adapted $\widetilde{H}$-valued processes. (See Sect. 4.5.)

In particular, $C_W([0, T]; L^2(\Omega, \widetilde{H}))$ is the space of adapted $\widetilde{H}$-valued continuous processes on $[0, T]$.

As seen earlier, the controlled system associated with (4.70) can be rewritten as a state-system

$$\begin{aligned} &\frac{dy}{dt} + \mathscr{A} y = u, \quad \forall t \geq 0, \\ &y(0) = x \end{aligned} \tag{4.72}$$

where $y : [0, \infty) \to \widetilde{H}$. (We take $x \in H$.)

In the following, we denote by the same symbol $|\cdot|$ the norm in H, $\widetilde{H}$ and $\mathbb{C}$. (The difference will be clear from the context.)

Now, we fix $\{\xi_k\}_{k=1}^M \subset \mathscr{O}$ and $\{\mu_k\}_{k=1}^M \subset \mathbb{C}$ such that

$$\left| \sum_{k=1}^M \mu_k \varphi_i^*(\xi_k) \right| > 0, \quad \forall i = 1, 2, \ldots, N. \tag{4.73}$$

Theorem 4.3 is the main result.

Theorem 4.3 *Under Assumptions* (4.73), *for* $|\eta|$ *sufficiently large, the feedback noise controller*

$$u(t) = \eta \sum_{j=1}^N (y(t), \varphi_j^*) \dot{\beta}_j(t) \sum_{k=1}^M \mu_k \delta(\xi_k) \tag{4.74}$$

weakly exponentially stabilizes in probability the state-system (4.72). *More precisely, the solution* y *to the closed-loop system*

$$\begin{aligned} &dy(t) + \mathscr{A} y(t) dt = \eta \sum_{j=1}^N (y(t), \varphi_j^*(t)) d\beta_j(t) \sum_{k=1}^M \mu_k \delta(\xi_k), \\ &y(0) = x, \end{aligned} \tag{4.75}$$

satisfies

$$\mathbb{P}\Big[\lim_{t\to\infty} e^{\gamma t}(X(t),\psi)=0\Big]=1,\quad \forall\psi\in(H^2(\mathcal{O})\cap H_0^1(\mathcal{O}))^d\cap H. \tag{4.76}$$

Here, $\delta(\xi_k)$ is the Dirac measure concentrated in ξ_k.

Equation (4.75) is taken in Ito's sense in the dual space $((H^2(\mathcal{O})\cap H_0^1(\mathcal{O}))^d\cap H)'$. More precisely, the solution y to (4.75) is in the following "mild" sense (see Sect. 4.5)

$$y(t)=e^{-\mathcal{A}t}x+\eta\sum_{j=1}^{N}\sum_{k=1}^{M}\int_0^t (y(s),\varphi_j^*)e^{-\mathcal{A}(t-s)}(\delta(\xi_k))d\beta_j(s), \tag{4.77}$$

where $e^{-\mathcal{A}t}\delta(\xi_k)\in((H^2(\mathcal{O})\cap H_0^1(\mathcal{O}))^d\cap H)'$ is defined by

$$e^{-\mathcal{A}t}\delta(\xi_k)(\psi)=(e^{-\mathcal{A}t}\psi)(\xi_k),\quad \forall\psi\in(H^2(\mathcal{O})\cap H_0^2(\mathcal{O}))^d\cap H)=D(A).$$

(Here $'$ stands for the dual space.)

Since $e^{-\mathcal{A}t}\psi\in H^2(\mathcal{O})\subset C(\overline{\mathcal{O}})$, the latter makes sense and so, (4.75) has a solution $y\in C_W([0,T];L^2(\Omega,(D(A))'))$ on each interval $[0,T]$.

It should be emphasized that Feedback (4.74) is a distribution (as a matter of fact, a measure) on $\mathcal{O}$ with support in the set $\{\xi_k\}_{k=1}^N$.

More generally, if $\mu\in(\mathcal{M}(\mathcal{O}))^d$ is a bounded Radon measure on $\mathcal{O}$ such that

$$\mu(\varphi_i^*)\neq 0,\quad \forall i=1,\dots,N, \tag{4.78}$$

we have the following theorem.

Theorem 4.4 *For* $|\eta|$ *large enough, the feedback law*

$$u=\eta\mu\sum_{j=1}^{N}(y,\varphi_j^*)\dot\beta_j \tag{4.79}$$

stabilizes System (4.72) *in the sense of* (4.76).

For instance, one might take $\mu\in((H^2(\mathcal{O})\cap H_0^1(\mathcal{O}))^d\cap H)'$ of the form

$$\mu(\psi)=\int_\Gamma h(\xi)\psi(\xi)d\sigma_\xi,\quad \forall\psi\in(H^2(\mathcal{O})\cap H_0^1(\mathcal{O}))^d\cap H,$$

where Γ is a smooth surface (or manifold) of $\mathcal{O}$ and h is a continuous function on $\overline{\mathcal{O}}$.

In particular, by Theorem 4.3, it follows that, if $\xi_0\in\mathcal{O}$ is such that

$$|\varphi_i^*(\xi_0)|\neq 0,\quad \forall i=1,\dots,N,$$

then the feedback law

$$u = \eta\delta(\xi_0)\sum_{j=1}^{N}(y,\varphi_j^*)\dot{\beta}_j \tag{4.80}$$

stabilizes, for $|\eta|$ large enough, System (4.72) in the sense of (4.76).

Since, by the unique continuation property of the eigenfunctions to Stokes–Oseen operator $\mathscr{A}^*$, each φ_j^* is not identically zero on any open subset of $\mathscr{O}$, we may conclude therefore that for almost all $\xi_0 \in \mathscr{O}$ there is a noise controller of the form (4.80) which weakly stabilizes in probability System (4.70). It should be emphasized that the feedback controller (4.74) uses only a discrete set of points ξ_k, $k = 1, \dots, M$, for actuation. This means that the controlled velocity field consists of a steady-state impulse component

$$\widetilde{\mu} = \sum_{k=1}^{M}\mu_k\delta(\xi_k)$$

modulated by the unsteady feedback noise controller

$$u_0(t) = \sum_{j=1}^{N}(y(t),\varphi_j^*)\dot{\beta}_j(t). \tag{4.81}$$

Since the steady-state component of the controller is singular (in fact, it is a measure), the stabilization is in the weak topology only, that is, in the sense of distributions on $\mathscr{O}$. However, as we see later, this controller is robust with respect to small perturbations of the system.

4.2.1 Proof of Theorem 4.3

We consider the spaces

$$\mathscr{X}_u = \text{lin span}\{\varphi_j\}_{j=1}^N = P_N\widetilde{H}, \qquad \mathscr{X}_s = (I - P_N)\widetilde{H}.$$

The operator $\mathscr{A}$ leaves invariant both spaces and we set, as in the previous cases,

$$\mathscr{A}_u = \mathscr{A}|_{\mathscr{X}_u}, \qquad \mathscr{A}_s = \mathscr{A}|_{\mathscr{X}_s}.$$

Notice that $\sigma(\mathscr{A}_u) = \{\lambda_j\}_{j=1}^N$, $\sigma(\mathscr{A}_s) = \{\lambda_j\}_{j=N+1}^\infty$.

Moreover, since $\sigma(\mathscr{A}_s) \subset \{\lambda;\ \operatorname{Re}\lambda_j > \gamma\}$, we have, for some $\varepsilon > 0$,

$$\|e^{-\mathscr{A}_s t}\|_{L(\widetilde{H},\widetilde{H})} \le Ce^{-(\gamma+\varepsilon)t}, \quad \forall t > 0. \tag{4.82}$$

Next, we write System (4.75) as

$$
\begin{aligned}
& y = y^1 + y^2, \quad y^1 = \sum_{j=1}^{N} y_j \varphi_j, \\
& dy^1 + \mathscr{A}_u y^1 dt = \eta P_N \widetilde{\mu} \sum_{j=1}^{N} y_j d\beta_j, \\
& y^1(0) = P_N x,
\end{aligned}
\tag{4.83}
$$

$$
\begin{aligned}
& dy^2 + \mathscr{A}_s y^2 dt = \eta (I - P_N) \widetilde{\mu} \sum_{j=1}^{N} y_j d\beta_j, \\
& y^2(0) = (I - P_N) x,
\end{aligned}
\tag{4.84}
$$

where

$$
\widetilde{\mu} = \sum_{k=1}^{M} \mu_k \delta(\xi_k).
$$

The solution y^2 to (4.84) is taken in the "mild" sense (4.77), that is,

$$
y^2(t) = e^{-\mathscr{A}_s t}(I - P_N)x + \eta \sum_{j=1}^{N} \int_0^t y_j(s) e^{-\mathscr{A}_s (t-s)} (I - P_N) \widetilde{\mu} d\beta_j(s), \tag{4.85}
$$

where $e^{-\mathscr{A}_s t}(I - P_N)\widetilde{\mu} \in ((H^2(\mathscr{O}) \cap H_0^1(\mathscr{O}))^d \cap H)'$ is given by

$$
\begin{aligned}
& (e^{-\mathscr{A} t}(I - P_N)\widetilde{\mu})(\psi) = \sum_{k=1}^{N} \mu_k (I - P_N^*) e^{-\mathscr{A}_s^* t} \psi(\xi_k), \\
& \forall \psi \in (H^2(\mathscr{O}) \cap H_0^1(\mathscr{O}))^d \cap H.
\end{aligned}
\tag{4.86}
$$

Now, by virtue of (4.71), System (4.83) can be rewritten as

$$
\begin{aligned}
& dy_i + \lambda_i y_i dt = \eta \sum_{j=1}^{M} y_j \zeta_i d\beta_j, \quad i = 1, \ldots, N, \\
& y_i(0) = y_i^0 = (P_N x, \varphi_i^*),
\end{aligned}
\tag{4.87}
$$

where

$$
\zeta_i = \sum_{k=1}^{M} \mu_k \varphi_i^*(\xi_k), \quad i = 1, \ldots, N.
$$

We set $z_i = e^{\widetilde{\gamma}t} y_i$, where $\widetilde{\gamma} = \gamma + \varepsilon$ is such that $\operatorname{Re}\lambda_j > \gamma + \varepsilon$ for $j > N$, and rewrite (4.87) as

$$\begin{aligned} &dz_i + (\lambda_j - \widetilde{\gamma})z_i\,dt = \eta\zeta_i \sum_{j=1}^{N} z_j d\beta_j, \\ &z_i(0) = y_i^0. \end{aligned} \tag{4.88}$$

Applying Ito's formula, we obtain that

$$\begin{aligned} &\frac{1}{2} d|z_i|^2 + (\operatorname{Re}\lambda_i - \widetilde{\gamma})|z_i|^2 dt \\ &\quad = \frac{1}{2}\eta^2|\zeta_i|^2 \sum_{j=1}^{N} |z_j|^2 dt \\ &\qquad + \eta \sum_{j=1}^{N} (\operatorname{Re}(\zeta_i z_i)\operatorname{Re}, z_j + \operatorname{Im}(\zeta_i z_i)\operatorname{Im} z_j) d\beta_j^1 \\ &\qquad + \eta \sum_{j=1}^{N} (\operatorname{Re}(\zeta_i z_i)\operatorname{Im} z_j - \operatorname{Im}(\zeta_i z_i)\operatorname{Re} z_j) d\beta_j^2, \quad i = 1, \dots, N. \end{aligned} \tag{4.89}$$

(This is the complex version of Ito's formula (4.163), which follows by taking separately the real and imaginary part of (4.88).)

Now, we apply Ito's formula with $\varphi(r) = r^\delta$, where $0 < \delta < \frac{1}{2}$. (Of course, this function is not of class C^2, but the computation below can be made rigorous by replacing the function φ by $\varphi_\varepsilon(r) = (r^2 + \varepsilon)^{\frac{\delta}{2}}$ and letting $\varepsilon \to 0$.) (See the proof of Theorem 4.1.) We have

$$\varphi'(r) = \delta(r)^{\delta-1} r, \qquad \varphi''(r) = \delta(\delta - 1)r^{\delta-2}$$

and so, (4.89) yields via Ito's formula for the real stochastic equation (4.89) with Brownian motion system $\{\beta_j^1, \beta_j^2\}_{j=1}^N$

$$\begin{aligned} &d|z_i|^{2\delta} + 2\delta(\operatorname{Re}\lambda_i - \widetilde{\gamma})|z_i|^{2\delta} dt \\ &\quad = (2\delta - 1)\delta\eta^2|\zeta_i|^2|z_i|^{2(\delta-1)} \sum_{j=1}^{N} |z_j|^2 dt \\ &\qquad + 2\delta|z_i|^{2(\delta-1)} \operatorname{Re}\left(\sum_{j=1}^{N} (\zeta_i z_i \bar{z}_j) d\beta_j\right), \quad \mathbb{P}\text{-a.s., } i = 1, \dots, N. \end{aligned} \tag{4.90}$$

We set

$$|z|^{2\delta} = \sum_{j=1}^{N} |z_i|^{2\delta}.$$

Then (4.90) yields

$$|z(t)|^{2\delta} + \int_0^t H(s)ds = |y^0|^{2\delta} + M(t), \quad \mathbb{P}\text{-a.s., } t \geq 0, \tag{4.91}$$

where

$$H = \delta\eta^2 \sum_{i=1}^N (1-2\delta)(|\zeta_i|^2 |z_i|^{2(\delta-1)}|z|^2 + 2(\operatorname{Re}\lambda_i - \widetilde{\gamma})|z_i|^{2\delta}),$$

$$M(t) = 2\delta \operatorname{Re} \int_0^t \sum_{i=1}^N \sum_{j=1}^N |z_i|^{2(\delta-1)}(\zeta_i z_i \bar{z}_j)d\beta_j.$$

Since, by Assumption (4.73), $|\zeta_i| > 0$, we have for $0 < \delta < \frac{1}{2}$ and $|\eta|$ sufficiently large

$$H(t) \geq \rho |z(t)|^{2\delta}, \quad \mathbb{P}\text{-a.s., } \forall t \geq 0, \tag{4.92}$$

where $\rho > 0$.

Then, by the martingale convergence theorem (see Lemma 4.5), it follows by (4.91) and (4.92), as in the proof of Theorem 4.1, that

$$\lim_{t\to\infty} |z(t)|^{2\delta} < \infty, \quad \mathbb{P}\text{-a.s.}$$

and

$$\int_0^\infty E|z(t)|^{2\delta}dt < \infty.$$

Hence,

$$\lim_{t\to\infty} |z(t)| = \lim_{t\to\infty} |y^1(t)|e^{\widetilde{\gamma}t} = 0, \quad \mathbb{P}\text{-a.s.} \tag{4.93}$$

and

$$\int_0^\infty e^{2\widetilde{\gamma}t}|y^1(t)|^2 dt < \infty, \quad \mathbb{P}\text{-a.s.} \tag{4.94}$$

Now, we come back to the infinite-dimensional system (4.84). It can be, equivalently, written as

$$d(y^2 e^{\gamma t}) + (\mathscr{A}_s - \gamma)(y^2 e^{\gamma t})dt = \eta(I - P_N)\widetilde{\mu} \sum_{j=1}^N e^{\gamma t} y_j d\beta_j(t).$$

Then, for each $\psi \in D(A) = (H^2(\mathscr{O}) \cap H_0^1(\mathscr{O}))^d \cap H$, we have (see (4.85))

$$(y^2(t), \psi)e^{\gamma t} = e^{-(\mathscr{A}_s - \gamma)t}((I - P_N)x, \psi)$$

$$+\eta\sum_{j=1}^{N}\sum_{k=1}^{M}\mu_k\int_0^t e^{\gamma s}y_j(s)(I-P_N^*)e^{-(\mathscr{A}_s^*-\gamma)(t-s)}\psi(\xi_k)d\beta_j(s).$$

Since, as seen earlier,

$$\|e^{-\mathscr{A}_s t}(I-P_N)\|_{L(\widetilde{H},\widetilde{H})}\le Ce^{-\widetilde{\gamma}t},\quad \forall t\ge 0, \tag{4.95}$$

it remains to estimate the integral term

$$Z(t)=\eta\sum_{j=1}^{N}\sum_{k=1}^{M}\mu_k\int_0^t e^{\gamma s}y_j(s)(I-P_N^*)e^{-(\mathscr{A}_s^*-\gamma)(t-s)}\psi(\xi_k)d\beta_j(s),$$

$$\forall\psi\in(H^2(\mathscr{O})\cap H_0^1(\mathscr{O}))^d\cap H.$$

Let $z(t)$ be the solution to the stochastic differential equation

$$\begin{aligned}&dz(t)+(\mathscr{A}_s^*-\gamma)z(t)=\eta A\psi\sum_{j=1}^{N}e^{\gamma t}y_j(t)d\beta_j(t),\\&z(0)=0.\end{aligned} \tag{4.96}$$

Then, we have

$$Z(t)=\sum_{k=1}^{M}\mu_k A^{-1}z(t)(\xi_k),\quad \mathbb{P}\text{-a.s., } t>0. \tag{4.97}$$

Since, by (4.95), $e^{-(\mathscr{A}_2^*-\gamma)t}$ is exponentially stable in $\mathscr{X}_s^*$, it follows by Lyapunov's theorem, already invoked so far in similar situations, that there is a self-adjoint, continuous and positive definite operator Q on $\mathscr{X}_s^*$ (that is, $(Qz,z)>0$, $\forall z\ne 0$) such that

$$\operatorname{Re}((\mathscr{A}_2^*-\gamma)z,Qz)=\frac{1}{2}|z|^2,\quad \forall z\in\mathscr{X}_s^*. \tag{4.98}$$

(Here, $\mathscr{X}_s^*=(I-P_N^*)\widetilde{H}$.)

Then, applying Ito's formula in (4.96) to the function $z\to\frac{1}{2}(Qz,z)$, we obtain by (4.98) that

$$\begin{aligned}&\frac{1}{2}(Qz(t),z(t))+\frac{1}{2}\int_0^t|z(s)|^2ds\\&=\frac{1}{2}(Qz(0),z(0))+\frac{1}{2}\eta^2\sum_{j=1}^{N}\int_0^t e^{2\gamma s}|y_j(s)|^2(QA\psi,A\psi)ds\\&\quad+\eta\sum_{j=1}^{N}\operatorname{Re}\int_0^t e^{\gamma s}y_j(s)(Q\psi,z(s))d\beta_j(s).\end{aligned} \tag{4.99}$$

By (4.94) and (4.98), it follows once again by Lemma 4.5 that there exists

$$\lim_{t\to\infty}(Qz(t), z(t)) < \infty, \quad \mathbb{P}\text{-a.s.}$$

and

$$\int_0^\infty |z(s)|^2 ds < \infty, \quad \mathbb{P}\text{-a.s.}$$

Hence

$$\lim_{t\to\infty} |z(t)|^2 = 0, \quad \mathbb{P}\text{-a.s.}$$

Now, taking into account that $H^2(\mathscr{O}) \subset C(\overline{\mathscr{O}})$, we have

$$|A^{-1}z(t)| \le C|z(t)|_{C(\overline{\mathscr{O}})}, \quad \forall t \ge 0.$$

(Here, $C(\overline{\mathscr{O}})$ is the space of continuous functions on $\overline{\mathscr{O}}$.)

Therefore, we infer by (4.97) that

$$\lim_{t\to\infty} |Z(t)| = \lim_{t\to\infty} (|(y(t), \psi)| e^{\gamma t}) = 0, \quad \mathbb{P}\text{-a.s.},$$

$\forall \psi \in (H^2(\mathscr{O}) \cap H_0^1(\mathscr{O}))^d \cap H$, and along with (4.93) this implies (4.76), as claimed.

4.2.2 *Proof of Theorem 4.4*

It is identical with that of Theorem 4.3 and so, we omit it. We note only that, in this case, System (4.87) reduces to

$$dy_i + \lambda_i y_i\, dt = \eta \mu(\varphi_i^*) \sum_{j=1}^N y_j\, d\beta_j, \quad i = 1, \ldots, N,$$

that is, $\zeta_i = \mu(\varphi_i^*)$, and so, the conclusions of the theorem follow as above.

4.2.3 *The Robustness of the Noise Feedback Controller*

We show here that the feedback controller (4.74) is robust to small structural perturbation of System (4.70). Indeed, if Condition (4.73) holds, then it still remains true for small perturbations of System (4.70) and, more precisely, of its eigenfunctions system. In fact, by the spectral stability of the Stokes–Oseen operator $\mathscr{A}$ (see, e.g.,

[59], Chap. 4), a small variation of magnitude ε in X_e leads to a new eigenfunction system $\{\varphi^*_{j,\varepsilon}\}_{j=1}^N$ for which we still have

$$\sup_{0\le\varepsilon\le\varepsilon_0}\left|\sum_{k=1}^{M}\mu_k\varphi^*_{j,\varepsilon}(\xi_k)\right| > \rho > 0, \quad \forall j=1,\dots,N. \tag{4.100}$$

This implies that, at the level of unstable modes, the system has a gain stability margin independent of ε. In other words, the solution $y^1 = y^1_\varepsilon$ to the corresponding system (4.83) satisfies (4.94) uniformly in ε and so, (4.96), (4.98) imply that

$$\lim_{t\to\infty}(y_\varepsilon(t),\psi)e^{\gamma t} = 0, \quad \mathbb{P}\text{-a.s.}, \quad \forall\psi\in(H^2(\mathcal{O})\cap H_0^1(\mathcal{O}))^d\cap H,$$

uniformly for $0<\varepsilon<\varepsilon_0$. (Here, y_ε is the solution to the perturbed system (4.75).)

The main conclusion from this brief analysis is that the noise controller has a robust stabilizing effect which, as seen earlier, is not always the case with the deterministic stabilizing feedback controllers of the form (4.74).

Remark 4.5 In the absence of Assumption (J1), Theorem 4.3 (as well as Theorem 4.4) remains true for a feedback controller u of the form

$$u(t) = \eta\sum_{j=1}^{N}(y(t),\phi_j)\dot{\beta}_j(t)\sum_{k=1}^{M}\delta(\xi_k). \tag{4.101}$$

Here, $\{\phi_j\}_{j=1}^N$ is obtained by $\{\varphi_j\}$ by Schmidt's orthogonalization algorithm. Then,

$$\mathcal{X}_u = \text{lin span}\{\phi_j\}_{j=1}^N$$

and, if one assumes that

$$\left|\sum_{j=1}^{N}\mu_k\phi_i(\xi_k)\right| < 0, \quad \forall i=1,\dots,N, \tag{4.102}$$

then it follows by the same argument that u is weakly stabilizable in the sense of Theorem 4.3. The details are omitted.

We have designed here for the linearized Navier–Stokes equation a stochastic stabilizing feedback controller with the support in an arbitrary finite set of points ξ_k in the spatial domain $\mathcal{O}$. The design of this feedback controller involves the knowledge and actuation in the points ξ_k of a finite system of eigenfunctions of the dual Stokes–Oseen operator corresponding to unstable eigenvalues and it is robust to small structural perturbations of the system. Theoretically, this is a substantial reduction in computation over existing Riccati-based methods. The stabilization is, however, in probability and in a weak distributional sense and this is the price paid for the "inconsistency" of the controller u which, having a discrete support, should be taken in a distribution space on $\mathcal{O}$.

4.2.4 Deterministic Impulse Controller

The above design of the stabilizable noise controller suggests the consideration of feedback deterministic controllers of the same form, that is (see, also, Sect. 3.3),

$$\widetilde{u}(t) = \eta \sum_{j=1}^{N} (y(t), \varphi_j^*) \sum_{k=1}^{M} \mu_k \delta(\xi_k). \tag{4.103}$$

In order to stabilize System (4.72) by a feedback controller of this form, one must impose the condition that $\sum_{k=1}^{M} \mu_k \varphi^*(\xi_k)$ is real and not zero, and this implies as in the proof of Theorem 4.3 that Controller (4.103) stabilizes exponentially System (4.72) weakly in $(H^2(\mathscr{O}) \cap H_0^1(\mathscr{O}))^d \cap H$ for η of the form

$$\eta_i = -\lambda \operatorname{sign} \sum_{k=1}^{M} \mu_k \varphi_i^*(\xi_k),$$

where $\lambda > |\operatorname{Re} \lambda_j - \gamma|,\ j = 1, \ldots, N$. Indeed, in this case, (4.72) can be written as

$$\frac{dy^1}{dt} + \mathscr{A}_u y^1 = -\lambda \left(\sum_{k=1}^{M} \mu_k \varphi_i^*(\xi_k) \right) y^1,$$

$$\frac{dy^2}{dt} + \mathscr{A}_s y^2 = -\eta (I - P_N) \left(\sum_{k=1}^{M} \mu_k \delta(\xi_k) \right) y^1,$$

and the proof of the weak asymptotic stability is quite immediate under condition (4.73).

We have, therefore, obtained a deterministic stabilizable impulse controller with support in a given set of points $\{\xi_k\}_{k=1}^{M} \subset \mathscr{O}$. However, though this controller is stabilizable, it is not robust for the reasons explained in Chaps. 2 and 3. This is the main advantage of the noise controller given by Formula (4.74).

4.3 The Tangential Boundary Stabilization by Noise

We keep the notation of Sects. 3.3 and 4.1, respectively. We come back to the Stokes–Oseen system with boundary controller, that is,

$$\begin{aligned} & X_t - \nu \Delta X + (X \cdot \nabla) y_e + (y_e \cdot \nabla) X = \nabla p && \text{in } (0, \infty) \times \mathscr{O}, \\ & \nabla \cdot X = 0 && \text{in } (0, \infty) \times \mathscr{O}, \\ & X \cdot n = 0, \quad X = u && \text{on } (0, \infty) \times \partial \mathscr{O}, \\ & X(0) = x && \text{in } \mathscr{O}. \end{aligned} \tag{4.104}$$

Our purpose here is to stabilize the null solutions to (4.104) by a noise boundary controller u of the form

$$u = \eta \sum_{i=1}^{N} \frac{\partial \widetilde{\phi}_i}{\partial n} (X, \varphi_i^*) \dot{\beta}_i, \tag{4.105}$$

where N is, as above, the number of eigenvalues λ_j of the operator $\mathscr{A}$ with $\mathrm{Re}\,\lambda_j \le \gamma$ and $\widetilde{\phi}_i$ is defined below. As in the previous case, φ_j^* are the eigenfunctions of $\mathscr{A}^*$ corresponding to $\overline{\lambda}_j$ and $\{\beta_i\}_{i=1}^N$ is an independent system of real Brownian motions in $\{\Omega, \mathbb{P}, \mathscr{F}, \mathscr{F}_t\}$.

Here, we assume that Hypothesis (J1) holds and also that

(J2) *The system* $\{\frac{\partial \varphi_i^*}{\partial n}\}_{i=1}^N$ *is linearly independent in* $(L^2(\partial\mathscr{O}))^d$.

As mentioned in Sect. 3.1, one might suspect that this property is generic in the class of equilibrium solutions y_e, as might be the case with the following weaker version of (J2).

Each $\frac{\partial \varphi_i^*}{\partial n}$ *is not identically zero on* $\partial\mathscr{O}$.

We set

$$\mathscr{L}y = -\nu\Delta y + (y\cdot\nabla)y_e + (y_e\cdot\nabla)y \quad \text{in } \mathscr{O}.$$

Then, as seen in Sect. 3.4, the Stokes–Oseen system

$$\begin{aligned}
&X_t + \mathscr{L}X = \nabla p && \text{in } (0,\infty)\times\mathscr{O},\\
&\nabla\cdot X = 0 && \text{in } (0,\infty)\times\mathscr{O},\\
&X\cdot n = 0,\ \ X = u && \text{on } (0,\infty)\times\partial\mathscr{O},\\
&X(0) = x && \text{in } \mathscr{O},
\end{aligned}$$

can be, equivalently, written as

$$\begin{aligned}
&\frac{d}{dt}X(t) + \widetilde{\mathscr{A}}X(t) = \widetilde{\mathscr{A}_k}Du(t), \quad t \ge 0,\\
&X(0) = x,
\end{aligned} \tag{4.106}$$

where $y = Du$ is the solution to the equation

$$\begin{aligned}
&ky + \mathscr{L}y = \nabla p && \text{in } \mathscr{O},\\
&\nabla\cdot y = 0 && \text{in } \mathscr{O},\\
&y = u,\ \ y\cdot n = 0 && \text{on } \partial\mathscr{O},
\end{aligned}$$

and $k > 0$ is fixed and sufficiently large. ($D : (L^2(\partial\mathscr{O}))^d \to H$ is the Dirichlet map associated with the operator $\mathscr{L} + kI$ and $\widetilde{\mathscr{A}_k} = \mathscr{A} + kI$.)

As seen earlier, $\widetilde{\mathscr{A}} : \widetilde{H} \to (D(A))'$ is the extension by transposition of $\mathscr{A} = P\mathscr{L}$ to all of $\widetilde{H}$ and with values in $(D(A))'$, defined by

$$\widetilde{\mathscr{A}} y(\psi) = \int_{\mathscr{O}} y \overline{\mathscr{A}}^* \psi \, d\xi = (y, \mathscr{A}^* \psi), \quad \forall \psi \in D(\mathscr{A}^*), \ y \in \widetilde{H}.$$

Here, $(D(A))' = (D(\mathscr{A}^*))'$ is the dual of the space $D(A)$ endowed with the graph norm in pairing induced by $\widetilde{H}$ as pivot space; we have $D(A) \subset \widetilde{H} \subset (D(A))'$ algebraically and topologically.

It should be noticed that, in this formulation which is standard in the boundary control theory, the right-hand side of (4.106) is an element of $(D(A))' = (D(\mathscr{A}^*))'$, that is, roughly speaking, it is a "pure" distribution on $\mathscr{O}$ which incorporates the boundary control u. We recall also that the dual $D^* \mathscr{A}_k^*$ of $\mathscr{A}_k D$ is given by (see (3.100))

$$D^* \mathscr{A}_k^* \psi = -\nu \frac{\partial \psi}{\partial n}, \quad \forall \varphi \in D(A).$$

Our aim here is to insert into the controlled system (4.106) a stochastic boundary controller of the form (4.105). Of course, as in the previous closed-loop systems with stochastic noise controller, the expression (4.105) is symbolic and its exact meaning is given by the stochastic differential equation

$$\begin{aligned} &dX(t) + \widetilde{\mathscr{A}} X(t) dt = \eta \sum_{i=1}^{N} \widetilde{\mathscr{A}_k} D \left(\frac{\partial \widetilde{\phi}_i}{\partial n} \right) (X(t), \varphi_i^*) d\beta_i(t), \quad t \geq 0, \\ &X(0) = x. \end{aligned} \tag{4.107}$$

Here, the system $\{\widetilde{\phi}_i\}_{i=1}^N$ is given by $\widetilde{\phi}_i = \sum_{j=1}^N \alpha_{ij} \varphi_j^*$, where α_{ij} are chosen in such a way that

$$\sum_{i=1}^{N} \alpha_{ij} \left(\frac{\partial \varphi_i^*}{\partial n}, \frac{\partial \varphi_k^*}{\partial n} \right)_1 = \delta_{jk}, \quad j, k = 1, \ldots, N.$$

Here, $|\cdot|_1 = |\cdot|_{(L^2(\partial \mathscr{O}))^d}$ and $(u, v)_1 = \int_{\partial \mathscr{O}} \frac{\partial u}{\partial n} \frac{\partial \bar{v}}{\partial n} d\xi$. By Assumption (J2), it is clear that the system $\{\widetilde{\phi}_i\}_{i=1}^N$ is well-defined and

$$(\varphi_i^*, \widetilde{\phi}_j)_1 = \delta_{ij}, \quad i, j = 1, \ldots, N.$$

A "mild" solution X to (4.107) is defined by

$$X(t)=e^{-\mathscr{A}t}x+\eta\int_0^t\sum_{i=1}^N e^{-\widetilde{\mathscr{A}}(t-s)}\left(\widetilde{\mathscr{A}_k}D\left(\frac{\partial\widetilde{\phi}_i}{\partial n}\right)\right)(X(s),\varphi_i^*)d\beta_i(s). \quad (4.108)$$

Equation (4.108) has a unique mild solution $X=X(t)$. This is an $(D(A))'$-valued continuous process which can be viewed as solution to the problem

$$\begin{aligned}
&X_t-\nu\Delta X+(X\cdot\nabla)y_e+(y_e\cdot\nabla)X=\nabla p && \text{in } (0,\infty)\times\mathscr{O},\\
&\nabla\cdot X=0 && \text{in } (0,\infty)\times\mathscr{O},\\
&X(0,\xi)=x(\xi) && \text{in } \mathscr{O},\\
&X=\sum_{i=1}^N\frac{\partial\widetilde{\phi}_i}{\partial n}(X,\varphi_i^*)\dot{\beta}_i && \text{on } (0,\infty)\times\partial\mathscr{O}.
\end{aligned} \quad (4.109)$$

In other words, the boundary controller $u=X|_{\partial\mathscr{O}}$ is a white noise on $\partial\mathscr{O}$. Moreover, since $(\frac{\partial\widetilde{\phi}_i}{\partial n}\cdot n)\cdot n=0$ on $\partial\mathscr{O}$, this stochastic controller is tangential, that is, $X\cdot n=0$ on $(0,\infty)\times\partial\mathscr{O}$.

Theorem 4.5 *Assume that Hypotheses* (J1) *and* (J2) *are satisfied. Then, for* $|\eta|$ *large enough, we have for the solution* X *to* (4.107) (*equivalently,* (4.108))

$$\mathbb{P}\left[\lim_{t\to\infty}e^{\gamma t}\|X(t)\|_{(D(A))'}\right]=0. \quad (4.110)$$

In particular, we have

$$\lim_{t\to\infty}e^{\gamma t}(X(t),\psi)=0,\quad \mathbb{P}\text{-a.s.},\ \forall\psi\in D(A). \quad (4.111)$$

Proof We argue as in the proof of Theorem 4.1. Namely, as in the previous case, we decompose System (4.107) in two parts,

$$\begin{aligned}
&dX_u+\mathscr{A}_uX_udt=\eta\widetilde{P}_N\sum_{i=1}^N\widetilde{\mathscr{A}_k}\left(\frac{\partial\widetilde{\phi}_i}{\partial n}\right)(X,\varphi_i^*)d\beta_i,\quad \mathbb{P}\text{-a.s.},\\
&X_u(0)=P_Nx,
\end{aligned} \quad (4.112)$$

$$\begin{aligned}
&dX_s+\mathscr{A}_sX_sdt=\eta(I-\widetilde{P}_N)\sum_{i=1}^N\widetilde{\mathscr{A}_k}\left(\frac{\partial\widetilde{\phi}_i}{\partial n}\right)(X,\varphi_i^*)d\beta_i,\quad \mathbb{P}\text{-a.s.},\\
&X_s(0)=(I-P_N)x.
\end{aligned} \quad (4.113)$$

Here,

$$\widetilde{P}_N:(D(A))'\to\mathscr{X}_u=\text{lin span}\{\varphi_i\}_{i=1}^N$$

is the projector on $\mathscr{X}_u$ and $\mathscr{A}_u=\widetilde{P}_N\mathscr{A}|_{\mathscr{X}_u}$, $\mathscr{A}_s=(I-\widetilde{P}_N)\mathscr{A}|_{\mathscr{X}_s}$, $\mathscr{X}_s=(I-\widetilde{P}_N)\widetilde{H}$. The operator $\widetilde{\mathscr{A}_s}$ is the extension of $\mathscr{A}_s$ to all of $\widetilde{H}$.

We represent the solution X_u to (4.111) as $X_u = \sum_{j=1}^{N} y_j \varphi_j$ and we obtain for $\{y_j\}_{j=1}^{N}$ the finite-dimensional stochastic system

$$\begin{aligned} &dy_j + \lambda_j y_j dt = \eta y_j d\beta_j, \quad j = 1, \dots, N, \\ &y_j(0) = y_j^0, \end{aligned}$$

which has the solution

$$y_j(t) = e^{-(\lambda_j + \frac{1}{2}\eta^2)t + \eta \int_0^t d\beta_j(s)}, \quad \mathbb{P}\text{-a.s., } \forall t \geq 0.$$

This yields for $\eta^2 > \gamma - 2 \operatorname{Re} \lambda_j$, $j = 1, \dots, N$, that

$$\begin{aligned} &\lim_{t \to \infty} e^{\gamma t} |y(t)| = 0, \quad \mathbb{P}\text{-a.s.,} \\ &\int_0^\infty e^{2\gamma t} |y(t)|^2 dt < \infty, \quad \mathbb{P}\text{-a.s.} \end{aligned}$$

where

$$|y|^2 = \sum_{j=1}^{N} y_j^2, \quad y = X_u.$$

Now, coming back to System (4.113), we write it as

$$dX_s + \widetilde{\mathscr{A}_s} X_s dt = \eta \sum_{i=1}^{N} Y_i(t) d\beta_i, \quad t \geq 0,$$

where

$$Y_i(t) = (I - \widetilde{P}_N)\widetilde{\mathscr{A}_s} D \left(\frac{\partial \widetilde{\phi}_i}{\partial n} \right) (X_u(t), \varphi_i^*), \quad i = 1, \dots, N.$$

Since $\sigma(\mathscr{A}_s) \subset \{\lambda_j;\ \operatorname{Re} \lambda_j > \gamma\}$, it follows that

$$\|e^{-\mathscr{A}_s t} x\|_{(D(A))'} \leq C e^{-\gamma t} \|x\|_{(D(A))'}, \quad \forall x \in (D(A))',$$

and so, by the Lyapunov theorem, there is a self-adjoint, continuous and positive operator $Q = L((D(A))', (D(A))')$ such that

$$\operatorname{Re}(Qx, \widetilde{\mathscr{A}_s} x)_* = \gamma (Qx, x)_* + \frac{1}{2} \|x\|^2_{(D(A))'}, \quad \forall x \in (D(A))',$$

where $(\cdot, \cdot)_*$ is the natural scalar product in $(D(A))'$.

Applying Ito's formula in the above system, we obtain that

$$\frac{1}{2} d(QX_s(t), X_s(t))_* + \frac{1}{2} \|X_s(t)\|^2_{(D(A))'} dt + \gamma (QX_s(t), X_s(t))_* dt$$

$$= \frac{1}{2}\eta^2 \sum_{i=1}^{N} (QY_i(t), Y_i(t)) + Hdt$$

$$+ \eta \sum_{i=1}^{N} ((\mathrm{Re}(QX_s), Y_i)_* + (\mathrm{Im}(QX_s), \mathrm{Im}\, Y_i)_*) d\beta_i.$$

This yields

$$e^{2\gamma t}(QX_s(t), X_s(t))_* + \int_0^t e^{2\gamma s} \|X_s(s)\|^2_{(D(A))'} ds$$

$$= (Q(I - P_N)x, (I - P_N)x)_* + 2\eta \sum_{i=1}^{N} \int_0^t e^{2\gamma s}(QY_i(s), T_i(s))_* ds$$

$$+ 2\eta \sum_{i=1}^{N} \int_0^t e^{2\gamma t}((\mathrm{Re}(QX_s), \mathrm{Im}\, Y_i)_* + (\mathrm{Im}(QX_s), \mathrm{Im}\, Y_i)_*) d\beta_i.$$

Now, we apply Lemma 4.5 to the processes Z, I, I_1, M defined below

$$Z(t) = e^{2\gamma t}(QX_s(t), X_s(t))_*,$$

$$I(t) = \int_0^t e^{2\gamma s} |X_s(s)|^2_{\widetilde{H}} ds,$$

$$I_1(t) = \eta^2 \sum_{i=1}^{N} \int_0^t e^{2\gamma t}(QY_i, Y_i) ds,$$

$$M(t) = 2\eta \sum_{i=1}^{N} \int_0^t e^{2\gamma s}((\mathrm{Re}(QX_s(s), \mathrm{Re}\, Y_i(s))$$

$$+ (\mathrm{Im}(QX_i(s), \mathrm{Im}\, Y_i(s)), \mathrm{Im}\, Y_i(s))) d\beta_j(s), \quad \mathbb{P}\text{-a.s., } t \geq 0.$$

(We note that $M(t)$ is a local martingale and $t \to I(t)$, $t \to I_1(t)$ are increasing processes and so, $t \to e^{2\gamma t}(QX_s(t), X_s(t))_*$ is a semimartingale as the sum of a local martingale and of an adapted process with finite dimension.)

Since, as seen above, $I_1(\infty) < \infty$, we conclude, therefore, that

$$\mathbb{P}\Big[\lim_{t\to\infty} (QX_s(t), X_s(t))_* e^{2\gamma t} = 0\Big] = 1$$

and, since $(Qx, x)_* = 0$ implies $x = 0$, we infer that

$$\mathbb{P}\Big[\lim_{t\to\infty} e^{\gamma t} |X_s(t)|_{(D(A))'} = 0\Big] = 1,$$

as claimed. This completes the proof. □

Remark 4.6 Compared with the boundary stabilization Theorem 3.7, a distinct feature of Theorem 4.4 is that the dissipation mechanism induced by the feedback controller (4.105) is not in the space H but in a much weaker distribution space $(D(A))'$. (We met this also in the case of inpulse noise controllers in Sect. 4.2.) This fact is not necessarily due to the stochastic nature of the controller as to its structure. In fact, the deterministic feedback controller acting in Theorem 3.7 has a regularity margin in t which compensates its singularity with respect to spatial variable. It is exactly what is missing from the stochastic controller (4.105) and this feature is also present in other boundary stabilization problem discussed here.

4.4 Stochastic Stabilization of Periodic Channel Flows by Noise Wall Normal Controllers

We come back here to the laminar flow model in a two-dimensional chan nel with the walls located at $y = 0, 1$ studied in Sect. 3.6.

Recall that the dynamic of flow is governed by the incompressible 2-D Navier–Stokes equation

$$\begin{aligned}
&u_t - \nu\Delta u + uu_x + vu_y = p_x, \quad x \in R,\ y \in (0,1),\\
&v_t - \nu\Delta v + uv_x + vv_y = p_y, \quad x \in R,\ y \in (0,1),\\
&u_x + v_y = 0,\\
&u(t,x,0) = u(t,x,1) = 0,\\
&v(t,x,0) = 0, \quad v(t,x,1) = v^*, \quad \forall x \in R,\\
&u(t,x+2\pi,y) \equiv u(t,x,y), \quad \forall x \in R,\ y \in (0,1),\\
&v(t,x+2\pi,y) \equiv v(t,x,y), \quad y \in (0,1),
\end{aligned} \tag{4.114}$$

and consider a steady-state flow governed by (4.114) with zero vertical velocity component, that is,

$$(U(y),0), U(y) = C(y^2 - y), \quad \forall y \in (0,1),$$

where $C = -\frac{a}{2\nu}$, $a \in R^+$. Recall that we have shown in Theorem 3.12 the boundary stabilization of the linearized system associated with (4.114) and U with a normal feedback controller on the wall $y = 1$.

Our aim here is the stabilization of this flow profile by a noise boundary controller $v(t,x,1) = v^*(t,x)$, $t \geq 0$, $x \in R$.

As seen earlier, the linearization of (4.114) around steady-state flow $(U(y), 0)$ leads to the following system (see (3.166))

$$\begin{aligned}
&u_t - \nu\Delta u + u_x U + vU' = p_x, \quad y \in (0,1),\ x, t \in R,\\
&v_t - \nu\Delta v + v_x U = p_y,\\
&u_x + v_y = 0, \qquad u(t,x,0) = u(t,x,1) = 0,\\
&v(t,x,0) = 0, \qquad v(t,x,1) = v^*(t,x),\\
&u(t, x+2\pi, y) \equiv u(t,x,y),\\
&v(t, x+2\pi, y) \equiv v(t,x,y).
\end{aligned} \tag{4.115}$$

Here, the actuator v^* is a normal velocity boundary controller on the wall $y = 1$. However, there is no actuation in $x = 0$ or inside the channel.

The main result here is that the exponential stability with probability 1 can be achieved by using a finite number M of Fourier modes and a stochastic feedback controller

$$\begin{aligned}
&v^*(t,x) = \sum_{|k|\le M} v_k^*(t) e^{ikx}, \quad t \ge 0,\ x \in R,\\
&v_k^*(t) = -\eta \sum_{j=1}^N \left(\int_0^{2\pi} \int_0^1 (v_{yy}(t,x,y) - k^2 v(t,x,y)) e^{-ikx} (\varphi_j^k)^*(y)\,dx\,dy \right) \dot\beta_j.
\end{aligned} \tag{4.116}$$

Here, $\{\varphi_j^*\}_{j=1}^N$ is a system of functions in $L^2(0,1)$ and $\beta_j(t) = \beta_j^1(t) + i\beta_j^2(t)$ are independent complex Brownian motions in a probability space $\{\Omega, \mathbb{P}, \mathcal{F}, \mathcal{F}_t\}$.

If we adopt the Fourier functional setting of Sect. 3.5, we can represent the solution (u, v) to System (4.115) as $u = \sum_{k\ne 0} u_k(t,y)e^{ikx}$, $v = \sum_{k\ne 0} v_k(t,y)e^{ikx}$, where

$$\begin{aligned}
&(u_k)_t - \nu u_k'' + (\nu k^2 + ikU)u_k + U'v_k = ikp_k, \quad \text{a.e. in } (0,1),\\
&(v_k)_t - \nu v_k'' + (\nu k^2 + ikU)v_k = p_k',\\
&iku_k + v_k' = 0, \quad \text{a.e. on } (0,1),\ k \ne 0,\\
&u_k(t,0) = u_k(t,1) = 0,\\
&v_k(t,0) = 0, \qquad v_k(t,1) = v_k^*(t).
\end{aligned} \tag{4.117}$$

Here,

$$p = \sum_{k\ne 0} p_k(t,y)e^{ikx}, \qquad u = \sum_{k\ne 0} u_k(t,y)e^{ikx}, \qquad v = \sum_{k\ne 0} v_k(t,y)e^{ikx}.$$

This yields (see (3.168))

$$\begin{aligned}
&(v_k'' - k^2 v_k)_t - \nu v_k^{\mathrm{iv}} + (2\nu k^2 + ikU)v_k'' \\
&\qquad - k(\nu k^3 + ik^2 U + iU'')v_k = 0, \quad t \ge 0,\ y \in (0,1), \\
&v_k'(t,0) = v_k'(t,1) = 0, \\
&v_k(t,0) = 0, \qquad v_k(t,1) = v_k^*(t).
\end{aligned} \tag{4.118}$$

In the following, we denote one more time by H the complexified space $L^2(0,1)$ with the norm $|\cdot|$ and the product scalar denoted by $(\cdot,\cdot)$. We denote by $H^m(0,1)$, $m = 1, 2, 3$, the standard Sobolev spaces on $(0,1)$ and

$$H_0^1(0,1) = \{v \in H^1(0,1);\ v(0) = v(1) = 0\},$$

$$H_0^2(0,1) = \{v \in H^2(0,1) \cap H_0^1(0,1);\ v'(0) = v'(1) = 0\}.$$

We set $\mathscr{H} = H^4(0,1) \cap H_0^2(0,1)$ and denote by $\mathscr{H}'$ the dual of $\mathscr{H}$ in the pairing with pivot space H, that is $\mathscr{H} \subset H \subset \mathscr{H}'$ algebraically and topologically. Denote by $(H^2(0,1))'$ the dual of $H^2(0,1)$ and by $H^{-1}(0,1)$ the dual of $H_0^1(Q)$ with the norm $\|\cdot\|_{-1}$. Denote also by $H_\pi^{-1}(Q)$ the space $L^2(0,2\pi; H^{-1}(0,1))$ with the norm $\|\cdot\|_{H_\pi^{-1}(Q)}$.

We recall some notation and results from Sect. 3.5 related to (4.118).

For each $k \in R$, we denote by $L_k : D(L_k) \subset H \to H$ and $F_k : D(F_k) \subset H \to H$ the operators

$$L_k v = -v'' + k^2 v, \quad v \in D(L_k) = H_0^1(0,1) \cap H^2(0,1), \tag{4.119}$$

$$\begin{aligned}
&F_k v = \nu v^{\mathrm{iv}} - (2\nu k^2 + ikU)v'' + k(\nu k^3 + ik^2 U + iU'')v, \\
&\qquad \forall v \in D(F_k) = H^4(0,1) \cap H_0^2(0,1),
\end{aligned} \tag{4.120}$$

$$\mathscr{F}_k v = \nu v^{\mathrm{iv}} - (2\nu k^2 + ikU)v'' + k(\nu k^3 + ik^2 U + iU'')v,$$

and consider the solution V_k of the equation

$$\begin{aligned}
&\theta V_k + \mathscr{F}_k V_k = 0, \quad y \in (0,1), \\
&V_k'(0) = V_k'(1) = 0, \qquad V_k(0) = 0, \qquad V_k(1) = v_k^*(t).
\end{aligned} \tag{4.121}$$

As seen earlier, for θ positive and sufficiently large, there is a unique solution V_k to (4.121). Then, subtracting (4.118) and (4.121), we obtain that

$$(L_k v_k)_t + F_k(v_k - V_k) - \theta V_k = 0, \quad t \ge 0.$$

Equivalently,

$$(L_k(v_k - V_k))_t + F_k(v_k - V_k) = \theta V_k - (L_k V_k)_t, \quad v_k - V_k \in D(F_k). \tag{4.122}$$

(Here, $L_k V_k$ is a distribution on $(0,1)$, which is made precise below.)

We consider the operator $\mathscr{A}_k : D(\mathscr{A}_k) \subset H \to H$ defined by

$$\mathscr{A}_k = F_k L_k^{-1}, \qquad D(\mathscr{A}_k) = \{u \in H;\ L_k^{-1} u \in D(F_k)\}. \tag{4.123}$$

We have by Lemma 3.6 that $-\mathscr{A}_k$ generates a C_0-analytic semigroup on H and, for each $\lambda \in \rho(-\mathscr{A}_k)$, $(\lambda I + \mathscr{A}_k)^{-1}$ is compact. Moreover, one has, for each $\gamma > 0$,

$$\sigma(-\mathscr{A}_k) \subset \{\lambda \in \mathbb{C};\ \operatorname{Re}\lambda \le -\gamma\},$$

$$\forall |k| \ge M = \frac{1}{\sqrt{2\nu}}\left(\gamma + 1 + \frac{a}{\sqrt{2\nu}}\right)^{\frac{1}{2}}. \tag{4.124}$$

In particular, it follows that, for $|k| \ge M$, we have

$$\|e^{-\mathscr{A}_k t}\|_{L(H,H)} \le C e^{-\gamma t}, \quad \forall t \ge 0,$$

and by (3.178)

$$\int_0^1 (|v_k'(t,y)|^2 + k^2 |v_k(t,y)|^2)dy$$
$$\le C e^{-\nu k^2 t} \int_0^1 (|v_k'(0,y)|^2 + k^2 |v_k(0,y)|^2)dy, \quad t \ge 0, \tag{4.125}$$

for $|k| \ge M$.

Now, coming back to System (4.122), we set

$$\tilde{z}_k(t) = L_k(v_k(t) - V_k(t)) \tag{4.126}$$

and write it as (see (3.180))

$$\begin{aligned}\tilde{z}_k(t) &= e^{-\mathscr{A}_k t}\tilde{z}_k(0) + \int_0^t e^{-\mathscr{A}_k(t-s)}(\theta V_k(s) - (L_k V_k(s))_s)ds \\ &= e^{-\mathscr{A}_k t} z_k(0) - L_k V_k(t) + e^{-\mathscr{A}_k t} L_k V_k(0) \\ &\quad + \int_0^t e^{-\mathscr{A}_k(t-s)}(\theta V_k(s) + \widetilde{F}_k V_k(s))ds,\end{aligned} \tag{4.127}$$

where $\widetilde{F}_k : H \to \mathscr{H}'$ is the extension of F_k to all of H defined by

$${}_{\mathscr{H}'}\langle \widetilde{F}_k v, \psi\rangle_{\mathscr{H}} = \int_0^1 v(y) F_k^* \psi(y) dy, \quad \forall \psi \in D(F_k^*),$$

and

$$F_k^* = \nu\psi^{\mathrm{iv}} - ((\nu k^2 - ik) \cup \psi)'' + (k - ik^2 U - iU'')\psi,$$
$$D(F_k^*) = H^4(0,1) \cap H_0^2(0,1).$$

Then, we have

$$\frac{d}{dt} z_k(t) + \widetilde{\mathscr{A}}_k z_k(t) = (\theta + \widetilde{F}_k) V_k(t), \quad t \geq 0. \tag{4.128}$$

For each $u \in R$, we denote by $V = D_k u \in H^4(0,1)$ the solution to the equation

$$\begin{aligned} &\theta V + \mathscr{F}_k V = 0, \quad \forall y \in (0,1), \\ &V'(0) = V'(1) = 0, \qquad V(0) = 0, \qquad V(1) = u. \end{aligned} \tag{4.129}$$

It is easily seen that the dual $((\theta + \widetilde{F}_k) D_k)^*$ is given by (see (3.183))

$$((\theta + \widetilde{F}_k) D_k)^* \varphi = \nu \varphi'''(1), \quad \forall \varphi \in D(F_k), \tag{4.130}$$

and so, (4.128) can be rewritten as

$$\frac{d}{dt} z_k(t) + \widetilde{\mathscr{A}}_k z_k(t) = (\theta + \widetilde{F}_k) D_k v_k^*(t), \quad \forall t \geq 0. \tag{4.131}$$

4.4.1 Feedback Stabilization

Let $\gamma > 0$ and let $k \in R$, $|k| \leq M$ given by (4.124). Then, the operator $-\mathscr{A}_k$ has a finite number $N = N_k$ of the eigenvalues $\lambda_j = \lambda_j^k$ with $\operatorname{Re} \lambda_j \geq -\gamma$. (In the following, we omit the index k from $\mathscr{A}_k$ and λ_j^k.)

We denote by $\{\varphi_j^k\}_{j=1}^N$ the corresponding eigen functions and repeat each λ_j according to its algebraic multiplicity m_j. We have $\mathscr{A} \varphi_j^k = -\lambda_j^k \varphi_j^k$, $j = 1, \dots, N$, and recall that the geometric multiplicity of λ_j is the dimension of the eigenfunction space corresponding to λ_j. The algebraic multiplicity of λ_j^k is the dimension of the range of the projection operator

$$P_j = \frac{1}{2\pi i} \int_{\Gamma_j} (\lambda I + \mathscr{A})^{-1} d\lambda,$$

where Γ_j is a smooth closed curve encircling λ_j^k. We assume, as in the previous cases, that the following assumption holds.

(J1) *All the eigenvalues $\lambda_j = \lambda_j^k$ with $1 \leq j \leq N$ are semisimple.*

If we denote by $(\varphi_j^k)^*$ the eigenfunctions to the dual operator $-\mathscr{A}^*$, that is,

$$\mathscr{A}^* (\varphi_j^k)^* = -\bar{\lambda}_j^k (\varphi_j^k)^*, \quad j = 1, \dots, N,$$

the system $\{(\varphi_j^k)^*\}$ can be chosen in such a way that

$$\left\langle \varphi_\ell^k, (\varphi_j^k)^* \right\rangle = \delta_{\ell j}, \quad \ell, j = 1, \dots, N. \tag{4.132}$$

We denote by $\mathscr{X}_u$ the space generated by $\{\varphi_j^k\}_{j=1}^N$ and $\mathscr{X}_s = (I - P_N)H$. We consider in (4.131) (equivalently (4.128)) the feedback controller

$$v_k^*(t) = \eta \sum_{\ell=1}^{N} \left\langle L_k v_k(t), (\varphi_j^k)^* \right\rangle \dot{\beta}_j(t), \quad t \geq 0, \tag{4.133}$$

where $\{\beta_j\}_{j=1}^N$ are independent complex Brownian and $\dot{\beta}_j$ is the white noise associated with β_j. More precisely, we take $\beta_j = \beta_j^1 + i\beta_j^2$, where $\{\beta_j^\ell\}_{j=1}^N$ are independent real Brownian motions. Then, we are lead to the stochastic closed-loop system

$$\begin{aligned} &d(L_k v_k(t)) + \widetilde{\mathscr{A}}_k(L_k v_k(t))dt = \eta \sum_{j=1}^{N} (\theta + \widetilde{F}_k) D_k \left\langle L_k v_k(t), (\varphi_j^k)^* \right\rangle d\beta_j, \\ &(L_k v_k)(0) = L_k v_k^0, \end{aligned} \tag{4.134}$$

which represents the abstract and rigorous formulation of the boundary closed-loop stochastic system

$$\begin{aligned} &d(L_k v_k(t)) + \mathscr{F}_k v_k(t)dt = 0, \quad t \geq 0, \\ &v_k(t,1) = \eta \sum_{j=1}^{N} \left\langle L_k v_k(t), (\varphi_j^k)^* \right\rangle \dot{\beta}_j(t). \end{aligned} \tag{4.135}$$

The feedback controller (4.133) can be, equivalently, expressed in term of normal velocity v as (see (4.116))

$$\begin{aligned} &v_k(t,1) = -\eta \sum_{j=1}^{N} \left(\int_0^{2\pi} \int_0^1 (v_{yy}(t,x,y) - k^2 v(t,x,y)) e^{-ikx} (\varphi_j^k)^*(y) dx\, dy \right) \dot{\beta}_j(t), \\ &|k| \leq M, \end{aligned} \tag{4.136}$$

$$v_k(t,1) = 0 \quad \text{for } |k| > M.$$

Equation (4.134) should be viewed in the following mild sense,

$$\begin{aligned} v_k(t) &= L_k^{-1} e^{-\mathscr{A}_k t} (L_k v_k^0) \\ &\quad + \eta \sum_{j=1}^{N} L_k^{-1} \int_0^t e^{-\mathscr{A}_k(t-s)} (\widetilde{F}_k + \theta) D_k \left\langle L_k v_k(s), (\varphi_j^k)^* \right\rangle d\beta_j(s) \end{aligned}$$

and has a unique solution $v_k \in (C(0,\infty); L^2(\Omega, H_0^1(0,1)))$ which is adapted to the filtration $\mathscr{F}_t$ (see Sect. 4.5).

We have the following result.

Theorem 4.6 *For $|k| \leq M$ and $|\eta|$ sufficiently large, we have, for $0 < \delta < \frac{1}{2}$,*

$$\begin{aligned} &\mathbb{P}\Big[\lim_{t\to\infty} e^{\gamma t}\|v_k(t)\|_{-1} = 0\Big] = 1, \\ &E\int_0^\infty e^{2\gamma\delta t}|v_k(t)|_{-1}^2 dt \leq C|v_k(0)|_{-1}^2. \end{aligned} \tag{4.137}$$

Taking into account (4.117) and (4.125), we obtain by Theorem 4.6 the exponential stabilization of System (4.115) with the feedback controller (4.136).

Theorem 4.7 *Under the assumptions of Theorem* 4.6, *the solution*

$$\begin{aligned} u(t,x,y) &= \sum_{k\neq 0} u_k(t,y)e^{ikx}, \\ v(t,x,y) &= \sum_{k\neq 0} v_k(t,y)e^{ikx}, \quad t\geq 0,\ x\in R,\ y\in(0,1), \end{aligned} \tag{4.138}$$

to (4.115) *with the boundary feedback controller* (4.136) *is exponentially stable with probability one. Namely, one has*

$$\mathbb{P}\Big[\lim_{t\to\infty} e^{\gamma t}(\|u(t)\|_{H_\pi^{-1}(Q)} + \|v(t)\|_{H_\pi^{-1}(Q)}) = 0\Big] = 1, \tag{4.139}$$

and

$$\begin{aligned} &E\int_0^\infty e^{2\gamma\delta t}\left(\|u(t)\|_{H_\pi^{-1}(\mathcal{O})}^{2\delta} + \|v(t)\|_{H_\pi^{-1}(\mathcal{O})}^{2\delta}\right)dt \\ &\quad\leq C\left(\|u(0)\|_{H_\pi^{-1}(\mathcal{O})}^{2\delta} + \|v(0)\|_{H_\pi^{-1}(\mathcal{O})}^{2\delta}\right). \end{aligned}$$

Everywhere below, we omit the exponent k from φ_j^k and $(\varphi_j^k)^*$.

4.4.2 Proof of Theorem 4.6

We set $y = L_k v_k$ and represent it as

$$y = P_N y + (I - P_N)y, \quad P_N y = \sum_{j=1}^N y_j\varphi_j.$$

Then, by virtue of (4.132), System (4.134) can be rewritten as

$$\begin{aligned} &dy_\ell + \lambda_\ell y_\ell dt = \eta((\widetilde{F}_k+\theta)D_k)^*\varphi_\ell^* \sum_{j=1}^N y_j d\beta_j, \\ &y_\ell(0) = \langle P_N L_k v_k(0), \varphi_\ell^*\rangle = y_\ell^0, \quad \ell = 1,\dots,N, \end{aligned} \tag{4.140}$$

and

$$dy^s + \widetilde{\mathscr{A}_k^s}\, y^s dt = \eta(I - P_N)\sum_{j=1}^{N}(\widetilde{F}_k + \theta)D_k(y_j)d\beta_j,$$
$$y^s(0) = (I - P_N)L_k v_k(0), \tag{4.141}$$

where $y^s = (I - P_N)y$, $\mathscr{A}_k^s = (I - P_N)\mathscr{A}_k$ and $\widetilde{\mathscr{A}_k^s}$ is the extension of $\mathscr{A}_k^s$ to all of H. (When there is no danger of confusion, we omit $\sim$.) Taking into account that

$$\overline{\lambda}_\ell L_k\varphi_\ell^* + F_k^*\varphi_\ell^* = 0, \quad \ell = 1, \dots, N, \tag{4.142}$$

and, therefore, $\varphi_\ell^* \in D(F_k^*) = D(F_k)$, we see by (4.130) that

$$\zeta_\ell = ((\widetilde{F}_k + \theta)D_k)^*\varphi_\ell^* = \nu(\varphi_\ell^*)'''(1), \quad \ell = 1, \dots, N.$$

Then, by virtue of (4.142), we have that

$$|\zeta_\ell| \geq \rho > 0, \quad \forall \ell = 1, 2, \dots, N. \tag{4.143}$$

Indeed, by (4.142), we see that

$$(\varphi_\ell^*)'''(1)\varphi(1) - (\varphi_\ell^*)''(1)\varphi'(1) = 0,$$

for all the solutions $\varphi \in H^4(0, 1)$ to the equation

$$\mathscr{F}_k\varphi + \lambda(-\varphi'' + k^2\varphi) = 0,$$
$$\varphi(0) = \varphi'(0) = 0,$$

and taking into account that

$$\varphi_\ell^*(0) = \varphi_\ell^*(1) = 0,$$
$$(\varphi_\ell^*)'(0) = (\varphi_\ell^*)'(1) = 0,$$

it follows that $(\varphi_\ell^*)''(1) = 0$ and, therefore, $(\varphi_\ell^*)'''(1) \neq 0$, unless $\varphi_\ell^* \equiv 0$. Since the latter is absurd, we have (4.143).

We rewrite (4.140) as

$$dy_\ell + \lambda_\ell y_\ell dt = \eta\zeta_\ell \sum_{j=1}^{N} y_j d\beta_j, \quad \ell = 1, \dots, N,$$
$$y_\ell(0) = y_\ell^0. \tag{4.144}$$

Lemma 4.4 *For $|\eta|$ sufficiently large, we have*

$$\lim_{t\to\infty}\left(\sum_{\ell=1}^{N}|y_\ell(t)|e^{\gamma t}\right) = 0, \quad \mathbb{P}\text{-a.s.} \tag{4.145}$$

$$\int_0^\infty e^{2\gamma t} \sum_{\ell=1}^N |y_\ell(t)|^2 dt < \infty, \quad \mathbb{P}\text{-a.s.}, \tag{4.146}$$

$$E \int_0^\infty e^{2\gamma\delta t} \sum_{\ell=1}^N |y_\ell(t)|^{2\delta} dt \le C \sum_{\ell=1}^N |y_\ell(0)|^{2\delta}. \tag{4.147}$$

Proof We prove (4.145) and (4.146) for $\gamma = 0$ because the general case follows from this by substituting y_e into (4.144) by $y_e e^{\gamma t}$ taking into account that $\operatorname{Re}\lambda_\ell \le -\gamma$ for $\ell = 1, \dots, N$.

We apply in (4.144) Ito's formula to the function $y \to \frac{1}{2}|y|^2$. We get

$$\begin{aligned}
&\frac{1}{2} d|y_\ell(t)|^2 + \operatorname{Re}\lambda_\ell |y_\ell(t)|^2 dt \\
&= 2\eta^2 |\zeta_\ell|^2 \sum_{j=1}^N |y_j|^2 dt \\
&\quad + \eta \sum_{j=1}^N (\operatorname{Re}(\zeta_\ell y_\ell) \operatorname{Re} y_j + \operatorname{Im}(\zeta_\ell y_\ell) \operatorname{Im} y_j) d\beta_j^1 \\
&\quad + \eta \sum_{j=1}^N (\operatorname{Re}(\zeta_\ell y_\ell) \operatorname{Im} y_j - \operatorname{Im}(\zeta_\ell y_\ell) \operatorname{Re} y_j) d\beta_j^2, \quad \ell = 1, \dots, N.
\end{aligned}$$

Equivalently,

$$\begin{aligned}
&dz_\ell(t) + 2 \operatorname{Re}\lambda_\ell z_\ell(t) dt \\
&= \eta^2 |\zeta_\ell|^2 \sum_{j=1}^N z_j dt \\
&\quad + 2\eta \sum_{j=1}^N ((\operatorname{Re}\zeta_\ell y_\ell) \operatorname{Re} R y_j + \operatorname{Im}(\zeta_\ell y_\ell) \operatorname{Im} y_j) d\beta_j^1 \\
&\quad + \eta \sum_{j=1}^N (\operatorname{Re}(\zeta_\ell y_\ell) \operatorname{Im} y_j - \operatorname{Im}(\zeta_\ell y_\ell) \operatorname{Re} y_j) d\beta_j^2,
\end{aligned} \tag{4.148}$$

where $z_\ell = |y_\ell|^2$. In (4.148), we apply once again Ito's formula to the function $\psi(z) = z^\delta$, where $0 < \delta < \frac{1}{2}$. We have

$$\begin{aligned}
&dz_\ell^\delta(t) + 2\delta \operatorname{Re}\lambda_\ell z_\ell^\delta(t) dt \\
&= 2\delta\eta^2 |\zeta_\ell|^2 \sum_{j=1}^N z_j(t) z_\ell^{\delta-1}(t) dt
\end{aligned}$$

$$+\delta(\delta-1)\eta^2|\zeta_\ell|^2\sum_{j=1}^N z_j(t)z_\ell^{\delta-1}(t)dt$$

$$+2\eta\delta\sum_{j=1}^M (M_{j\ell}^1(t)d\beta_j^1(t)+M_{j\ell}^2(t)d\beta_j^2(t)),\quad \ell=1,\dots,N, \tag{4.149}$$

where

$$M_{j\ell}^1=\mathrm{Re}(\zeta_\ell y_\ell)\,\mathrm{Re}\,y_j+\mathrm{Im}(\zeta_\ell y_\ell)\,\mathrm{Im}\,y_j,$$
$$M_{j\ell}^2=\mathrm{Re}(\zeta_\ell y_\ell)\,\mathrm{Im}\,y_j-\mathrm{Im}(\zeta_\ell y_\ell)\,\mathrm{Re}\,y_j.$$

(The previous calculation is somewhat formal because ψ is not of class C^2. However, it can be made rigorous, as in the proof of Theorem 4.1, if we replace ψ by $\psi_\varepsilon(z)=(\varepsilon+z)^\delta$ and let $\varepsilon\to 0$.)

Then, (4.149) yields, for all $\ell=1,\dots,N$,

$$d|y_\ell(t)|^{2\delta}+2\delta\,\mathrm{Re}\,\lambda_\ell|y_\ell(t)|^{2\delta}dt+\delta\eta^2|\zeta_\ell|^2(1-2\delta)\sum_{j=1}^N|y_j(t)|^2|y_\ell(t)|^{2(\delta-1)}dt$$

$$=2\eta\delta\sum_{j=1}^N(M_{j\ell}^1(t)d\beta_j^1(t)+M_{j\ell}^2(t)d\beta_j^2(t))|y_\ell(t)|^{2(\delta-1)}.$$

Finally,

$$d\sum_{\ell=1}^N|y_\ell(t)|^{2\delta}+2\delta\sum_{\ell=1}^N\mathrm{Re}\,\lambda_\ell|y_\ell(t)|^{2\delta}dt$$
$$+\delta\eta^2(1-2\delta)\sum_{j=1}^N|y_j(t)|^2\sum_{\ell=1}^N|\zeta_\ell|^2|y_\ell(t)|^{2(\delta-1)}dt$$
$$=2\eta\delta\sum_{\ell=1}^N\sum_{j=1}^N(M_{j\ell}^1(t)d\beta_j^1(t)+M_{j\ell}^2(t)d\beta_j^2(t))|y_\ell(t)|^{2(\delta-1)}. \tag{4.150}$$

Recalling (4.143) and that $0<\delta<\frac{1}{2}$, we see by (4.150) that

$$E\sum_{\ell=1}^N|y_\ell(t)|^{2\delta}+2\delta E\sum_{\ell=1}^N\mathrm{Re}\,\lambda_\ell\int_0^t|y_\ell(s)|^{2\delta}ds$$
$$+\delta\eta^2(1-2\delta)\rho^2E\int_0^t\sum_{\ell=1}^N|y_\ell(s)|^{2(\delta-1)}\sum_{j=1}^N|y_j(s)|^2ds$$

$$\le \sum_{\ell=1}^{N} |y_\ell(0)|^{2\delta}, \quad \mathbb{P}\text{-a.s.}, \ t \ge 0,$$

and, therefore, for

$$\eta^2 > \frac{2}{(1-2\delta)\rho^2} \max_{1\le\ell\le N} \{\operatorname{Re}\lambda_\ell\},$$

we have

$$E\sum_{\ell=1}^{N} |y_\ell(t)|^{2\delta} + \gamma_0 E \int_0^t \sum_{\ell=1}^{N} |y_\ell(s)|^{2\delta} ds \le \sum_{\ell=1}^{N} |y_\ell(0)|^{2\delta} \tag{4.151}$$

where $\gamma_0 > 0$ and, therefore,

$$E\int_0^\infty \sum_{\ell=1}^{N} |y_\ell(s)|^{2\delta} ds < \sum_{\ell=1}^{N} |y_\ell(0)|^{2\delta}. \tag{4.152}$$

We set

$$Z(t) = \sum_{\ell=1}^{N} |y_\ell(t)|^{2\delta},$$

$$I(t) = \delta\eta^2(1-2\delta) \int_0^t \sum_{j=1}^{N} |y_j(s)|^2 \sum_{\ell=1}^{N} |\zeta_\ell|^2 |y_\ell(s)|^{2(\delta-1)} ds,$$

$$I_1(t) = -2\delta \sum_{\ell=1}^{N} \operatorname{Re}\lambda_\ell \int_0^t |y_\ell(s)|^{2\delta} ds,$$

$$M(t) = 2\eta\delta \int_0^t \sum_{\ell=1}^{N}\sum_{j=1}^{N} (M_{j\ell}^1(s)d\beta_j^1(s) + M_{j\ell}^2(s)d\beta_j^2(s))|y_\ell(s)|^{2(\delta-1)},$$

and rewrite (4.150) as

$$Z(t) + I(t) = Z(0) + I_1(t) + M(t), \quad \mathbb{P}\text{-a.s.} \tag{4.153}$$

Since $I(t)$ and $I_1(t)$ are nondecreasing and $M(t)$ is a semimartingale, it follows by the martingale convergence theorem (see Lemma 4.5) that there is

$$\lim_{t\to\infty} Z(t) < \infty, \quad \mathbb{P}\text{-a.s.}$$

Then, by (4.152), we see that

$$\lim_{t\to\infty} Z(t) = 0, \quad \mathbb{P}\text{-a.s.}$$

This completes the proof of Lemma 4.4. □

Proof of Theorem 4.6 (continued). Coming back to (4.141), we note that, since $\sigma(-\mathscr{A}_k^s) \subset \{\lambda;\ \operatorname{Re}\lambda \le -\gamma\}$, we have that

$$\|e^{-\mathscr{A}_k^s t}\|_{L(H,H)} \le Ce^{-\gamma t}, \quad \forall t \ge 0. \tag{4.154}$$

We have

$$\begin{aligned} y^s(t) &= e^{-\mathscr{A}_k^s t}(I - P_N)L_k v_k^0 \\ &\quad + \eta \sum_{j=1}^N \int_0^t e^{-\mathscr{A}_k^s (t-s)}(I - P_N)(\widetilde{F}_k + \theta)D_k[y_j(s)])d\beta_j(s), \quad \forall t \ge 0. \end{aligned} \tag{4.155}$$

Recalling that $\mathscr{A}_k = F_k L_k^{-1}$, we have by (4.155) that

$$\begin{aligned} y^s(t) &= e^{-\mathscr{A}_k^s t}(I - P_N)L_k v_k^0 \\ &\quad + \eta \mathscr{A}_k^s \sum_{j=1}^N \int_0^t e^{-\mathscr{A}_k^s (t-s)}(I - P_N)L_k D_k(y_j(s))d\beta_j(s) \\ &\quad + \eta\theta \sum_{j=1}^N \int_0^t e^{-\mathscr{A}_k^s (t-s)}(I - P_N)D_k(y_j(s))d\beta_j(s). \end{aligned}$$

Hence

$$\begin{aligned} (\theta + \mathscr{A}_k)^{-1} y^s(t) &= (\theta + \mathscr{A}_k)^{-1} e^{-\mathscr{A}_k^s t}(I - P_N)L_k v_k^0 \\ &\quad + \eta \sum_{j=1}^N \int_0^t e^{-\mathscr{A}_k^s (t-s)}(I - P_N)L_k(D_k(y_j(s)))d\beta_j(s). \end{aligned} \tag{4.156}$$

(We may take θ sufficiently large such that $(\theta + \mathscr{A}_k)^{-1} \in L(H,H)$.)

We set $X_s(t) = (\theta + \mathscr{A}_k)^{-1} y^s(t)$ and rewrite (4.156) as

$$\begin{aligned} &dX_s(t) + \mathscr{A}_k^s X_s(t)dt = \eta(I - P_N)\sum_{j=1}^N L_k(D_k y_j)d\beta_j(t), \\ &X_s(0) = (\theta + \mathscr{A}_k)^{-1}(I - P_N)L_k v_k^0. \end{aligned} \tag{4.157}$$

Since, as seen earlier, the operator $-\mathscr{A}_k^s$ generates a γ-exponentially stable C_0-semigroup on H and in the space $H^{-1}(0,1) = H^{-1}$ too, which we endow with the scalar product $\langle y, z\rangle_{-1} = \langle L_k^{-1} y, z\rangle$ and with the corresponding norm $|\cdot|_{-1}$. Then, by the Lyapunov theorem there is $Q \in L(H^{-1}, H^{-1})$, $Q = Q^* \ge 0$, such that

$$\operatorname{Re}\langle Qx, \mathscr{A}_k^s x - \gamma x\rangle_{-1} = \frac{1}{2}|x|_{-1}^2, \quad \forall x \in D(\mathscr{A}_k^s).$$

We note that Q is positively definite in the sense that $\inf\{\langle Qx,x\rangle;\ |x|=1\}>0$.

Applying Ito's formula in (4.156) to the function $\varphi(x)=\frac{1}{2}\langle Qx,x\rangle_{-1}$, we obtain

$$\frac{1}{2}d\langle QX_s(t),X_s(t)\rangle_{-1}+\frac{1}{2}|X_s(t)|_{-1}^2dt+\gamma\langle QX_s(t),X_s(t)\rangle_{-1}dt$$
$$=\frac{1}{2}\eta^2\sum_{j=1}^N\langle QY_j(t),Y_j(t)\rangle_{-1}dt+\eta dM_0(t),$$

where $Y_j(t)=(I-P_N)L_k(D_ky_j(t))$ and

$$dM_0(t)=\sum_{j=1}^N\left(\left\langle\operatorname{Re}QX_s(t),\operatorname{Re}Y_j(t)\right\rangle_{-1}+\left\langle\operatorname{Im}QX_s(t),\operatorname{Im}Y_j(t)\right\rangle_{-1}\right)d\beta_j^1(t)$$
$$+\sum_{j=1}^N\left(\left\langle\operatorname{Re}QX_s(t),\operatorname{Im}Y_j(t)\right\rangle_{-1}-\left\langle\operatorname{Im}QX_s(t),\operatorname{Re}Y_j(t)\right\rangle_{-1}\right)d\beta_j^2(t).$$

By Lemma 4.5, it follows that $\int_0^\infty|Y_j(t)|_{-1}^2e^{2\gamma t}dt<\infty$, $\mathbb{P}$-a.s. This yields

$$e^{2\gamma t}\langle QX_s(t),X_s(t)\rangle_{-1}+\int_0^t e^{2\gamma s}|X_s(s)|_{-1}^2ds$$
$$=\langle Q(I-P_N)x,(I-P_N)x\rangle_{-1}$$
$$+\eta^2\sum_{j=1}^N\int_0^t e^{2\gamma s}\left\langle QY_j(s),Y_j(s)\right\rangle_{-1}ds+2\eta\sum_{j=1}^N\int_0^t e^{2\gamma s}dM_0(s),$$

$\mathbb{P}$-a.s., $t\geq 0$.

Then, once again by Lemma 4.5, where

$$Z(t)=e^{2\gamma t}\langle QX_s(t),X_s(t)\rangle_{-1},$$
$$I(t)=\int_0^t e^{2\gamma s}|X_s(s)|_{-1}^2ds,\qquad I_1(t)=\eta^2\sum_{j=1}^N\int_0^t e^{2\gamma s}\left\langle QY_j,Y_j\right\rangle_{-1}ds,$$
$$M(t)=2\eta\sum_{j=1}^N\int_0^t e^{2\gamma s}dM_0(s),\quad \mathbb{P}\text{-a.s., } t\geq 0,$$

we infer that $\lim_{t\to\infty}e^{2\gamma t}\langle QX_s(t),X_s(t)\rangle_{-1}=0$, $\mathbb{P}$-a.s. This yields

$$\lim_{t\to\infty}(e^{\gamma t}|X_s(t)|_{-1})=0,\quad \mathbb{P}\text{-a.s.}$$

and, therefore,

$$\lim_{t\to\infty}(e^{\gamma t}|(\theta+\mathscr{A}_k)^{-1}L_kv_k(t)|_{-1})=0,\quad \mathbb{P}\text{-a.s.}\tag{4.158}$$

Taking into account that, by (4.123), $(\theta - \mathscr{A}_k)^{-1}L_k$ is an isomorphism from $H^{-1}(0,1)$ to itself, we infer by (4.152), (4.155) that (4.137) holds, thereby completing the proof of Theorem 4.6. □

Remark 4.7 As in the case of Theorem 4.3 the feedback controller designed here is robust with respect to deterministic perturbations of the system as well as to small stochastic perturbations. As in the case of other stabilizable noise controllers, this property is due to its stochastic construction whose contribution to the resulting system is a linear positive definite term which induces a robust dissipative mechanism. It should be also said that, as in the case of the impulse distributed controller considered in the previous section, the boundary noise controller implies weak stabilization only, and the motivation here is also that the boundary noise controller, likewise the impulse noise controllers, are "meager controllers" compared with normal boundary controllers designed in Theorem 4.4, which are stabilizable in the strong topology.

4.5 Stochastic Processes

A probability space is a triple $(\Omega, \mathscr{F}, \mathbb{P})$, where Ω is the set of "events", $\mathscr{F}$ is a σ-algebra of subsets of Ω and $\mathbb{P} : \mathscr{F} \to [0, \infty)$ is a probability measure that is a measure taking values in the interval $[0, 1]$, which assigns "probabilities" to events of Ω and $\mathbb{P}(\Omega) = 1$.

A random variable $X : \Omega \to H$, where H is a Banach space, is a measurable function on Ω, that is, $X^{-1}(B) \subset \mathscr{F}$ for each Borelian set B of H. A family $\{X(t);\ t \geq 0\}$ of random variables is called a stochastic process. We call $E(X) = \int_\Omega X\, d\mathbb{P}$, the expectation of the random variable X.

If $\mathscr{V}$ is a σ-algebra, $\mathscr{V} \subset \mathscr{F}$, then for each integrable random variable X (with respect to the measure $d\mathbb{P}$), one defines and denotes by $E(X|\mathscr{V})$ a random variables on Ω such that

(a) $E(X|\mathscr{V})$ is $\mathscr{V}$-measurable
(b) $\int_A X\, d\mathbb{P} = \int_A E(X|\mathscr{V}) d\mathbb{P}$, $\forall A \in \mathscr{V}$

and calls it the *conditional expectation* of X with respect to σ-algebra $\mathscr{V}$.

Definition 4.1 Let $X(t)$ be a stochastic process such that $E|X(t)| < \infty$ for all $t \geq 0$. It is called a *martingale* if

$$X(s) = E(X(t)|\mathscr{U}(s)), \quad \mathbb{P}\text{-a.s., for } t \geq s > 0,$$

and a *submartingale* (*supermartingale*)

$$X(s) \leq E(X(t)|\mathscr{U}(s)), \quad \mathbb{P}\text{-a.s., for all } t \geq s > 0,$$

respectively,

$$X(s) \geq E(X(t)|\mathscr{U}(s)), \quad \forall s \leq t.$$

Here, $\mathscr{U}(s) = \sigma(X(\tau); 0 \le \tau \le s)$ is the σ-algebra generated by the random variables $X(\tau)$ for $0 \le \tau \le s$.

A collection of σ-algebra $\{\mathscr{F}_t\}$ satisfying $\mathscr{F}_s \le \mathscr{F}_t \le \mathscr{F}$ for all $s \le t$ is called *filtration* and a stochastic process $X(t)$ is said to be adapted to filtration $\{\mathscr{F}_t\}$ if $X(t)$ is $\mathscr{F}_t$-measurable for all $t \ge 0$.

A random variable τ with values in $[0, \infty)$ is an $\{\mathscr{F}_t\}$-*stopping time* if

$$\{\tau \le t\} \subset \mathscr{F}_t, \quad \forall t \ge 0.$$

The stochastic process $X(t)$ is said to be a *local martingale* if there is a sequence of stopping times $\{\tau_n\}$ such that $\tau_n \to \infty$, $\mathbb{P}$-a.s., as $n \to \infty$ and, for each n, $X(t) = X(t, \Lambda\tau_n)$ is a martingale.

Finally, the stochastic process $X(t)$ is $\{\mathscr{F}_t\}$-*semimartingale* if $X = M + Y$, where M is a local martingale with respect to $\{\mathscr{F}_t\}$ and Y is an $\{\mathscr{F}_t\}$-adapted finite variation process, that is, for each $t > 0$, $\operatorname{asup} \sum_i |Y(t_{i+1}) - Y(t_i))| < \infty$, where the supremum is taken over all the partitions of $[0, t]$.

Lemma 4.5 is related to the martingale convergence theorem and plays an important role to obtain convergence in probability of stochastic processes. For the proof, we refer to [63].

Lemma 4.5 *Let I and I_1 be nondecreasing adapted processes, Z be a nonnegative semimartingale and M a local martingale such that $E(Z(t)) < \infty, \forall t \ge 0$, $I_1(\infty) < \infty$, $\mathbb{P}$-a.s., and $Z(t) + I(t) = Z(0) + I_1(t) + M(t)$, $\forall t \ge 0$. Then, there is*

$$\lim_{t\to\infty} Z(t) < \infty, \quad \mathbb{P}\text{-a.s.} \quad \text{and} \quad I(\infty) < \infty, \quad \mathbb{P}\text{-a.s.}$$

Definition 4.2 A real-valued stochastic process $\beta(\cdot)$ is called a *Brownian motion* or a *Wiener process* if

(i) $\beta(0) = 0$, $\mathbb{P}$-a.s.,
(ii) $\beta(t) - \beta(s)$ is $\mathscr{N}(0, t-s)$ Gaussian distributed for all $t \ge s \ge 0$,
(iii) for all times $0 < t_1 < t_2 < \cdots < t_n$, the random variables $\beta(t_1), \beta(t_2) - \beta(t_1), \ldots, \beta(t_n) - \beta(t_{n-1})$ are independent ("independent increments").

A real-valued random variable Y is said to be $\mathscr{N}(0, q)$ Gaussian distributed if

$$\mathbb{P}[a \le Y \le b] = \frac{1}{\sqrt{2\pi q}} \int_0^b e^{-\frac{x^2}{2q}}\, dx, \quad \text{for all } -\infty < a < b < \infty.$$

Lemma 4.6 *Let $\beta(t)$, $t \ge 0$, be a real Brownian motion in some probability space $(\Omega, \mathscr{F}, \mathbb{P})$. Then, for each $\lambda > 0$, we have*

$$\mathbb{P}\left(\sup_{t>0} e^{\beta(t)-\lambda t} \ge r\right) = \mathbb{P}\left(e^{\sup_{t>0}(\beta(t)-\lambda t)} \ge r\right)$$
$$= \mathbb{P}\left(\sup_{s>0}(\beta(s) - \lambda s) \ge \log r\right) = r^{-2\lambda}. \quad (4.159)$$

Proof Fix $T > 0$. By Girsanov's theorem, $\widetilde{\beta}(t) := \beta(t) - \lambda t$, $t \le T$, is a Brownian motion in $(\Omega, \mathscr{F}, \widetilde{\mathbb{P}})$, where $d\widetilde{\mathbb{P}} = e^{\lambda\beta(T) - \frac{1}{2}\lambda^2 Y}\, d\mathbb{P}$. We have

$$\mathbb{P}\left(\sup_{0\le t\le T} e^{\beta(t)-\lambda t} \ge r\right) = \mathbb{P}\left(\sup_{0\le t\le T} e^{\widetilde{\beta}(t)} \ge r\right).$$

Setting

$$M_T = \sup_{0\le t\le T} e^{\widetilde{\beta}(t)},$$

we have $\mathbb{P}(M_T \ge r) = \int_\Omega \mathbf{1}_{[r,+\infty)}(M_T) d\mathbb{P} = \int_\Omega \mathbf{1}_{[r,+\infty)}(M_T) e^{-\lambda\beta(T)+\frac{1}{2}\lambda^2 T}\, d\widetilde{\mathbb{P}}$.

Replacing in the latter the identity $\beta(t)$ by $\widetilde{\beta}(t) + \lambda t$, it yields

$$\mathbb{P}(M_T \ge r) = \int_\Omega \mathbf{1}_{[r,+\infty)}(M_T) e^{-\lambda\widetilde{\beta}(T)-\frac{1}{2}\lambda^2 T}\, d\widetilde{\mathbb{P}}.$$

Because $\widetilde{\beta}$ is a Brownian motion with respect to $\widetilde{\mathbb{P}}$, we can compute the above integral by using the well-known expression of the law of $(M_t, \widetilde{\beta}(t))$. We obtain that

$$\mathbb{P}(M_T \ge r) = \frac{2}{\sqrt{2\pi T^3}} \int_0^\infty db \int_{-\infty}^b (b-a) e^{-\lambda a - \frac{1}{2}\lambda^2 T} e^{-\frac{(2b-a)^2}{2T}}\, da.$$

It follows that

$$\mathbb{P}(M_T \ge r) = \frac{1}{2} e^{-2\lambda r} \operatorname{Erfc}\left(\frac{r-\lambda T}{\sqrt{2T}}\right) + \frac{1}{2} e^{2\lambda r} \operatorname{Erfc}\left(\frac{r+\lambda T}{\sqrt{2T}}\right),$$

where

$$\operatorname{Erfc}(x) = \frac{2}{\sqrt{\pi}} \int_x^{+\infty} e^{-t^2}\, dt.$$

For $T \to \infty$, we obtain (4.159). □

The Brownian motion β is nowhere differentiable on $(0, \infty)$. However, its formal derivative

$$\dot{\beta}(t) = \frac{d\beta}{dt}(t)$$

is called *white noise* because, heuristically, one may view $\eta(t) = \dot{\beta}(t)$ as a Gaussian process which satisfies $E(\eta(t)\eta(s)) = \delta(t-s)$ for $t \ge s$.

If U is a separable Hilbert space with the scalar product $(\cdot,\cdot)_U$, then a standard way to define a Wiener process on U is by the formula

$$W(t) = \sum_{j=1}^\infty \sqrt{\lambda_j}\, \beta_j(t) e_j, \tag{4.160}$$

where β_j are real-valued Brownian motions linearly independent in $(\Omega, \mathscr{F}, \mathbb{P})$ and $\{e_j\}$ is an orthonormal basis in H given by eigenfunctions e_j of a self-adjoint continuous positive operator Q with finite trace $\operatorname{Tr} Q$, that is, $Qe_j = \lambda_j e_j$, $j = 1, \dots, \infty$.

If H is another separable Hilbert space and $\Phi(t)$, $t \in (0, T)$, is a $L(U, H)$-valued stochastic process of the form

$$\Phi(t) = \sum_{j=1}^{m} \chi_j(t)\Phi(t_j),$$

where χ_j is the characteristic function of interval $[t_j, t_{j+1}]$, $0 < t_1 < \cdots < t_m < t$, then one defines the stochastic integral

$$\int_0^t \Phi(s)dW(s) = \sum_{j=0}^{m} \Phi(t_j)(W(t_{m+1} \wedge t) - W(t_m \wedge t)$$

and this definition extends in a standard way (see [44], pp. 1–4) to the adapted processes $\Phi : [0, T] \to L(U, H)$ such that

$$\mathbb{P}\left(\int_0^t \|\Phi(s)\|_2^2 ds < +\infty,\ t \geq 0\right) = 1,$$

where $\|\cdot\|_2$ is the Hilbert–Schmidt norm in $L(U, H)$.

In these terms, the solution $X = X(t)$ of the stochastic differential equation

$$\begin{aligned} dX(t) &= \Phi(t)dW(t) + f(t)dt, \quad t \in (0, T), \\ X(0) &= x, \end{aligned} \tag{4.161}$$

is defined as the process given by

$$X(t) = x + \int_0^t f(s)ds + \int_0^t \Phi(s)dW(s), \quad \forall t \in [0, T]. \tag{4.162}$$

Theorem 4.8 *Let $F : [0, T] \times H \to R$ be a function which is uniformly continuous along with partial derivatives F_t, F_x, F_{xx} on each bounded subset of $[0, T] \times H$. Then, if X is the solution to* (4.162), *we have*

$$\begin{aligned} dF(t, X(t)) &= (F_x(t, X(t)), \Phi(t)dW(t)) \\ &\quad + F_t(t, X(s))dt + (F_x(t, X(t)), f(t))dt \\ &\quad + \frac{1}{2}\operatorname{Tr}[F_{xx}(t, X(t))(\Phi(t)Q^{\frac{1}{2}})(\Phi(t)Q^{\frac{1}{2}})^*]dt, \quad t \in [0, T], \end{aligned} \tag{4.163}$$

that is,

$$F(t,X(t)) = F(0,x) + \int_0^t (F_s(s,X(s)) + (F_x(s,X(s)), f(s)))ds$$
$$+ \int_0^t (F_x(s,X(s)), \Phi(s))ds, \quad t \in (0,T).$$

This chain differentiation stochastic formula is the famous *Ito's formula* (for the proof, see [44], p. 115). In this special case, where $X(t) = \{X_j(t)\}_{j=1}^n$ is an R^n-valued stochastic process such that

$$dX(t) = f(t)dt + dW(t), \tag{4.164}$$

where $W(t)$ is an m-dimensional Wiener process $W(t) = \sum_{j=1}^m g_{ij}(t)\beta_j(t)$ ($\{\beta_j\}_{j=1}^m$ are independent Brownian motions), Formula (4.163) yields

$$dF(t,X(t)) = (f(t) + dW(t), F_x(t,X(t))dt + F_t(t,X(t))dt$$
$$+ \frac{1}{2}\sum_{i,j=1}^n F_{x_i x_j} \sum_{k=1}^m g_{ik}g_{kj}\, dt.$$

Given two separable Hilbert spaces, H and U with the norms $|\cdot|$ and $|\cdot|_U$, and $\{\beta_j\}$ a sequence of mutually independent Brownian motions on a probability space $(\Omega, \mathcal{F}, \mathbb{P})$, we consider in the following the filtration $\mathcal{F}_t$ generated by the σ-algebra generated by all $\{\beta_j(s),\ s \le t\}_{j=1}^\infty$. Consider, as above, a Wiener process $W(t)$ defined by (4.160) and denote by $L^2(\Omega, \mathcal{F}, \mathbb{P}, H)$ the space of H-valued random variable X with $E|X|_H^2 < \infty$.

We denote also by $C_W([0,T]; L^2(\Omega, \mathcal{F}, \mathbb{P}, H))$ the space of all the continuous functions $u : [0,T] \to L^2(\Omega, \mathcal{F}, \mathbb{P}, H)$, which are adapted to the filtration $\{\mathcal{F}_t\}$. Consider the stochastic differential equation

$$\begin{aligned} &dX(t) + AX(t)dt = f(t)dt + B(X(t))dW(t), \\ &X(0) = x, \end{aligned} \tag{4.165}$$

where $-A$ is the infinitesimal generator of a C_0-semigroup e^{-At} on H, and B is a continuous operator from U to $L(U,H)$.

The adapted process $X(t)$ is said to be a "mild" solution to (4.165) if

$$X(t) = e^{-At}x + \int_0^t e^{-A(t-s)} f(s)ds + \int_0^t e^{-A(t-s)} B(X(s))dW(s),$$
$$t \in [0,T]. \tag{4.166}$$

Theorem 4.9 *Assume that* $f \in L^2(0,T;H)$ *and*

$$\|S(t)B(x) - S(t)B(y)\|_{HS} \le \gamma|x-y|, \quad \forall t \in [0,T],\ x,y \in H,$$

where $\|\cdot\|_{HS}$ *is the Hilbert–Schmidt norm. Then,* (4.165) *has a unique "mild" solution* $X \in C_W([0,T]; L^2(\Omega, \mathcal{F}, \mathbb{P}, H))$.

(For the proof, see [45], p. 67.)

Theorem 4.9 extends to nonlinear differential equations of the form

$$dX + AXdt + F(X)dt = fdt + B(X)dW$$

for Lipschitzian mappings $F : H \to H$, but there are, however, few results on more general nonlinearities F.

For the stochastic Navier–Stokes equation

$$\begin{aligned} &dX + \mathscr{A}X\,dt + SX\,dt = dW, \\ &X(0) = x, \end{aligned} \tag{4.167}$$

where $\mathscr{A} = \nu A + A_0$ is the Stokes–Oseen operator in the complexified space $\widetilde{H}$ and $SX = \mathbb{P}((X \cdot \nabla)X$, it follows the existence of a unique mild solution in 2-D (see, e.g., Theorem 1.5.31 in [45]).

Now, if we consider the stochastic Stokes–Oseen equation

$$\begin{aligned} &dX(t) + \mathscr{A}X(t)dt = \sum_{j=1}^{N}(X(t), \varphi_j)\psi_j\,d\beta_j, \\ &X(0) = x, \end{aligned} \tag{4.168}$$

where $\varphi_j, \psi_j \in \widetilde{H}$, then it has a unique mild solution $X \in C_W([0,T]; L^2(\Omega, \mathscr{F}, \mathbb{P}, \widetilde{H})$, that is,

$$X(t) = e^{-\mathscr{A}t}x + \sum_{j=1}^{N}\int_0^t e^{-\mathscr{A}(t-s)}(X(s), \varphi_j)\psi_j d\beta_j(s). \tag{4.169}$$

The existence follows in this case by a standard fixed-point arguments we do not reproduce here.

Consider the Navier–Stokes equation (4.166) with multiplicative noise of the form (4.168), that is,

$$\begin{aligned} &dX(t) + AX(t)dt + SX(t)dt = \sum_{j=1}^{N}(X(t), \varphi_j)\psi_j\,d\beta_j, \\ &X(0) = x. \end{aligned} \tag{4.170}$$

We set

$$U(t) = \prod_{j=1}^{N} e^{\beta_j(t)\Gamma_j}, \quad t \ge 0, \quad \Gamma_j x = (x, \varphi_j)\psi_j, \quad j = 1, \dots, N,\ x \in \widetilde{H},$$

$$\mathscr{A}_\Gamma = \mathscr{A} + \frac{1}{2}\sum_{j=1}^{N}\Gamma_j^2, \qquad D(\mathscr{A}_\Gamma) = D(\mathscr{A}).$$

We associate with (4.170) the random differential equation

$$\begin{aligned}&\frac{dy}{dt}(t)+U^{-1}(t)\mathscr{A}_\Gamma U(t)y(t)+U^{-1}(t)S(U(t)y(t))=0,\\&y(0)=x.\end{aligned}\qquad(4.171)$$

We have the following theorem (see [44], p. 127).

Theorem 4.10 *If the process* $y:[0,\omega)\times\Omega\to\widetilde{H}$ *is adapted and, for each* $\omega\in\Omega$, $y(t,\omega)$ *is of class* C^1 *and satisfies* (4.171) $\mathbb{P}$-*a.s., then the process* $X(t)=U(t)y(t)$ *is a solution to* (4.170).

4.6 Comments on Chap. 4

The results of Sects. 4.1.1–4.1.2 are taken from Barbu [14] and those of Sect. 4.1.3 on stochastic internal stabilization of Navier–Stokes system are taken from the work [20] of Barbu and Da Prato. The main results of Sect. 4.2 are taken from the work of Barbu [18], while Theorem 4.5 on the stochastic boundary tangential stabilization was established in Barbu [14]. The results of Sect. 4.4 on the normal stabilization by noise were established in Barbu [17]. There is an extensive literature on the stabilization by noise of ODEs and PDEs and the reference list mentions a few works in this direction (see [5, 6, 38–40, 46, 64]). In this context, we mention also the works of Duan and Fursikov [48] on the stochastic stabilization of Stokes–Oseen equations and also [58, 72]. However, there are few connections and no overlap between these works and the results presented in this chapter, which refer to the internal and boundary noise stabilization.

Chapter 5
Robust Stabilization of the Navier–Stokes Equation via the H^∞-Control Theory

Since most of the fluid dynamic systems are subject to uncertainties and external disturbances, a major problem is the design of feedback controllers which achieve asymptotic stability not only for a nominal system (which is only partially known) but also for an entire set of systems covering a neighborhood of the given system. Such a control is called *robust* and the H^∞-control theory provides an efficient and popular approach to this question. We discuss in some details the H^∞-control problem for the stabilization problems studied in Chap. 3. We begin with a general presentation and some basic results on the H^∞-control problem for linear infinite-dimensional systems.

5.1 The State-space Formulation of the H^∞-Control Problem

We consider the abstract input–output system described by the equations

$$\begin{aligned} y' &= Ay + B_2 u + B_1 w \quad \text{in } R^+ = (0, \infty), \\ z &= C_1 y + D_{12} u, \end{aligned} \tag{5.1}$$

in a real Hilbert space X with the norm denoted $|\cdot|$ and the scalar product $(\cdot, \cdot)$.

Here, A is the infinitesimal generator of a C_0-semigroup e^{At}, $B_1 \in L(W, X)$, $B_2 \in L(U, X)$, $C_1 \in L(X, Z)$ and $D_{12} \in L(U, Z)$, where W, U and Z are real Hilbert spaces with the norms $|\cdot|_W$, $|\cdot|_U$, $|\cdot|_Z$ and the scalar products $(\cdot, \cdot)_W$, $(\cdot, \cdot)_U$ and $(\cdot, \cdot)_Z$, respectively.

In System (5.1), $y : R^+ \to X$ is the state variable, $u \in L^2(R^+; U)$ is the control input, while $w \in L^2(R^+; W)$ is an exogeneous variable (disturbances) and z is the measurable output variable.

We denote by $\mathscr{F}$ the set of all the linear feedback controllers $F \in L(X, U)$ which stabilizes exponentially the system, that is,

$$\|e^{(A+B_2F)t}\|_{L(X,X)} \le Me^{-\delta t}, \quad \forall t > 0,$$

for some $\delta > 0$.

V. Barbu, *Stabilization of Navier–Stokes Flows*,
Communications and Control Engineering,
DOI 10.1007/978-0-85729-043-4_5,

For any such $F \in \mathscr{F}$, consider the closed-loop operator

$$S_F : L^2(R^+; W) \to L^2(R^+; Z)$$

defined by

$$S_F(w)(t) = C_1 y^w + D_{12} F y^w = (C_1 + D_{12} F) \int_0^t e^{(A+B_2F)(t-s)} B_1 w(s) ds, \qquad (5.2)$$

where y^w is the mild solution to the closed-loop system

$$y' = (A + B_2 F)y + B_1 w; \qquad y(0) = 0. \qquad (5.3)$$

The H^∞-control problem for System (5.1) can be formulated as:

Given $\gamma > 0$, find $F \in \mathscr{F}$ such that

$$\|S_F\| < \gamma. \qquad (5.4)$$

Here, $\|S_F\|$ is the norm of the operator $S_F \in L(L^2(R^+; W), L^2(R^+; Z))$.

If such an operator F exists, it is called a *γ-suboptimal solution* to the *H^∞-control problem*, and, if (5.4) holds, one says that *F has L_2-gain less than γ*.

This is the state-space formulation of the H^∞-problem due to Doyle et al. (see, also, [10, 77, 79] for the H^∞-control problem in infinite dimensions). There is an equivalent frequential formulation on this problem in terms of the transfer function associated to System (5.1).

If we denote by $G_F(i\tau)$ the operator $(i\tau - A - B_2 F)^{-1} B_1$, $\tau \in R$, then S_F can be, equivalently, expressed as

$$(S_F w)(i\tau) = (C_1 + D_{12} F) G_F(i\tau) \widehat{w}(i\tau), \qquad w \in L^2(R^+; W),$$

where $\widehat{w}$ is the Fourier transform of w. The norm of S_F is given by the Hardy H^∞-norm of $(C_1 + D_{12} F) G_F(i\tau)$, that is,

$$\|S_F\| = \|(C_1 + D_{12} F) G_F\|_\infty = \sup_{\tau \in R} \|(C_1 + D_{12} F) G_F(i\tau)\|,$$

and this explains why Problem (5.4) is called the H^∞-control problem.

The problem we consider here is one of the simplest in the H^∞-control theory, that is, that of control with state feedback which, as we see later on, has the great advantage of being easily extended to nonlinear control systems.

The exact meaning of the H^∞-problem (5.1), (5.4) is the following: The plant (5.1) subject to external disturbances w has an internal state y, which cannot be observed directly. However, a state measurement z is made and a stabilizable feedback controller $u = Fy$ is designed in such a way as to minimize the effect of disturbances w. As we shall see below, this problem leads naturally to a max-min problem or a differential game having (5.1) as state-system. Roughly speaking, the H^∞-control strategy to design robust stabilizable controllers is concerned with the treatment of the worst-case disturbance of System (5.1).

To deal with the H^∞-control problem (5.4), we assume that the following conditions are satisfied.

(k) *The pair (A, B_2) is stabilizable, that is, there exists $L \in L(H, U)$ such that $A + B_2 L$ generates an exponentially stable semigroup.*

(kk) *The pair* (A, C_1) *is exponentially detectable, that is, there exists* $\widetilde{K} \in L(Z, X)$ *such that* $A + \widetilde{K}C_1$ *generates an exponentially stable semigroup, that is,* (A^*, C_1^*) *is stabilizable.*

(kkk) $D_{12}^*[C_1, D_{12}] = [0, I]$.

The latter assumption can be, equivalently, written as

(kkk)′ $|C_1 y + D_{12}u|_Z^2 = |C_1 y|_Z^2 + |u|_U^2, \ \forall (y, u) \in X \times U.$

Now, we present the main result of this section.

Theorem 5.1 *Let* $\gamma > 0$ *be given and let Assumptions* (k) (kk) *and* (kkk) *be satisfied. Then, the* H^∞*-control problem* (5.4) *has a* γ*-suboptimal solution* $F \in \mathcal{F}$ *if and only if there exists* $P = P^* \geq 0$, $P \in L(X, X)$ *such that*

$$(Ax, Py) + (Ay, Px) + (P(\gamma^{-2}B_1B_1^* - B_2B_2^*)Px, y) + (C_1x, C_1y) = 0,$$
$$\forall x, y \in D(A) \tag{5.5}$$

and $A_P = A + (\gamma^{-2}B_1B_1^* - B_2B_2^*)P$ *generates an exponentially stable semigroup. In this case,* $\widetilde{F} = -B_2B_2^*P$ *belongs to* $\mathcal{F}$ *and is a* γ*-suboptimal solution to the* H^∞*-control problem. The algebraic Riccati equation* (5.5) *has at most one solution* $P = P^* \geq 0$, $P \in L(X, X)$ *with the property that* A_P *is exponentially stable.*

Readers familiar with the differential game theory will recognize in (5.5) an algebraic Riccati equation arising in the linear quadratic differential game theory with infinite time horizon. We see later that the H^∞-control problem described above admits, indeed, an equivalent formulation in terms of a two-person zero sum differential game governed by System (5.1).

As in the case of the linear quadratic control problem, we may reformulate Theorem 5.1 in terms of Hamiltonian systems.

Theorem 5.2 *Let* $\gamma > 0$ *be given. Then the* H^∞*-control problem has a* γ*-suboptimal solution* $F \in \mathcal{F}$ *if and only if the Hamiltonian system*

$$\begin{aligned} y' &= Ay + (B_2B_2^* - \gamma^{-2}B_1B_1^*)p, \quad t > 0, \\ p' &= -A^*p + C_1^*C_1y, \quad t > 0, \end{aligned} \tag{5.6}$$

has a positively invariant manifold $E = \{(y, p) \times X \times X;\ p + Py = 0\}$, *where* $P = P^* \geq 0$, $P \in L(X, X)$ *and the flow* $(y(t), p(t))$ *in* E *is exponentially stable. In this case,* $F = -B_2^*P$ *is a solution to the* H^∞*-control problem.*

Theorem 5.2 is important because very often in computation it is more convenient to find the invariant manifold E of the Hamiltonian system (5.6) than to solve the Riccati equation (5.5).

Before proceeding further with the proof, we first prove the equivalence of Theorems 5.1 and 5.2.

Lemma 5.1 *Riccati equation* (5.5) *and Hamiltonian System* (5.6) *are equivalent.*

Proof Let $P \in L(X,X)$, $P = P^* \geq 0$, be a solution to the algebraic Riccati equation (5.5) such that A_P generates an exponentially stable semigroup $e^{A_P t}$. For $y_0 \in X$, consider the mild solution to the Cauchy problem

$$\begin{aligned} &y' = Ay - (B_2B_2^* - \gamma^{-2}B_1B_1^*)Py, \quad t \geq 0, \\ &y(0) = y_0, \end{aligned} \tag{5.7}$$

and set $p(t) = -Py(t)$, $\forall t \geq 0$.

If $y_0 \in D(A)$, then $y \in C^1(R^+; X)$ and, for every $v \in D(A^*)$, we have

$$\frac{d}{dt}(p(t), v) = -(y'(t), Pv) = -(Ay(t) - (B_2B_2^* - \gamma^{-2}B_1B_1^*)Py(t), Pv), \quad \forall t \geq 0.$$

By virtue of (5.5), this yields

$$\frac{d}{dt}(p(t), v) = -(p(t), A^*v) + (C_1C_1^*y(t), v), \quad \forall v \in D(A^*),\ \forall t \geq 0,$$

which is equivalent with the fact that p is a mild solution to the backward equation

$$p' = A^*p + C_1C_1^*y, \quad t > 0. \tag{5.8}$$

By continuity, this extends to all $y_0 \in X$. Since $e^{A_P t}$ is exponentially stable, we have

$$|y(t)| + |p(t)| \leq Me^{-\delta t}|y_0|, \quad \forall t \geq 0, \tag{5.9}$$

where $\delta > 0$ and M is some positive constant.

Hence, $E = \{(y,p) \in X \times X'\ p + Py = 0\}$ is a positively invariant manifold of System (5.6) and the flow $(y(t), p(t)) \in E$ is exponentially stable.

Conversely, let $P \in L(X,X)$, $P = P^* \geq 0$, be such that $\{(x,p) \in X \times X;\ p + Px = 0\}$ is positively invariant for System (5.6) and the solution (x,p) to (5.6) satisfies (5.9). This means that the solution $y = y(t)$ to the closed-loop system

$$\begin{aligned} &y' = Ay - (B_2B_2^* - \gamma^{-2}B_1B_1^*)Py, \\ &y(0) = y_0, \end{aligned} \tag{5.10}$$

has exponential decay, that is,

$$|y(t)| = |e^{A_P t}y_0| \leq Me^{-\delta t}|y_0|.$$

On the other hand, $p = -Py$ satisfies (5.8), that is,

$$\frac{d}{dt}Py(t) = -A^*Py(t) + C_1C_1^*y(t), \quad \forall t \geq 0. \tag{5.11}$$

Let $y_0 \in D(A)$. Then, y is continuously differentiable on $[0,\infty)$ and, if we multiply (5.10) by $-p'(t) = Py'(t)$, (5.11) by $y'(t)$ and subtract the results, we get

$$(Ay(t), p(t)) - 2^{-1}(B_2B_2^* - \gamma^{-2}B_1B_1^*)Py(t), Py(t)) - 2^{-1}|C_1y(t)|_Z^2 \equiv C, \quad \forall t \geq 0.$$

Since

$$\lim_{t\to\infty} Ay(t) = \lim_{t\to\infty} e^{A_P t}Ay_0 = 0 \quad \text{and} \quad \lim_{t\to\infty} p(t) = 0,$$

the constant C is zero and, therefore,

$$2(Ay_0, Py_0) + |B_2^* P y_0|_U^2 - \gamma^{-2}|B_1^* P y_0|_W^2 + |C_1 y_0|^2 = 0, \quad \forall y_0 \in D(A).$$

Differentiating (Gâteaux) the latter equation, we see that P satisfies (5.5). This completes the proof of the equivalence between Riccati equation (5.5) and Hamiltonian System (5.6). □

Proof of Theorem 5.1 1. *The only if part.* This is the hard part of the proof and it relies on an equivalent two person game formulation of the H^∞-problem.

Let $F \in L(X, X)$ be such that $A + B_2 F$ generates an exponentially stable semigroup and $\|S_F\| < \gamma$. Consider the sup inf problem

$$\sup_{w \in \mathcal{W}} \inf_{u \in \mathcal{U}} K(u, w), \tag{5.12}$$

where $\mathcal{W} = L^2(R^+; W)$, $\mathcal{U} = L^2(R^+; U)$ and $K : \mathcal{U} \times \mathcal{W} \to [-\infty, +\infty)$ defined by

$$\begin{aligned} K(u, w) &= \int_0^\infty (|C_1 y + D_{12} u|_Z^2 - \gamma^2 |w|_W^2)\,dt \\ &= \int_0^\infty (|C_1 y|_Z^2 + |u|_U^2 - \gamma^2 |w|_W^2)\,dt \end{aligned} \tag{5.13}$$

and y is the solution to System (5.1) with initial condition $y(0) = y_0$ ($y_0 \in X$ arbitrary but fixed).

We study first the minimization problem

$$\inf\{K(u, w);\ u \in \mathcal{U}\}, \tag{5.14}$$

where $w \in \mathcal{W}$ is arbitrary but fixed.

This is a linear quadratic optimal control problem with the nonhomogeneous state-system

$$\begin{aligned} &y' = Ay + B_2 u + B_1 w, \quad t > 0, \\ &y(0) = y_0. \end{aligned} \tag{5.15}$$

Obviously, this problem has for each $w \in \mathcal{W}$ a unique solution $\bar{u} = \Gamma w$ because $\|S_F\| < \gamma$ and the finite cost condition for the quadratic control problem associated with (5.14) holds. (See, e.g., [9, 32].)

Lemma 5.2 *There is* $\bar{p} \in C(R^+; X) \cap L^2(R^+; X)$ *such that*

$$\begin{aligned} &\bar{p}' = -A^* \bar{p} + C_1^* C_1 L y, \quad t > 0, \\ &\bar{p}(\infty) = 0, \end{aligned} \tag{5.16}$$

$$B_2^* \bar{p}(t) = \bar{u}(t), \quad \text{a.e. } t > 0, \tag{5.17}$$

where y *is the solution to* (5.15) *with* $u = \bar{u}$. *Conversely, any function* $\bar{u}$ *given by* (5.17) *is optimal in* (5.14).

Proof By the general theory of linear quadratic control problems (see, e.g., [9, 32], we know that there is $P_0 \in L(X,X)$, $P_0 = P_0^* \ge 0$ such that $A_{P_0} = A - B_2B_2^*P_0$ is exponentially stable. Let $\bar{p} \in C(R^+;X) \cap L^2(R^+;X)$ be the solution to

$$\begin{aligned} \bar{p}' &= -A_{P_0}^*\bar{p} - P_0B_2\bar{u} + C_1C_1^*\bar{y},\\ \bar{p}(\infty) &= 0, \end{aligned} \tag{5.18}$$

that is,

$$\bar{p}(t) = -\int_t^\infty e^{A_{P_0}^*(s-t)}(C_1C_1^*\bar{y}(s) - P_0B_2\bar{u}(s))ds, \quad \forall t \ge 0.$$

On the other hand, we have

$$\int_0^\infty (|C_1y_\lambda|_Z^2 + |\bar{u} + \lambda v|_U^2)dt \ge \int_0^\infty (|C_1\bar{y}|_Z^2 + |\bar{u}|_U^2)dt, \quad \forall \lambda > 0,$$

where $v \in L^2(R^+;U)$ and

$$\begin{aligned} y_\lambda' &= Ay_\lambda + B_2(\bar{u} + \lambda v) + B_1w,\\ y_\lambda(0) &= y_0. \end{aligned}$$

This yields

$$\int_0^\infty ((C_1z, C_1\bar{y})_Z + (\bar{u}, v)_U)dt = 0, \quad \forall v \in L^2(R^+;U), \tag{5.19}$$

where

$$\begin{aligned} z' &= Az + B_2v, \quad t > 0,\\ z(0) &= 0. \end{aligned} \tag{5.20}$$

Of course, in (5.19) and (5.20) we must confine ourselves to those $z \in L^2(R^+;U)$ having the property that $C_1z \in L^2(R^+;Z)$. In particular, we may take $v = u - B_2^*P_0y$, where $u \in L^2(R^+;U)$ and y is the solution to

$$y' = A_{P_0}y + B_2u, \quad t > 0, \ y(0) = 0. \tag{5.21}$$

This yields,

$$\int_0^\infty ((C_1y, C_1\bar{y})_Z + (\bar{u}, u - B_2^*P_0y)_U)dt = 0, \tag{5.22}$$

for all $(y,u) \in C(R^+;X) \times L^2(R^+;U)$ which satisfy (5.21).

On the other hand, by (5.18) and (5.21), it follows that

$$\int_0^T (C_1C_1^*\bar{y}, y)dt = (\bar{p}(T), y(T)) - \int_0^T ((B_2^*\bar{p}, u)_U - (B_2^*P_0y, \bar{u})_U)dt.$$

Then, letting $T \to \infty$, we get

$$\int_0^\infty (C_1C_1^*\bar{y}, y)dt = -\int_0^\infty ((B_2^*\bar{p}, u)_U - (B_2^*P_0y, \bar{u})_U)dt,$$

and so, by (5.22), we see that

$$\int_0^\infty (\bar{u} - B_2^* \bar{p}, u)dt = 0, \quad \forall u \in L^2(R^+; U),$$

which implies (5.17), as claimed. □

It is readily seen that Systems (5.16) and (5.17) are also necessary for optimality in Problem (5.14). In particular, this implies that the map Γ is of the form

$$\Gamma w = \Gamma_0 w + f_0, \quad \forall w \in L^2(R^+; W),$$

where $\Gamma_0 \in L(\mathscr{W}, \mathscr{U})$ and $f_0 \in \arg\{\inf K(u,0);\ u \in \mathscr{U}\}$.

Now, consider the function $\varphi \to R$ defined by

$$\varphi(w) = -K(\Gamma w, w), \quad \forall w \in \mathscr{W}.$$

Clearly, φ is quadratic, that is, there is $D \in L(\mathscr{W}, \mathscr{W})$ such that

$$\varphi(w) = \|Dw\|^2_{\mathscr{W}} + (Dw, g)_{\mathscr{W}} + \eta, \quad \forall w \in \mathscr{W}.$$

Since $\|S_F\| < \gamma$ for some $F \in \mathscr{F}$, we have

$$\inf\left\{\int_0^\infty (|C_1 y|^2_Z + |u|^2_U)dt;\ y' = Ay + B_2 u + B_1 w,\ y(0) = y_0,\ u \in \mathscr{U}\right\}$$
$$\le (\gamma^2 - \varepsilon)\int_0^\infty |w|^2_W dt + \beta|y_0|^2, \quad \forall w \in \mathscr{W},$$

where $\varepsilon > 0$ and $\beta \ge 0$. We have, therefore,

$$\varphi(w) \ge \varepsilon\|w\|^2_{\mathscr{W}} - \beta|y_0|^2, \quad w \in \mathscr{W}.$$

This implies that DD^* is positive definite and so, φ attains its infimum in a unique point w^*. In other words,

$$w^* = \arg\inf\{\varphi(w);\ w \in \mathscr{W}\}$$

and so, $(u^* = w^*, w^*)$ is the solution to Problem (5.12), that is,

$$(u^*, w^*) = \arg \sup_{w \in \mathscr{W}} \inf_{u \in \mathscr{U}} K(u, w).$$

Lemma 5.3 *We have*

$$w^* = -\gamma^{-2} B_1^* p, \quad a.e., t > 0, \tag{5.23}$$

where $p \in C(R^+; X) \cap L^2(R^+; X)$ *is any mild solution to*

$$\begin{aligned} &p' = -A^* p + C_1 C_1^* y^*, \quad t > 0, \\ &p(\infty) = 0. \end{aligned} \tag{5.24}$$

Here, y^* is the solution to System (5.1), where $u = u^*$ and $w = w^*$.

We note that, by Lemma 5.2, we already know that System (5.24) has a mild solution $p \in L^2(R^+; X)$.

Proof of Lemma 5.3 We write φ as

$$\varphi(w) = \gamma^2 \|w\|^2_{\mathscr{W}} - \widetilde{\varphi}(w), \quad w \in \mathscr{W}, \tag{5.25}$$

where

$$\begin{aligned} \widetilde{\varphi}(w) &= \int_0^\infty (|C_1 y^w|_Z^2 + |\Gamma_0 w + f_0|_U^2) dt, \\ (y^w)' &= A y^w + B_2(\Gamma_0 w + f_0) + B_1 w, \quad t > 0, \\ y^w(0) &= y_0. \end{aligned}$$

We have, therefore,

$$(\nabla \widetilde{\varphi}(w), \bar{w}) = 2 \int_0^\infty ((\Gamma_0 w + f_0, \Gamma_0 \bar{w})_U + (C_1 y^w, C_1 z)_Z dt, \quad \forall \bar{w} \in \mathscr{W},$$

where

$$z' = Az + B_2 \Gamma_0 \bar{w} + B_1 \bar{w}; \qquad z(0) = 0. \tag{5.26}$$

Since $C_1 \in L^2(R^+; Z)$ and the pair (A, C_1) is detectable, it follows as above that $z \in L^2(R^+; Z)$ and $\lim_{t\to\infty} z(t) = 0$. Then, recalling that $y^{w^*} = y^*$, we get, by (5.24) and (5.26) that

$$\int_0^\infty (C_1 C_1^* y^*, z) dt = \int_0^\infty (p' + A^* p, z) dt = -\int_0^\infty (p, B_1 \bar{w} + B_2 \Gamma_0 \bar{w}) dt.$$

Hence,

$$\begin{aligned} (\nabla \widetilde{\varphi}(w^*), \bar{w}) &= -2 \int_0^\infty (B_1^* p + \Gamma_0^* B_2^* p, \bar{w})_W dt + 2 \int_0^\infty (\Gamma_0^* B_2^* p, \bar{w})_W dt \\ &= -2 \int_0^\infty (B_1^* p, \bar{w})_W dt, \quad \forall \bar{w} \in \mathscr{W}. \end{aligned}$$

Hence, $\nabla \widetilde{\varphi}(w^*) = -2 B_1^* p$ and, by (5.25), we get (5.23), as claimed.

To summarize, (u^*, w^*) is expressed as

$$u^* = B_2^* p, \quad w^* = -\gamma^{-2} B_1^* p, \quad \text{a.e. } t > 0, \tag{5.27}$$

where $p \in L^2(R^+; X)$ and

$$\begin{aligned} (y^*)' &= A x^* + B_2 u^* + B_1 w^*, \quad t > 0, \\ p' &= -A^* p + C_1 C_1^* y^*, \quad t > 0, \\ y^*(0) &= y_0, \qquad p(\infty) = 0. \end{aligned} \tag{5.28}$$

Hence, (y^*, p) is a solution to the Hamiltonian system (5.6).

If we write the first equation in (5.28) as

$$(y^*)' = (A + KC_1) y^* - KC_1 y^* + B_2 u^* + B_1 w^*,$$

where $K \in L(Z, X)$ is as in Hypothesis (kk), we see that

$$y^* \in L^2(R^+; X). \tag{5.29}$$

Let $P : X \to X$ be the mapping defined by

$$Py_0 = \{-p(0)\}, \tag{5.30}$$

where $p \in C(R^+; X) \cap L^2(R^+; X)$ is any mild solution to the equation

$$\begin{aligned} &p' = -A^*p + C_1C_1^*y^*, \quad t > 0,\\ &p(\infty) = 0. \end{aligned} \tag{5.31}$$

□

Lemma 5.4 *The map P is single-valued, linear, continuous and*

$$(Py_0, y_0) = \sup_{w\in\mathcal{W}} \inf_{u\in\mathcal{U}} K(u, w) \geq 0, \tag{5.32}$$

$$p(t) = -Py^*(t), \quad t \geq 0. \tag{5.33}$$

Proof If we multiply the first equation in (5.28) by $p(t)$, the second by $y^*(t)$ and use (5.27), we get

$$(p(T), y(T)) - (p(0), y_0) = \int_0^T (|Cy^*(t)|_Z^2 + |u^*(t)|_U^2 - \gamma^2|w^*(t)|_W^2)dt. \tag{5.34}$$

Then, letting $T \to \infty$,

$$-(p(0), y_0) = \int_0^\infty (|Cy^*(t)|_Z^2 + |u^*(t)|_U^2 - \gamma^2|w^*(t)|_W^2)dt. \tag{5.35}$$

(Without loss of generality, we may assume that y, p are differentiable; otherwise, we use as above the variation of constant formula for y and p.) In particular, Equality (5.35) shows that P is single-valued. Indeed, if $-Py_0$ would contain two elements $p_1(0), p_2(0)$, then $p_1(0) + \lambda(p_1(0) - p_2(0)) \in -Py_0$ for all $\lambda \in R$ (because $p_1 + \lambda(p_1 - p_2)$ is a solution to (5.3) if p_1, p_2 are) and so, by (5.35), y_0 must be 0. Hence, Py_0 is single-valued for $y_0 \neq 0$. For $y_0 = 0$, we take by definition $P(0) = 0$. It is readily seen that P is linear. Next, by (5.28) we get after some calculation that

$$(Py_0, z_0) = \int_0^\infty ((C_1y^*, C_1z^*)_Z + (u^*, v^*)_U - \gamma^2(w^*, \bar{w}^*)_W)dt,$$

where (z^*, v^*, w^*) is the solution to Problem (5.15) with the initial value z_0. This, clearly, implies that $P = P^*$.

Since, as shown earlier,

$$\begin{aligned} \varphi(w^*) &= -\int_0^\infty (|C_1y^*(t)|^2 + |u^*(t)|_U^2 - \gamma^2|w^*(t)|_W^2)dt\\ &\geq \varepsilon \int_0^\infty |w^*(t)|_W^2dt - \beta|y_0|^2, \end{aligned}$$

it follows by (5.35) that

$$(Py_0, y_0) \leq \beta|y_0|^2, \quad \forall y_0 \in X, \tag{5.36}$$

where β is independent of y_0. Hence, $P \in L(X, X)$. Moreover, by (5.32) we see that $(Py_0, y_0) \geq 0$, $\forall y_0 \in X$.

To prove (5.33), we note that, for every $t \geq 0$, (y^*, u^*, w^*) is the solution to the sup inf problem

$$\sup_{w\in L^2(t,\infty;W)} \inf_{u\in L^2(t,\infty;U)} \left\{ \int_t^\infty (|C_1 y|_Z^2 + |u|_U^2 - \gamma^2 |w|_W^2)dt; \right.$$
$$\left. y' = Ay + B_2 u + B_1 w \text{ in } (t,\infty);\ y(t) = y^*(t) \right\}. \tag{5.37}$$

Indeed, if

$$\Gamma(t, w, y_0) = \arg\inf_u \left\{ \int_0^\infty (|C_1 y|_Z^2 + |u|_Z^2)ds;\ y' = Ay + B_2 u + B_1 w;\ y(t) = y_0 \right\},$$

we have by the dynamic programming principle

$$\Gamma(t, w, y^w(t)) = \Gamma(0, w, y_0) \quad \text{on } R_t = (t, \infty), \tag{5.38}$$

where $(y^w)' = Ay^w + B_2 \Gamma w + B_1 w;\ y^w(0) = y_0$.

Next, we have

$$w^* = \arg \sup_{w\in\mathscr{W}} \left\{ \int_0^\infty (|C_1 y^w|_Z^2 + |\Gamma(0, w, y_0)|^2 - \gamma^2 |w|_W^2)dt; \right.$$
$$\left. (y^w)' = Ay^w + B_2 \Gamma(0, w, y_0) + B_1 w,\ y^w(0) = y_0 \right\}$$
$$= \arg\sup_w \left\{ \int_t^\infty (|C_1 y|_Z^2 + |\Gamma(s, w, y^w)|^2 - \gamma^2 |w|_W^2)ds; \right.$$
$$\left. y' = Ay + B_2 \Gamma(t, w, y^w) + B_1 w;\ y^w(t) = y^*(t) \right\}$$
$$= \arg\sup_w \inf_u \left\{ \int_t^\infty (|C_1 y|_Z^2 + |u|_U^2 - \gamma^2 |w|_W^2)ds; \right.$$
$$\left. y' = Ay + B_2 u + B_1 w;\ y(t) = y^*(t) \right\}$$

as claimed.

Hence, $p(t) = -Py^*(t)$, $\forall t \geq 0$, and so, the solution p to (5.31) is unique.

This implies that $\{(y, p) \times X \times X;\ p + Py = 0\}$ is a positively invariant manifold of the Hamiltonian system (5.6) and $y^*(t) = e^{A_P t} y_0$, $\forall t \geq 0$. (Note that, because $(B_2 B_2^* - \gamma^{-2} B_1 B_1^*)P$ is continuous, A_P generates a C_0-semigroup.) Since, as seen earlier, $y^* \in L^2(R^+; X)$, by Datko's theorem it follows that $e^{A_P t}$ is exponentially stable. We have, therefore, shown that P satisfies all the conditions of Theorem 5.2 (and, by Lemma 5.1, also of Theorem 5.1). This completes the proof of the *only if part*.

2. *The if part.* Assume that there is $P \in L(X, X)$, $P = P^* \ge 0$, which satisfies (5.5) and such that A_P is exponentially stable. Consider the system

$$\begin{aligned} y' &= (A - B_2 B_2^* P)y + B_1 w, \quad t \ge 0, \\ y(0) &= y_0, \end{aligned} \tag{5.39}$$

where $y_0 \in X$ and $w \in L^2(R^+; W)$.

If $y_0 \in D(A)$ and $w \in C^1(R^+; W)$, then the solution y to (5.39) is differentiable. Thus, multiplying this equation with $Py(t)$ and using (5.5), we get

$$\begin{aligned} \frac{d}{dt}(Py(t), y(t)) &= 2(Ay(t), Py(t)) - 2|B_2^* Py(t)|_U^2 + 2(w(t), B_1^* Py(t))_W \\ &= 2(w(t), B_1^* Py(t))_W - |C_1 y(t)|_Z^2 - |B_2^* Py(t)|_U^2 \\ &\quad - \gamma^{-2} |B_1^* Py(t)|_W^2 \\ &= -|C_1 y(t)|_Z^2 - |B_2^* Py(t)|_U^2 + \gamma^2 |w(t)|_W^2 \\ &\quad - \gamma^2 |w(t) - \gamma^{-2} B_1^* Py(t)|_W^2, \quad \forall t \ge 0. \end{aligned}$$

If $w \in L^2(R^+; W)$, this yields

$$\begin{aligned} &\int_0^\infty (|C_1 y(t)|_Z^2 + |B_2^* Py(t)|_U^2 - \gamma^2 |w(t)|_W^2) dt \\ &\quad \le (Py_0, y_0) - \gamma^2 \int_0^\infty |w(t) - \gamma^{-2} B_1^* Py(t)|_W^2 \, dt. \end{aligned} \tag{5.40}$$

For $w \equiv 0$, the latter implies that $C_1 y \in L^2(R^+; Z)$, $B_1^* Py \in L^2(R^+; U)$ and so, by the detectability hypothesis (kk), $y \in L^2(R^+; X)$ and, therefore, $e^{(A - B_2 B_2^* P)t}$ is exponentially stable.

Next, for $w \in L^2(R^+; W)$ and $y_0 = 0$, we set $\bar{w} = w - \gamma^{-2} B_1^* Py$ and rewrite (5.39) as

$$\begin{aligned} y' &= A_P y + B_1 \bar{w}, \quad t > 0, \\ y(0) &= 0. \end{aligned}$$

We have

$$\|\bar{w} + \gamma^{-2} B_1^* Py\|_{L^2(R^+; X)} \ge \delta \|\bar{w}\|_{L^2(R^+; W)},$$

and so, by (5.40) we see that

$$\begin{aligned} &\|S_F \bar{w}\|_{L^2(R^+; Z)}^2 \le \gamma^2 \|\bar{w}\|_{L^2(R^+; W)}^2 - \delta^2 \|\bar{w}\|_{L^2(R^+; W)}^2 = (\gamma^2 - \delta^2) \|\bar{w}\|_{L^2(R^+; W)}^2, \\ &\quad \forall w \in L^2(R^+; W), \end{aligned}$$

where δ is independent of w and $F = -B_2^* P$. Hence, $\|S_F\| < \gamma$, as claimed.

The uniqueness of P is the consequence of the obvious equality

$$(A_{P_1} y_0, (P_1 - P_2) z_0) + ((P_1 - P_2) y_0, A_{P_2} z_0) = 0, \quad \forall y_0, z_0 \in D(A),$$

which yields

$$\frac{d}{dt}(e^{A_{P_1} t} y_0, (P_1 - P_2) e^{A_{P_2} t} z_0) = 0, \quad \forall t \ge 0.$$

Since $e^{A_{P_1}t}, e^{A_{P_2}t}$ are exponentially stable, the latter implies that

$$((P_1 - P_2)y_0, z_0) = 0, \quad \forall y_0, z_0 \in D(A),$$

that is, $P_1 = P_2$, as claimed. □

5.2 The H^∞-Control Problem for the Stokes–Oseen System

We apply the above general scheme to the linear control system

$$\begin{aligned} &\frac{dy}{dt} + \mathscr{A}y = B_2u + B_1w, \quad t \geq 0, \\ &y(0) = y_0, \end{aligned} \tag{5.41}$$

$$z = C_1y + D_{12}u, \quad t \geq 0, \tag{5.42}$$

on the space $H = \{y \in (L^2(\mathscr{O}))^d;\ \nabla \cdot y = 0,\ y \cdot n = 0 \text{ on } \partial\mathscr{O}\}$ and to the Stokes–Oseen operator $\mathscr{A} = \nu A + A_0$, where

$$Ay = P(-\nu\Delta y), \qquad A_0y = P((y_e \cdot \nabla)y + (y \cdot \nabla)y_e),$$

$$D(\mathscr{A}) = D(A) = (H^2(\mathscr{O}))^d \cap (H_0^1(\mathscr{O}))^d \cap H.$$

P is the Leray projector, y is the state of the system, u is the input control force taken from a real Hilbert space U, w is the exogeneous force taken from a Hilbert space W (the disturbances space) and $B_2 \in L(U, H)$, $B_1 = L(W, H)$.

Finally, the measurement operator C_1 is in $L(H, Z)$ and $D_{12} \in L(U, Z)$, where Z is another real Hilbert space which is made precise below. The norm of H is denoted by $|\cdot|$, and its scalar product by $(\cdot,\cdot)$.

We assume everywhere in the following that Assumptions (k)–(kkk) hold for $X = H$ and $-\mathscr{A}$ in place of A. Then, by Theorem 5.1, we have the following result.

Theorem 5.3 *The H^∞-control problem for System* (5.41), (5.42) *has a γ-suboptimal solution $F \in L(H, U)$, that is, $u = Fy$ exponentially stabilizes System* (5.41) *with $w = 0$ and*

$$\int_0^\infty (|C_1y(t)|_Z^2 + |Fy(t)|_U^2)dt \leq (\gamma^2 - \varepsilon)\int_0^\infty |w(t)|_W^2 dt, \quad \forall w \in L^2(R^+; W),$$

for some $\varepsilon > 0$, if and only if the algebraic Riccati equation

$$\mathscr{A}^*\widetilde{P} + \widetilde{P}\mathscr{A} + \widetilde{P}(B_2B_2^* - \gamma^{-2}B_1B_1^*\widetilde{P}) - C_1^*C_1 = 0 \tag{5.43}$$

has a unique solution $\widetilde{P} \in L(H, H)$, $\widetilde{P} = \widetilde{P}^* \geq 0$. Moreover, in this case, the feedback $\widetilde{F} = -B_2^*\widetilde{P}$ is a γ-suboptimal solution to the H^∞-problem for (5.41), (5.42) *and $\mathscr{A} + (B_2B_2^* - \gamma^{-2}B_1B_1^*)\widetilde{P}$ is exponentially stable.*

As seen in Sect. 5.1, the H^∞-control problem for System (5.41) is closely related to the max-min problem (see (5.13), (5.14))

$$\max_{w \in L^2(R^+;W)} \min_{u \in L^2(R^+;U)} \int_0^\infty (|C_1y(t)|_Z^2 + |u|_U^2 - \gamma^2|w(t)|_W^2)dt$$

subject to (5.41), (5.42), and the optimal strategy (u^*, w^*) is under certain circumstances a saddle-point of this cost functional.

The operator C_1 may be selected in such a way that some physical properties of the fluid flow (for instance, kinetic energy or enstrophy) are minimized or kept within reasonable limits. We discuss below the existence of γ-suboptimal solutions of the H^∞-control problem associated with (5.41), (5.42) in some specific situations treated in Chap. 3. As a matter of fact, the problem we address here is whether the stabilizable feedback controllers designed there are robust in the sense of the H^∞-control theory.

5.2.1 *Internal Robust Stabilization with Regulation of Turbulent Kinetic Energy*

We place ourselves in the conditions of the internal stabilization Theorem 3.2 (or Theorem 3.3) and take $V = R^{M^*}$, $W = H$, $B_1 \in L(H, H)$ and

$$B_2 u = \sum_{j=1}^{M^*} P(m\psi_j)u_j, \quad u = \{u_j\}_{j=1}^{M^*} \in R^{M^*}, \tag{5.44}$$

$$C_1 y = P(m_1 y), \quad \forall y \in H, \tag{5.45}$$

$$D_{12} u = \sum_{j=1}^{M^*} P(m_2 \phi_j)u_j, \quad u \in R^{M^*}. \tag{5.46}$$

Here, $m = \mathbf{1}_{\mathcal{O}_0}$, $m_1 = \mathbf{1}_{\mathcal{O}_1}$, $m_2 = \mathbf{1}_{\mathcal{O}_2}$, where $\mathcal{O}_i$, $i = 0, 1, 2$, are open subsets of $\mathcal{O}$ such that $\mathcal{O}_1 \cap \mathcal{O}_2 = \emptyset$, ψ_j are chosen as in Theorem 3.2 and $\{\phi_j\}_{j=1}^{M^*}$ is an orthonormal system in $(L^2(\mathcal{O}_2))^d$. Then, all the assumptions in Theorem 5.3 are satisfied, including (kk) because the system $(\mathcal{A}, C_1)$ is detectable if and only if $(\mathcal{A}^*, C_1^*)$ is stabilizable and here this happens because the linear system

$$\frac{dy}{dt} + \mathcal{A}^* y = C_1^* v = P(m_1 v), \quad t \geq 0, \tag{5.47}$$

is exponentially stabilizable by virtue of Theorem 3.2.

Everywhere in the following, when the internal stabilization theorem from Chap. 3 is invoked, we mean the stabilizable feedback controller with decay of the form $e^{\delta t}$ for some $\delta > 0$. In other words, N is here the number of eigenvalues λ_j of the Stokes–Oseen operator $\mathcal{A}$ with $\operatorname{Re} \lambda_j \leq 0$, $j = 1, \dots, N$, and the dimension M^* of the stabilizable controller is that which corresponds to this choice. We keep here and everywhere in the following the symbol γ for the parameter arising in the definition of the H^∞-control problem.

We analyze below the performance of the feedback control (see (3.53))

$$u(t) = Fy(t) = -P\left(m \sum_{j=1}^{M^*} (Ry(t), \psi_i)_0 \psi_i\right) = -B_2^* R y(t) \tag{5.48}$$

in System (5.41). We have the following theorem.

Theorem 5.4 *Under the assumptions of Theorem* 3.3, *the feedback controller* (5.48) *stabilizes exponentially System* (5.41) *and has an L_2-gain less than* $2\|B_1^*\|_{L(H,H)} \times \|R\|_{L(U,H)}$. *In other words, F given by* (5.48) *is a γ-suboptimal solution to the H^∞-control problem with* $\gamma = 2\|B_1^*\|_{L(H,H)}\|R\|_{L(U,H)} + \delta$, *where δ is arbitrarily small.*

Proof We multiply the closed-loop system (5.41), where u is the feedback controller (5.48), by Ry, where R is the solution to the Riccati equation (3.52). Then, by (3.51) and (3.52) we find, after some calculations that

$$\frac{1}{2}(Ry(t), y(t)) + \frac{1}{2}\int_0^\infty (|A^{\frac{3}{4}}y(t)|^2 + |B_2^*Ry(t)|^2)dt$$

$$= \int_0^\infty (Ry(t), B_1w(t))dt \le \|B_1^*\|_{L(H,H)}\|R\|_{L(V,H)} \int_0^\infty |A^{\frac{1}{2}}y(t)|\,|w(t)|dt.$$

This yields

$$\int_0^\infty (|A^{\frac{3}{4}}y(t)|^2 + |B_2^*Ry(t)|^2)dt \le 4\|B_1^*\|^2_{L(H,H)}\|R\|^2_{L(V,H)} \int_0^\infty |w(t)|^2dt$$

and, therefore,

$$\int_0^\infty (|C_1y(t)|^2 + |u(t)|^2)dt \le (\gamma^2 - \varepsilon)\int_0^\infty |w(t)|^2dt, \quad \forall w \in L^2(R^+; W),$$

for some $\varepsilon > 0$. (We have taken here, for simplicity, the normalized estimates $|A^{\frac{1}{2}}y| \le |A^{\frac{3}{4}}y|$, $|y| \le |A^{\frac{1}{2}}y|$.) □

By the same argument, it follows that the low-gain feedback controller (3.76) is γ-suboptimal in the H^∞-control problem associated with (5.41).

Namely, one has the following theorem.

Theorem 5.5 *Under the assumptions of Theorem* 3.5, *the feedback controller* (3.76) *stabilizes exponentially System* (5.41) *and has L_2-gain less than* $2\|B_1^*\|_{L(H,H)} \times \|R_0\|_{L(U,H)}$.

Remark 5.1 The above theorems show that the stabilizable feedback controllers designed via Riccati equations (3.52) and (3.81) are robust with an L_2-gain of the order of $\|B_1^*\|_{L(H,H)}\|R\|_{L(V,H)}$, respectively $\|B_1^*\|_{L(H,H)}\|R_0\|_{L(H,H)}$. The above computation shows that this is not, however, the case with the stabilizable feedback controller given by (3.42).

5.2.2 Robust Internal Stabilization with the Regulation of Fluid Enstrophy

We consider here the input–output Stokes–Oseen system (5.41), where $W = H$, $U = R^{M^*}$, $B_1 \in L(H, H)$, $B_2 = L(U, H)$ given by (5.44), $D_{12} = 0$, $Z = V'$ and

$$C_1 y = P(\operatorname{curl} y), \quad \forall y \in V, \tag{5.49}$$

where $V = (H_0^1(\mathcal{O}))^d \cap H$ and V' is its dual in duality pairing with pivot space H. Then, $C_1 \in L(H, Z)$ and note also that Assumption (kk) is satisfied with the operator $\widetilde{K} \in L(V', H')$ defined by

$$\widetilde{K} = -\eta C_1^*,$$

where η is positive and sufficiently large. Indeed, it is easily seen that the linear system

$$\frac{dy}{dt} + \mathcal{A} y + \eta C_1 C_1^* y = 0, \quad \forall t \geq 0,$$

is exponentially stable for η large enough because there is $\omega > 0$ such that

$$\int_{\mathcal{O}} |C_1 y|^2 d\xi \geq \omega \|y\|_V^2.$$

We recall that $\operatorname{curl} y = \nabla \times y$ in 3-D and $\operatorname{curl} y = D_2 y_1 - D_1 y_2$, $y = (y_1, y_2)$ in 2-D. (See Sect. 3.8.)

We have the following theorem.

Theorem 5.6 *Under the assumptions of Theorem* 3.2, *the stabilizing feedback controller* (5.48) *has L_2-gain less than* $2\|B_1^*\|_{L(H,H)}\|R\|_{L(V,H)}$ *(up to a normalized constant) for* C_1, D_{12}, Z *given by* (5.42), (5.46) *and* (5.49), *respectively.*

Proof Arguing as in the proof of Theorem 5.4, we find that the solution y to the closed-loop system (5.41) with Controller (5.48) satisfies the estimate

$$\int_0^\infty (|A^{\frac{3}{4}} y(t)|^2 + |B_2^* R y(t)|^2) dt \leq 4\|B_1^*\|_{L(H,H)}^2 \|R\|_{L(V,H)}^2 \int_0^\infty |w|^2 dt.$$

Now, keeping in mind that

$$|A^{\frac{3}{4}} y| \geq |A^{\frac{1}{2}} y| \geq |\operatorname{curl} y| = |C_1 y|$$

(we have normalized the constants), we get

$$\int_0^\infty |C_1 y(t)|_Z^2 dt \leq 4\|B_1^*\|_{L(H,H)}^2 \|R\|_{L(V,H)}^2 \int_0^\infty |w(t)|^2 dt, \quad \forall w \in L^2(0, \infty; H),$$

as claimed. □

Remark 5.2 By Theorem 5.6 it follows that the high-gain feedback controller (5.48) has robust H^∞-performance with respect to distributed disturbances $w \in H$ in attenuation of the enstrophy output $\|\operatorname{curl} y\|_{L^2(0,\infty;V')}$. It is well-known that in fluid dynamics the enstrophy is related to energy of the fluid flow and has an important role in studying the turbulence. Thus, one might view the stabilizable feedback controller (5.48) as one which "keeps the turbulence" in certain limits if the disturbance norm $\|w\|_{L^2(0,\infty;H)} \le 1$. So, (5.48) can be seen as a high-gain robust controller. On the other hand, it is easily seen by the above computation that the low-gain feedback controller arising in (3.76) is no longer robust in this case. In other words, it has no attenuation effect on the enstrophy of the system in the presence of distributed disturbances w. This amounts to saying that it has a low H^∞-robustness performance than the high-gain controller.

Remark 5.3 The algebraic Riccati equation (5.43) provides an optimal stabilizable feedback law $u = -B_2^*\widetilde{P}y$ for the control system (5.41), (5.42) from the point of view of the corresponding differential game problem. Indeed, for a given γ this controller minimizes the effect of disturbances w and in the same time the cost of the enstrophy $|C_1 y|^2$. The optimal γ is that for which the Riccati equation (5.43) is solvable in the class of linear continuous, self-adjoint and positive operators in H. One might say, therefore, that the H^∞-problem for (5.41), (5.42) reduces to solvability of this Riccati equation.

On the other hand, as shown above in Theorem 5.1, Riccati equation (5.43) is equivalent with Hamiltonian System (5.6) corresponding to the associated differential game and this fact might be used to compute the solution $\widetilde{P}$ as in Sect. 5.1 (see (5.30)).

5.3 The H^∞-Control Problem for the Navier–Stokes Equations

The H^∞-control theory was extended in a natural way to nonlinear control systems with exogeneous inputs (disturbances) (see [10, 77, 78] for a few references on this subject). In the case of the Navier–Stokes equations, the H^∞-control problem is defined as follows.

Consider the input–output system governed by the Navier–Stokes system

$$\begin{aligned}
&\frac{\partial y}{\partial t} - \nu\Delta y + (y\cdot\nabla)y_e + (y_e\cdot\nabla)y + (y\cdot\nabla)y = \nabla p + B_2 u + B_1 w, \quad t\ge 0,\\
&\nabla\cdot y = 0 \quad \text{in } (0,\infty)\times\mathcal{O},\\
&y = 0 \quad \text{on } (0,\infty)\times\partial\mathcal{O},\\
&y(0) = y_0 \quad \text{in } \mathcal{O}.
\end{aligned} \tag{5.50}$$

Equivalently,

$$\begin{aligned}
&\frac{dy}{dt} + \mathscr{A}y + Sy = B_2 u + B_1 w,\\
&y(0) = y_0,
\end{aligned} \tag{5.51}$$

where $S = P((y\cdot\nabla)y)$, $\mathscr{A} = \nu A + A_0$, $B_2 \in L(U,H)$, $B_1 \in L(W,H)$. We have also a controlled output z given by (5.42), that is,

$$z = C_1 y + D_{12} u,$$

where $C_1 \in L(H,Z)$, $D_{12} \in L(U,H)$.

Everywhere in the following, we assume that Assumptions (k), (kk) and (kkk) hold.

In analogy with the linear H^∞-control theory we have the following definition.

Definition 5.1 Given $\gamma > 0$, $L \in \mathscr{L}(H, U)$ is a γ-suboptimal solution to the local H^∞-control problem for System (5.51) if there is a neighborhood $\mathscr{U}$ of the origin such that the closed-loop system

$$\begin{aligned} &\frac{d}{dt} y + \mathscr{A} y + Sy - B_2 L y = B_1 w \quad \text{in } R^+, \\ &y(0) = y_0, \end{aligned} \tag{5.52}$$

has locally L_2-gain less than γ in the sense that there are $\varepsilon > 0$ and $\mu : H \to R^+$, $\mu(0) = 0$, such that

$$\int_0^\infty (|C_1 y(t)|_Z^2 + |Ly(t)|_U^2) dt \le (\gamma^2 - \varepsilon) \int_0^\infty |w(t)|_W^2 dt + \mu(y_0), \tag{5.53}$$

for all $w \in L^2(R^+; W)$ and $y_0 \in \mathscr{U}$ such that the state-space trajectories y do not leave the neighborhood $\mathscr{U}$. Moreover, the state feedback $u = Ly$ stabilizes asymptotically the closed-loop system,

$$\frac{d}{dt} y + \mathscr{A} y + Sy - B_2 L y = 0, \qquad y(0) = y_0 \in H.$$

The main result of this paper, Theorem 5.7, is a local H^∞-control result for (5.51) and amounts to saying that if the H^∞-control problem for the linearized Navier–Stokes system has a γ-suboptimal solution, then the H^∞-control problem for System (5.51) has a γ-suboptimal solution in a neighborhood of the origin.

Theorem 5.7 *Assume that the H^∞-control problem for System* (5.41), (5.42) *has a γ-suboptimal solution $L \in L(U, H)$. Then, there is a neighborhood $\Sigma_\rho = \{y \in V; \|y\| \le \rho\}$, $\rho > 0$ and a unique map $G \in C^1(\Sigma_\rho, V)$ such that*

$$2(\mathscr{A} x, G(x)) - \gamma^{-2} |B_1^* G(x)|_W^2 + B_2^* G(x)|_U^2 - |C_1 x|_Z^2 = 0, \quad \forall x \in \Sigma_\rho, \tag{5.54}$$

$$G(0) = 0; \qquad \nabla G(0) = \widetilde{P}, \tag{5.55}$$

$$G(x) = \nabla \varphi(x), \quad \forall x \in \Sigma_\rho, \tag{5.56}$$

where $\varphi \in C^1(\Sigma_\rho)$. Moreover, the feedback controller

$$u = -B_2^* G(y) \tag{5.57}$$

stabilizes asymptotically System (5.51) *on Σ_ρ and has locally L_2-gain less than or equal to γ^*. That is, there is $C \ge 0$ such that*

$$\int_0^\infty (|C_1 y|_Z^2 + |B_2^* G(y)|_U^2) dt \le C |y_0|^2 + \gamma^2 \int_0^\infty |w|_W^2 dt, \tag{5.58}$$

$\forall w \in L^2(R^+; W)$ and all $y_0 \in \Sigma_\rho$ such that $\{y(t);\ t \in R^+\}$ does not leave Σ_ρ.

It turns out also by (5.54), (5.55) that $\widetilde{P}$ is the solution to the algebraic Riccati equation (5.43) corresponding to the H^∞-control problem for the linearized system (5.41), (5.42).

Corollary 5.1 *Under the assumptions of Theorem* 5.4 (*respectively, Theorem* 5.5) *there is a neighborhood* $\Sigma_\rho = \{y \in V;\ \|y\| \le \rho\}$ *and* $G \in C^1(\Sigma_\rho, V)$, *which is the unique solution to the Hamilton–Jacobi equation* (5.54)∼(5.55).

Moreover, the feedback controller (5.57) *stabilizes exponentially System* (5.51) *on* Σ_ρ *and has locally* L_2-*gain less or equal than* $2\|B_2^*\|_{L(H,H)}\|R\|_{L(U,H)}$ (*respectively, less than* $2\|B_1^*\|_{L(H,H)}\|R_0\|_{L(U,H)}$).

The main conclusion of Theorem 5.7 is that if the linearized Navier–Stokes equation around an equilibrium solution y_e has a robust linear feedback (in the sense of H^∞-theory), then also the Navier–Stokes equation has a robust H^∞-controller in a neighborhood of y_e. In the specific situations treated in Theorems 5.4, 5.5 and 5.6 (see, e.g., (5.48)), the operator B_2^* is of the form

$$B_2^* p = P\left(m\sum_{j=1}^{M^*}(p,\psi_i)_0\psi_j\right)$$

and so, has a finite-dimensional structure (in the space H). However, the infinite-dimensional Hamilton–Jacobi equation (5.54) still remains a very complex object and its solvability is a hard problem since, in general, it has not a global classical solution. As in the stabilization theory, one might ask about the size of the neighborhood Σ_ρ, where the local H^∞-control problem is solvable. One might expect that on a larger domain the Hamilton–Jacobi equation (5.54) has a generalized solution (a viscosity solution, for instance) which can provide a robust feedback controller, but this problem still remained open.

5.3.1 Proof of Theorem 5.7

As seen in Sect. 5.1, the H^∞-control problem can be redefined as a differential game on the product space $L^2(R^+;U) \times L^2(R^+;W)$. We consider, therefore, the max-min control problem (see (5.12))

$$\begin{aligned}
&\sup_w \inf_u \int_0^\infty (|z(t)|_Z^2 - \gamma^2|w(t)|_W^2)\,dt,\\
&\qquad \text{subject to } u \in L^2(R^+;U),\ w \in L^2(R^+;W),\\
&\frac{dy}{dt} + \nu Ay + A_0 y = B_1 w + B_2 u, \quad t > 0,\\
&y(0) = y_0,\\
&z = C_1 y + D_{12} u.
\end{aligned}$$

As seen earlier, the formal optimality system for this problem is expressed by

$$u = B_2^* p, \qquad w = -\gamma^{-2} B_1^* p,$$

where (y, p) is the solution to the Hamiltonian system

$$\frac{dy}{dt} + \nu A y + A_0 y + S y = (B_2 B_2^* - \gamma^{-2} B_1 B_1^*) p, \tag{5.59}$$

$$\begin{aligned} &\frac{dp}{dt} - \nu A p - A_0^* p - (\nabla S(y))^* p = C_1 C_1^* y, \quad t \geq 0, \\ &y(0) = y_0, \qquad p(\infty) = 0. \end{aligned} \tag{5.60}$$

Let φ be the value function associated with the above max-min problem, that is

$$\varphi(y_0) = \sup_w \inf_u \int_0^\infty (|z(t)|_Z^2 - \gamma^2 |w(t)|_W^2) dt. \tag{5.61}$$

In analogy with the linear H^∞-control theory presented in Sect. 5.1, one might suspect there is a close connection between the Hamiltonian system (5.59), (5.60), the Hamilton–Jacobi equation (5.54) and the optimal value function φ.

Indeed, a subset $\Gamma \subset H \times H$ is a C^1 asymptotically stable *invariant manifold* for the Hamiltonian system (5.59)–(5.60) provided for any $(y_0, p_0) \in \Gamma$ and $t_0 > 0$, there is a unique solution $(y(t), p(t))$ for (5.59)–(5.60) with $y(t_0) = y_0$, $p(t_0) = p_0$ such that $(y(t), p(t)) \in \Gamma$ for all $t \geq t_0$ and $\lim_{t\to\infty}(y(t), p(t)) = 0$ in $H \times H$.

If such an invariant manifold exists, then it can be represented as $\Gamma = \{(y, p);\ p + G(y) = 0\}$, where $G \in C^1$. Moreover, in this case, $\varphi = \nabla G$ and G is solution to (5.54). By this heuristic argument, it follows that the existence of an invariant manifold for the Hamiltonian system (5.59), (5.60) is equivalent with the existence theory for the Hamilton–Jacobi equation (5.54). This formal approach, which resembles the classical characteristic methods in first-order PDEs, is perhaps the simplest way to study (5.54). For a general Hamiltonian system of the form

$$\frac{dy}{dt} = -\nabla_y \mathscr{H}(y, p), \qquad \frac{dp}{dt} = \nabla_p \mathscr{H}(y, p),$$

this is true under certain strong regularity and growth conditions on the Hamiltonian function $\mathscr{H}$.

We pursue this approach here and our effort for the proof of Theorem 5.7 is oriented toward the existence of an asymptotically stable C^1-invariant manifold for (5.59)–(5.60) in sufficiently small neighborhood Σ_ρ of the origin. A key element of this approach is Proposition 5.1.

Proposition 5.1 *There is* $\rho > 0$ *such that, for all* $y_0 \in \Sigma_\rho$, *System* (5.59)–(5.60) *has a unique solution*

$$(y, p) \in (C([0, \infty); V) \cap L^2(R^+; D(A)) \cap W^{1,2}([0, \infty); H))^2.$$

Here, $W^{1,2}([0, \infty); H) = \{y \in L^2(R^+; H),\ y' \in L^2(R^+; H)\}$, where y' is the derivative of $y : R^+ \to H$ in the sense of H-valued distributions (equivalently, y is absolutely continuous, $y \in L^2(R^+; H)$ and $\frac{dy}{dt} = y' \in L^2(R^+; H)$). V is the space $(H_0^1(\mathscr{O}))^d \cap H$ with the norm $\|y\| = |A^{\frac{1}{2}} y|$.

First, we study the linearized Hamilton–Jacobi system corresponding to System (5.59)–(5.60).

Lemma 5.5 *Let $y_0 \in H$ and $f, g \in L^2(R^+; H)$ be given. Then, the Hamiltonian system*

$$\zeta'(t) + \nu A\zeta(t) + A_0\zeta(t) = (B_2B_2^* - \gamma^{-2}B_1B_1^*)q(t) + f(t), \quad a.e.\ t > 0, \tag{5.62}$$

$$\begin{aligned} q'(t) - \nu Aq(t) - A_0^* q(t) = C_1C_1^*\zeta(t) + g(t), \quad a.e.\ t > 0, \\ \zeta(0) = y_0, \qquad q(\infty) = 0, \end{aligned} \tag{5.63}$$

has a unique solution

$$(\zeta, q) \in (C([0,\infty); H) \cap L^2(R^+; V) \cap W^{1,1}([\delta,\infty); H))^2, \quad \forall \delta > 0,$$

which satisfies the estimates

$$\|\zeta\|_{L^2(R^+;V)} + \|q\|_{L^2(R^+;V)} \le C(|y_0| + \|f\|_{L^2(R^+;H)} + \|g\|_{L^2(R^+;H)}), \tag{5.64}$$

$$\begin{aligned} |\zeta(t)| + |q(t)| \le e^{-\alpha t}|\zeta_0| + C\int_0^t e^{-\alpha(t-s)}|\bar{f}(s)|ds \\ + \int_t^\infty e^{-\alpha(s-t)}|\bar{g}(s)|ds, \end{aligned} \tag{5.65}$$

where

$$\|\bar{f}\|_{L^2(R^+;H)} + \|\bar{g}\|_{L^2(R^+;H)} \le C(|y_0| + \|f\|_{L^2(R^+;H)} + \|g\|_{L^2(R^+;H)}). \tag{5.66}$$

If $y_0 \in V$, then

$$(\zeta, q) \in (C([0,\infty); V) \cap L^2(R^+; V) \cap W^{1,2}([0,\infty); H))^2$$

and

$$\begin{aligned} &\|A\zeta\|_{L^2(R^+;H)} + \|Aq\|_{L^2(R^+;H)} \\ &\quad \le C(\|\zeta_0\| + \|f\|_{L^2(R^+;H)} + \|g\|_{L^2(R^+;H)}), \end{aligned} \tag{5.67}$$

$$\begin{aligned} &\|\zeta(t)\| + \|q(t)\| \\ &\quad \le \left(e^{-\alpha t}\|y_0\| + \int_0^t e^{-\alpha(t-s)}|\bar{f}(s)|ds + \int_t^\infty e^{-\alpha(s-t)}|\bar{g}(s)|ds\right). \end{aligned} \tag{5.68}$$

Here, A_0^* is the adjoint of A_0, C is a positive constant independent of y_0, f, g and $\alpha > 0$.

Proof Since the H^∞-control problem for the linear system (5.41), (5.42) has a γ-suboptimal solution $u = L\zeta$, $L \in L(H, H)$, we have

$$\begin{aligned}&\int_0^\infty (|C_1\zeta|_Z^2+|L\zeta|_U^2+2(g,\zeta))dt\\&\quad\le(\gamma^2-2^{-1}\varepsilon)\int_0^\infty |w|_W^2dt\\&\qquad+C(|y_0|^2+\|f\|^2_{L^2(R^+;H)}+\|g\|^2_{L^2(R^+;H)}),\end{aligned}\tag{5.69}$$

for all $w\in L^2(R^+;W)$ and $\zeta\in C([0,\infty);H)\cap L^2(R^+;V)\cap W^{1,2}([\delta,\infty);H)$, $\forall\delta>0$, the solution to the closed-loop system

$$\begin{aligned}&\zeta'+\nu A\zeta+A_0\zeta+B_2L\zeta=B_1w+f,\quad \text{a.e. } t>0,\\&\zeta(0)=y_0.\end{aligned}$$

We establish (5.69) in the following way. Since the linearized H^∞-control problem has a γ-suboptimal solution, we have

$$\int_0^\infty (|C_1\tilde\zeta q|_Z^2+|L\tilde\zeta|_U^2)dt\le(\gamma^*-2^{-1}\varepsilon)\int_0^\infty |w|_W^2dt,\tag{5.70}$$

for some $\varepsilon>0$, where

$$\tilde\zeta(t)=\int_0^t S_{\nu A+A_0-B_2L}(t-r)B_1w(r)dr.$$

We note that $S_{\nu A+A_0-B_2L}(t)=\exp(-t(\nu A+A_0-B_2L))$ is an exponential stable semigroup. Now, we write the solution of the inhomogeneous linearized problem as

$$\begin{aligned}\zeta(t)=S_{\nu A+A_0-B_2L}(t)y_0&+\int_0^t S_{\nu A+A_0-B_2L}(t-r)f(r)dr\\&+\int_0^t S_{\nu A+A_0-B_2L}(t-r)B_1w(r)dr,\end{aligned}\tag{5.71}$$

that is,

$$\zeta(t)=\tilde\zeta(t)+\stackrel{\approx}{\zeta}(t),\tag{5.72}$$

with

$$\stackrel{\approx}{\zeta}(t)=S_{\nu A+A_0-B_2L}(t)y_0+\int_0^t S_{\nu A+A_0-B_2L}(t-r)f(r)dr.\tag{5.73}$$

Now, using the exponential stability of the semigroup and also Young's inequality for the integral, and then using the fact that $C_1\in\mathscr{L}(H,Z)$ and $L\in\mathscr{L}(H,U)$, we obtain

$$\int_0^\infty (|C_1\stackrel{\approx}{\zeta}(t)|_Z^2+|L\stackrel{\approx}{\zeta}(t)|_U^2)dt\le C\left(|y_0|^2+\int_0^\infty |f(t)|^2dt\right).\tag{5.74}$$

Note also that

$$\int_0^\infty (g,\zeta)dt=\int_0^\infty (g,\tilde\zeta)dt+\int_0^\infty (g,\stackrel{\approx}{\zeta})dt.$$

Using (5.71), we get

$$\left|\int_0^\infty (g,\zeta)dt\right| \le \varepsilon_1 \int_0^\infty |w|^2 dt + C_{\varepsilon_1}\left(|y_0|^2 + \int_0^\infty |f(t)|^2 dt + \int_0^\infty |g(t)|^2 dt\right). \tag{5.75}$$

Hence, combining (5.70), (5.74) and (5.75), we get (5.69).

Now, let us show that the sup-inf problem

$$\sup_{w\in L^2(R^+;H)} \inf_{u\in L^2(R^+;U)} \left\{\int_0^\infty (|C_1\zeta|_Z^2 + |u|_U^2 + 2(g,\zeta) - \gamma^2|w|_W^2)dt;\right.$$
$$\left.\zeta' + \nu A\zeta + A_0\zeta = B_2u + B_1w + f;\ y(0) = y_0\right\} \tag{5.76}$$

has at least one solution

$$(\zeta, u, w) \in (C([0,\infty); H)\cap L^2(R^+; V)) \times L^2(R^+; U) \times L^2(R^+; W),$$
$$\zeta \in W^{1,2}([\delta,\infty); H)\cap L^2((\delta,\infty); D(A)), \quad \forall\delta > 0.$$

It should be noted that (5.62), (5.63) is just the Hamiltonian system corresponding to the differential game (5.76). In fact, if we write the Hamiltonian $\mathscr{H}$ of the linearized system as

$$\mathscr{H}(\zeta,q) := -\langle \nu A\zeta + A_0\zeta, q\rangle - \frac{1}{2}\{|C_1\zeta|_Z^2 - |B_2^*q|_U^2 + \gamma^{-2}|B_1^*q|_W^2\}$$
$$+ \langle f, g\rangle - \langle g,\zeta\rangle,$$

then (5.62), (5.63) can be written as

$$\frac{d\zeta}{dt} = \nabla_q\mathscr{H} \quad\text{and}\quad \frac{dq}{dt} = -\nabla_\zeta\mathscr{H}.$$

Moreover, the substitution of the linear invariant manifold representation $q = -\widetilde{P}\zeta$ in the above Hamiltonian system (with $f = g = 0$) will immediately produce the operator–Riccati equation (5.43) for $\widetilde{P}$. (See Lemma 5.1.)

We set, for any $w \in L^2(R^+; W)$,

$$\psi(w) := \inf_{u\in L^2(R^+;U)} \left\{\int_0^\infty (|C_1\zeta|_Z^2 + |u|_U^2 + 2(g,\zeta))dt;\right.$$
$$\left.\zeta' + \nu A\zeta + A_0\zeta = B_2u + B_1w + f;\ \zeta(0) = w_0\right\}. \tag{5.77}$$

Note that the function

$$J(w) := \gamma^2 \int_0^\infty |w|_W^2 dt - \psi(w)$$

is convex and coercive. In fact, coerciveness can be seen from

$$J(w) = \gamma^2 \int_0^\infty |w|^2 dt - \int_{u\in L^2(R^+;U)} \left\{\int_0^\infty (|C_1\zeta|_Z^2 + |u|_U^2 + 2(g,\zeta))dt\right\}$$
$$\ge \gamma^2 \int_0^\infty |w|^2 dt - \int_0^\infty (|C_1\zeta|_Z^2 + |u|_U^2 + (g,\zeta))dt.$$

Now, using Estimate (5.69), we get

$$J(w) \geq \frac{1}{2}\varepsilon \int_0^\infty |w|_W^2 dt - C\left(|y_0|^2 + \int_0^\infty |f|^2 dt + \int_0^\infty |g|^2 dt\right).$$

That is, for $y_0 \in H$ and $f, g \in L^2(R^+; H)$, we have

$$J(w) \geq \frac{1}{2}\varepsilon \int_0^\infty |w|_W^2 dt - C_0, \quad \forall w \in L^2(R^+; W),$$

which is the coerciveness property. The convexity of J is immediate because J is a quadratic function. The problem

$$\sup_{w \in L^2(R^+; W)} \left\{ \psi(w) - \gamma^2 \int_0^\infty |w|_W^2 dt \right\} \tag{5.78}$$

has a unique solution w^*. Hence, the inf-sup problem (5.76) has a unique solution $(u^*, w^*) \in L^2(R^+; U) \times L^2(R^+; W)$.

For each $w \in L^2(R^+; W)$, let us denote by $\bar{u} = \Gamma w$ the solution to Problem (5.77), that is,

$$\bar{u} = \Gamma w = \arg \inf_{u \in L^2(R^+; U)} \left\{ \int_0^\infty (|C_1 \zeta|_Z^2 + |u|_U^2 + 2(g, \zeta))dt; \right.$$
$$\left. \zeta' + \nu A \zeta + A_0 \zeta = B_2 u + B_1 w + f;\ \zeta(0) = y_0 \right\}.$$

We then get, from the LQG theory (see, e.g., [9, 32]), $\exists \bar{q} \in L^2(R^+; X) \cap C(R^+; X)$, such that

$$\begin{aligned} &-\bar{q}_t + (\nu A + A_0^*)\bar{q} = C_1 C_1^* \bar{\zeta} + g, \quad t > 0, \\ &\bar{q}(\infty) = 0, \quad \bar{u} = B_2 \bar{q}, \quad \forall t \geq 0. \end{aligned} \tag{5.79}$$

Now, assuming stabilizability, we write

$$\begin{aligned} &\bar{q}_t + (\nu A + A_0 + B_2 D)^* \bar{q} = D^* \bar{u} + C_1 C_1^* \bar{\zeta} + g, \quad t > 0, \\ &\bar{q}(\infty) = 0. \end{aligned} \tag{5.80}$$

Since $\nu A + A_0 + B_1 D$ is exponentially stable, we deduce that the solution $\bar{q}$ to (5.80) is unique and

$$|\bar{q}(t)| \leq C \int_t^\infty \exp(-\alpha(s-t))(|\bar{u}(s)| + |C_1 \bar{\zeta}(s)| + |g(s)|)ds, \quad t \geq 0, \tag{5.81}$$

where $\alpha > 0$. Hence, $\bar{q} \in L^2(R^+; H)$.

Note also that

$$\nabla \psi(w) = -2B_1^* \bar{q}, \quad \text{a.e. } t > 0, \tag{5.82}$$

where $\bar{q}$ is the solution to the above equation. In fact, if ζ^w is the solution of

$$\begin{aligned} &\zeta_t^w + (\nu A + A_0)\zeta^w = B_2 \Gamma w + B_1 w + f, \\ &\zeta^w(0) = y_0, \end{aligned} \tag{5.83}$$

then we have, $\forall \widehat{w} \in L^2(R^+; W)$,

$$\begin{aligned}
&\psi(w) - \psi(\widehat{w}) \\
&= \int_0^\infty (|C_1\zeta^w|_Z^2 - |C_1\zeta^{\widetilde{h}}|_Z^2)dt \\
&\quad + \int_0^\infty (|\Gamma w|_W^2 - |\Gamma \widetilde{h}|_W^2)dt + 2\int_0^\infty (g, \zeta^w - \zeta^{\widetilde{w}})dt \\
&\le 2\int_0^\infty [(C_1C_1^*\zeta^w, \zeta^w - \zeta^{\widetilde{w}}) + (\Gamma w, \Gamma(w - \widehat{w})) + (g, \zeta^w - \zeta^{\widehat{w}})]dt.
\end{aligned}$$

Then, using (5.78) and integrating by parts, we get

$$\begin{aligned}
\psi(w) - \psi\widehat{w}) &\le -2\int_0^\infty [(q, B_1(w - \widehat{w}) + (B_2^*q, \Gamma(w - \widehat{w})) \\
&\quad + (\Gamma w, \Gamma(w - \widehat{w}))]dt \\
&= -2\int_0^\infty (B_1^*q, w - \widehat{w})dt.
\end{aligned}$$

Hence, (5.82) holds, as claimed. Moreover, from (5.78), we have

$$\nabla\psi(w^*) - 2\gamma w^* = 0,$$

and, hence, comparing with (5.82), we get

$$w^* = -\gamma B_1^*q.$$

Since w^* solves (5.78), we see by (5.80), (5.82) that (u^*, w^*) solves (5.76). Since $C_1\zeta^* \in L^2(R^+; Z)$, it follows by detectability that $\zeta^* \in L^2(R^+; H)$.

Similarly, applying Young's inequality to (5.85), we see that $q \in L^2(R^+; H)$ and

$$\lim_{t\to\infty} q(t) = q(\infty) = 0.$$

We may conclude, therefore, that there is a solution (ζ, q) to (5.62), (5.63) such that

$$(\zeta, q) \in (C([0, T]; H) \cap L^2(0, T; V) \cap W^{1,2}([\delta, T]; H))^2, \quad \forall T > 0.$$

If $\zeta_0 \in V$, we have that

$$\zeta, q \in C([0, T]; V) \cap L^2(0, T; D(A)) \cap W^{1,2}([0, T]; H), \quad \forall T > 0.$$

(This follows from the regularity properties of solutions to linear evolution equations with principal part self-adjoint operators.) Moreover, inequality is satisfied with q instead of $L\zeta$. Thus, for $f = g$, it follows by (5.69) that

$$\int_0^\infty (|C_1\zeta|_Z^2 + |B_2^*q|_U^2 - \gamma^{-2}|B_1^*q|_W^2)dt \le -2^{-1}\varepsilon \int_0^\infty |B_1^*q|_W^2 dt.$$

Since, by virtue of (5.62), (5.63), we have

$$\int_0^\infty (|C_1\zeta|_Z^2 + |B_2^*q|_U^2 - \gamma^{-2}|B_1^*q|_W^2)dt = 0,$$

we infer that $B_1^*q = 0$, $C_1\zeta = 0$, $B_2^*q = 0$ and, therefore, $\zeta = 0$, $q = 0$. Hence, the solution (ζ, q) is unique. In general, we have by (5.69), (5.76) and (5.62), (5.63)

$$\int_0^\infty (|C_1\zeta|_Z^2 + |B_2^*q|_U^2 - \gamma^{-2}|B_1^*q|_W^2 + 2(g,\zeta))dt$$
$$\leq -2^{-1}\varepsilon \int_0^\infty |B_1^*q|_W^2 dt + C(|y_0|^2 + \|f\|_{L^2(R^+;H)}^2 + \|g\|_{L^2(R^+;H)}^2)$$

and

$$\int_0^\infty (|C_1\zeta|_Z^2 + |B_2^*q|_U^2 - \gamma^{-2}|B_1^*q|_W^2 + (g,\zeta) + (f,q))dt = 0.$$

We get, therefore,

$$\int_0^\infty (|C_1\zeta|_Z^2 + |B_2^*q|_U^2 + |B_1^*q|_W^2)dt$$
$$\leq C\left(\int_0^\infty ((f,g) + (g,\zeta))dt + |y_0|^2 + \|f\|_{L^2(R^+;H)}^2 + \|g\|_{L^2(R^+;H)}^2\right).$$

By Hypotheses (k) and (kk), we may write System (5.62), (5.63) as

$$\zeta' + \nu A\zeta + A_0\zeta + KC_1\zeta = (B_2B_2^* - \gamma^{-2}B_1B_1^*)q + KC_1y + f \quad (5.84)$$

and

$$q' - \nu Aq - A_0^*q - L^*B_2^*q = C_1C_1^*\zeta - L^*B_2^*q + g, \quad (5.85)$$

where $u = L\zeta$ is a stabilizable feedback controller. We represent the solution to (5.84) as

$$\zeta(t) = S_{\nu A+A_0+KC_1}(t)y_0 + \int_0^t S_{\nu A+A_0+KC_1}(t-r)\psi(r)dr,$$

where $S_{\nu A+A_0+KC_1}(t)$ is the exponentially stable semigroup generated by $\nu A + A_0 + KC_1$ and ψ is the right-hand side of (5.84). Then, we have

$$|\zeta(t)| \leq |y_0|\exp(-\alpha t) + C\int_0^t \exp(-\alpha(t-r))|\bar{f}(r)|dr,$$

where

$$|\bar{f}| = |B_1^*q| + |B_2^*q| + |C_1\zeta| + |f|.$$

Similarly, we also get

$$|q(t)| \leq C\int_t^\infty \exp(-\alpha(r-t))|\bar{g}(r)|dr,$$

where

$$|\bar{g}| = |C_1\zeta| + |B_2^*q|.$$

Here, from the earlier estimate, we have

$$\begin{aligned}\|\bar{f}\|_{L^2(R^+;H)} + \|\bar{g}\|_{L^2(R^+;H)} &\le \varepsilon(\|q\|^2_{L^2(R^+;H)} + \|\zeta\|^2_{L^2(R^+;H)})\\ &\quad + C_\varepsilon(\|y_0\|^2 + \|f\|_{L^2(R^+;H)} + \|g\|^2_{L^2(R^+;H)}).\end{aligned}$$

Now,

$$\begin{aligned}|\zeta(t)| + |q(t)| \le |y_0|\exp(-\alpha t) + C\int_0^t \exp(-\alpha(t-r))|\bar{f}(r)|dr\\ + C\int_t^\infty \exp(-\alpha(r-t))|\bar{g}(r)|dr.\end{aligned}$$

Applying Young's inequality to the integrals on the right-hand side, we get

$$\begin{aligned}\|\zeta\|^2_{L^2(R^+;H)} + \|q\|^2_{L^2(R^+;H)} &\le C(|y_0|^2 + \|\bar{f}\|^2_{L^2(R^+;H)} + \|\bar{g}\|^2_{L^2(R^+;H)})\\ &\le \varepsilon(\|q\|^2_{L^2(R^+;H)} + \|\zeta q\|^2_{L^2(R^+;H)})\\ &\quad + C_\varepsilon(|y_0|^2 + \|f\|^2_{L^2(R^+;H)} + \|g\|^2_{L^2(R^+;H)}).\end{aligned}$$

Hence,

$$\|\zeta\|^2_{L^2(R^+;H)} + \|q\|^2_{L^2(R^+;H)} \le C_1(|y_0|^2 + \|f\|^2_{L^2(R^+;H)} + \|g\|^2_{L^2(R^+;H)}).$$

Using this again on the estimates for $\bar{f}$ and $\bar{g}$, we get (5.64), (5.65). Now, recalling that

$$\begin{aligned}&(\zeta' + \nu A\zeta, \zeta) = 2^{-1}(|\zeta|^2)' + \nu\|\zeta\|^2, \quad \text{a.e., in } R^+,\\ &(A_0\zeta, \zeta) \ge -\alpha_0|\zeta|^2, \quad \forall \zeta \in V,\end{aligned}$$

by (5.84), (5.85) and (5.64), (5.65), Estimate (5.64) follows. Finally, if $y_0 \in V$, we have by (5.62), (5.63) that

$$\begin{aligned}(\|\zeta(t)\|^2)' + 2\nu|A\zeta(t)|^2 \le C(&|A\zeta(t)|\,\|\zeta(t)\| + |A\zeta(t)|(|\zeta(t)|\,\|\zeta(t)\|)^{\frac{1}{2}}\\ &+ |A\zeta(t)|\,|q(t)| + |f(t)|)), \quad \text{a.e. } t > 0,\end{aligned}$$

and a similar inequality for $p(t)$. Integrating from 0 to t and using the Holder inequality and the previous estimates, we get (5.65) and (5.66), thereby completing the proof of Lemma 5.5. □

Proof of Proposition 5.1 Let $y_0 \in V$ be arbitrary but fixed. Denote by

$$\Gamma : L^2(R^+; H) \times L^2(R^+; H) \to L^2(R^+; H) \times L^2(R^+; H)$$

the operator defined by

$$\Gamma(f, g) = (\zeta, q), \tag{5.86}$$

where (ζ, q) is the solution to System (5.62), (5.63). By Lemma 5.5 we know that Γ is Lipschitzian from $L^2(R^+; H) \times L^2(R^+; H)$ to $(C([0,\infty); V) \cap L^2(R^+; D(A)))^d$

or to $(C([0,\infty);H)\times L^2(R^+;V))^d$ if $y_0\in V$. In terms of the mapping Γ, we may rewrite the solution to the nonlinear system (5.59)–(5.60) as

$$(y,p)=\Gamma(-S(y),(\nabla S(y))^*p). \tag{5.87}$$

We set

$$\mathscr{E}(y,p)=(-S(y),(\nabla S(y))^*p),\qquad \mathscr{G}=\Gamma\circ\mathscr{E}.$$

By Proposition 1.7, we have

$$|b(y,z,w)|\le C|A^{\frac34}y|\,\|z\|\,|w|\le C|Ay|^{\frac12}\|y\|^{\frac12}\|z\|\,|w|\quad \text{if } d=2,$$
$$|b(y,z,w)|\le C|A^{\frac34}y|\,|A^{\frac32}z|\,|w|\le C|Ay|^{\frac12}\|y\|^{\frac12}\,|Az|^{\frac12}\|z\|^{\frac12}|w|\quad \text{if } d=3,$$

where $\|y\|=|A^{\frac12}y|$.

For simplicity, we work below in the case $d=2$ only, since the case $d=3$ can be treated completely similarly.

This yields

$$\begin{aligned}
|S(y)|&\le C(\|y\||Ay|)^{\frac12}\|y\|,\quad \forall y\in D(A).\\
|S(y)-S(z)|&\le C((\|y\||Ay||y-z|\|y-z\|)^{\frac12}\\
&\quad+(|z|\|z\|\|y-z\||A(y-z)|)^{\frac12},\quad \forall y,z\in D(A).\\
|(\nabla S(y))^*)p))&\le C((\|y\||Ay||p|\|p\|)^{\frac12}\\
&\quad+(|y|\|y\||p||Ap|)^{\frac12},\quad \forall y,p\in D(A).
\end{aligned} \tag{5.88}$$

$$\begin{aligned}
|(\nabla S(y))^*(p)-(S(z))^*(r)|&\le C(\|y\||Ay||p-r|\|p-r\|)^{\frac12}\\
&\quad+(\|y-z\||A(y-z)||r|\|r\|)^{\frac12}\\
&\quad+(|y|\|y\|\|p-r\||A(p-r)|)^{\frac12}\\
&\quad+(|y-z|\|y-z\|\|p\||Ap|)^{\frac12}),\\
&\quad\forall y,z,p,r\in D(A).
\end{aligned}$$

We have, therefore,

$$\begin{aligned}
&\|\mathscr{E}(y,p)-\mathscr{E}(z,r)\|_{L^2(R^+;H)\times L^2(R^+;H)}\\
&\quad\le C(\|y-z\|_{C([0,\infty);V)}(\|Ay\|_{L^2(R^+;H)}+\|Ap\|_{L^2(R^+;H)})\\
&\qquad+\|p-r\|_{C([0,\infty);V)}\|Ay\|_{L^2(R^+;H)}+\|(A(p-r)\|_{L^2(R^+;H)}\|y\|_{C([0,\infty);V)}\\
&\qquad+\|A(y-z)\|_{L^2(R^+;H)}(\|p\|_{C([0,\infty);V)}+\|z\|_{C([0,\infty);V)}),
\end{aligned} \tag{5.89}$$

for all $y,p,z,r\in L^2(R^+;D(A))\cap C([0,\infty);V)$.

$$\begin{aligned}
\|\mathscr{E}(y,p)\|_{L^2(R^+;H)}&\le C(\|y\|_{C([0,\infty);V)}\|Ay\|_{L^2(R^+;H)}\\
&\quad+\|p\|_{C([0,\infty);V)}\|Ay\|_{L^2(R^+;H)}\\
&\quad+\|y\|_{C([0,\infty);V)}\|Ap\|_{L^2(R^+;H)}),
\end{aligned} \tag{5.90}$$

for all $y,p\in L^2(R^+;D(A))\cap C([0,\infty);V)$.

In the space $X = (L^2(R^+; H) \cap C([0,\infty); V))^2$, consider the subset

$$\mathscr{K} = \{(y,p) \in X;\ \|y\|_{C([\infty,0);V)} + \|p\|_{C([0,\infty);V)} \le \rho,\quad \|y\|_{L^2(R^+;D(A))} + \|p\|_{L^2(R^+;D(A))} \le \rho\}. \tag{5.91}$$

By Lemma 5.5 and estimates (5.89), (5.90), it follows that, for $\|y_0\|$ and ρ sufficiently small, the operator $\mathscr{G}$ maps $\mathscr{K}$ into itself and is a contraction on X. Hence, System (5.59), (5.60) has a unique solution $(y,p) \in X$ for a sufficiently small $\|y_0\|$. Moreover, by the regularity theory of linear evolution equations with self–adjoint operators, it follows that this solution also satisfies the desired condition. This completes the proof. □

Now, we show that there is an invariant manifold for the Hamiltonian system and that this leads to C^1 solutions to the Hamilton–Jacobi–Isaac equation in a small neighborhood. Consider the map $G : \Sigma_\rho \to V$ defined by

$$G(y_0) = -p(0), \tag{5.92}$$

where (y,p) is the solution to System (5.59), (5.60). By uniqueness in (5.59), (5.60), it follows that

$$p(t) = -G(y(t)), \quad \forall t \ge 0, \tag{5.93}$$

for all the solutions (y,p) having the property that $y(t) \in \Sigma_\rho$, $\forall t \ge 0$. (This happens, for instance, for all the solutions with $\|y_0\|$ sufficiently small.) In other words, $\{(y,p);\ p + G(y) = 0;\ y \in \Sigma_\rho\}$ is a positively invariant manifold for the Hamiltonian system (5.59), (5.60). Moreover, by Lemma 5.5 and estimates (5.89), (5.90), it follows that G is Lipschitzian on Σ_ρ and $G(0) = 0$.

Next, consider the value function $\varphi : \Sigma_\rho \to R$,

$$\varphi(y_0) = 2^{-1} \int_0^\infty (|C_1 y|_Z^2 + |B_2^* p|_U^2 - \gamma^{-2} |B_2^* p|_W^2)\,dt, \tag{5.94}$$

where (y,p) is the solution to (5.59), (5.60). It is readily seen that the map $y_0 \to (y,p)$ denoted Φ is Fréchet differentiable from Σ_ρ to $L^2(R^+; H \times L^2(R^+; H)$ and, for all $h \in V$,

$$\nabla\Phi(y_0)(h) = (z,q),$$

where (z,q) is the solution to

$$\begin{aligned} z' + \nu Az + A_0 z(y) z &= (B_2 B_2^* - \gamma^{-2} B_1 B_1^*) q \quad \text{in } R^+, \\ q' - \nu Aq - A_0^* q &= C_1 C_1^* z + (\nabla^2 B(z))^* q \quad \text{in } R^+, \\ z(0) = h, &\qquad q(\infty) = 0. \end{aligned} \tag{5.95}$$

Here, $(\nabla^2 S(z))^* p \in H$ is defined by

$$((\nabla^2 S(z))^* q, w) = b(z,w,q) + b(w,z,q), \quad \forall w \in H.$$

Arguing as in the proof of Proposition 5.1, it follows via Lemma 5.5 that System (5.95) has a unique solution

$$(z,q) \in (C([0,\infty); V) \cap L^2(R^+; D(A)) \cap W^{1,2}([0,\infty); H))^2.$$

This implies that G is differentiable on Σ_ρ and

$$\nabla G(y_0)h = -q(0),$$

where (z, q) is the solution to (5.95). In fact, defining the linear operator Γ_0 as

$$(\Gamma_0 \circ \Phi)(y_0) = (y_0, -G(y_0)),$$

we note that

$$\nabla(\Gamma_0 \circ \Phi)(y_0)(h) = \Gamma_0 \circ (\nabla\Phi(y_0))(h) = (h, -\nabla G)(y_0)(h).$$

In particular, for $y_0 = 0$ (since the corresponding solution y in (5.59), (5.60) is zero), it follows that $\nabla G(y_0)(h) = -q(0)$, where (z, q) is the solution to the system

$$\begin{aligned} z' + \nu Az + A_0 z &= (B_2B_2^* - \gamma^{-2}B_1B_1^*)q, \quad t > 0,\\ q' - \nu Aq - A_0^* q &= C_1C_1^* z,\\ z(0) = h,& \qquad q(\infty) = 0. \end{aligned} \tag{5.96}$$

This is, precisely, the Hamiltonian system corresponding to the H^∞-control problem for the linear system (5.41), (5.42). Hence, $\nabla G(0) = \widetilde{P}$, where $\widetilde{P}$ is the solution to the algebraic Riccati equation (5.43). Next, by (5.59), (5.60), (5.94), (5.95), we see that

$$\varphi(y(t)) = 2^{-1} \int_t^\infty (|C_1y(s)|_Z^2 + |B_2^*p|_U^2 - \gamma^{-2}|B_1^*p|_W^2), \quad \forall t \in R^+. \tag{5.97}$$

This means that

$$\nabla\varphi(y_0) = G(y_0), \quad \forall y_0 \in \Sigma_\rho.$$

Then, by (5.93) and (5.94), we have that

$$(\varphi(y(t)))' = (y'(t), Gy(t)) = -(y'(t), p(t)), \quad \text{a.e. } t \in R^+,$$

and

$$\varphi(y(t)) = 2^{-1} \int_t^\infty (|C_1y(s)|_Z^2 + |B_2^*p|_U^2 - \gamma^{-2}|B_1^*p|_W^2)dt, \quad \forall t \in R^+. \tag{5.98}$$

This yields

$$\begin{aligned} &((B_2B_2^* - \gamma^{-2}B_1B_2^*)p(t), p(t)) - ((\nu Ay(t) + A_0y(t) + S(y(t)), p(t))\\ &\quad - 2^{-1}(|C_1y(t)|_Z^2 + |B_2^*p(t)|_W^2 - \gamma^{-2}|B_1^*p(t)|_W^2) = 0, \quad \forall t \geq 0, \end{aligned}$$

that is,

$$\begin{aligned} &|B_2^*p(t)|_U^2 - \gamma^{-2}|B_1^*p(t)|_W^2 - 2(\nu Ay(t) + A_0y(t)\\ &\quad + Sy(t), p(t)) - |C_1y(t)|_Z^2 = 0, \quad \forall t \geq 0. \end{aligned}$$

Taking into account (5.93), for $t = 0$, we get the Hamilton–Jacobi–Isaac equation (5.54), as claimed.

For $\|y_0\| < \rho$ and $w \in L^2(R^+; W)$ the closed-loop system

$$y' + \nu Ay + A_0 y + Sy + B_2 B_2^* G(y) = B_1 w \quad \text{in } R^+, \qquad y(0) = y_0, \tag{5.99}$$

has a unique local solution

$$y \in C([0,T]; V) \cap L^2(R^+; D(A)) \cap W^{1,2}([0,T]; H)$$

such that $y(t) \in \Sigma_\rho$, $\forall t \in [0,T]$ for some $T > 0$. This follows by usual fixed-point arguments taking into account that G is Lipschitzian in V on Σ_ρ. Let us assume that this solution is global, that is, it exists on all of R^+. Then, if we multiply (5.99) by $G(y(t))$, we see by (5.54) that

$$\begin{aligned}2(\varphi(y(t)))' &+ |B_2^* G(y(t)|_U^2 + |C_1 y(t)|_Z^2 + \gamma^{-2}|B_1^* G(y))|_W^2 \\ &= 2(B_1^* G(y(t)), w(t)).\end{aligned}$$

This yields

$$\begin{aligned}&\int_0^T (|C_1 y|_Z^2 + |B_2^* G(y)|_U^2 - \gamma^2 |w|_W^2)dt \\ &\quad = 2(\varphi(y_0) - \varphi(y(T))) - \int_0^T |\gamma^{-1} B_1^* G(y) - \gamma w|_W^2 dt \\ &\quad \le C\|y_0\|^2 - \varphi(y(T)), \quad \forall T > 0,\end{aligned} \tag{5.100}$$

because by (5.54) it follows that $\varphi(y) \le C\|y\|^2$, $\forall y \in \Sigma_\rho$. Letting T tend to ∞, we get Inequality (5.58), that is,

$$\int_0^\infty (|C_1 y|_Z^2 + |B_2^* G(y)|_U^2 - \gamma^2 |w|_W^2)dt \le C\|y_0\|^2, \quad \forall w \in L^2(R^+; W).$$

In particular, for $w = 0$, the latter implies that the function $h = B_2 B_2^* G(y)$ is in $L^2(R^+; H)$. Consider System (5.99) (with $w = 0$)

$$y' + \nu Ay + A_0 \eta + B_2 B_2^* G(y) = 0, \qquad y(0) = y_0.$$

Let $K \in \mathscr{L}(Z, H)$ be such that $\nu A + A_0 + KC_1$ generates an exponentially stable semigroup $S_{\nu A + A_0 + KC_1}(t)$ (such a K exists because of the detectability hypothesis).

Now, we write the nonhomogeneous System (5.99) as

$$\frac{d}{dt} y + (\nu A + A_0 + KC_1)y = KC_1 y + h - S(y).$$

Note that the inertial term has the estimate

$$\|S(y)\|_{D(A^{-s})} \le C|y|\|y\|,$$

with $s = \frac{1}{2}$ in two-dimensions and $s \ge \frac{3}{4}$ for three-dimensions (we may take $s = 1 - \varepsilon$ for the arguments below to work for three-dimensions). Then, we have

$$\|S(y)\|_{D(A^{-s})} \le C\rho|y|, \quad \forall y \in \Sigma_\rho.$$

Thus, writing

$$\begin{aligned}
y(t) = S_{\nu A+A_0+KC_1}(t)y_0 &- \int_0^t S_{\nu A+A_0+KC_1}(t-r)A^{\frac{1}{2}}A^{-\frac{1}{2}}S(y(r))dr \\
&+ \int_0^t S_{\nu A+A_0+KC_1}(t-r)h(r)dr \\
&+ \int_0^t S_{\nu A+A_0+KC_1}(t-r)KC_1 y(r)dr,
\end{aligned}$$

we estimate

$$\begin{aligned}
|y(t)| \le \exp(-\alpha t)|y_0| &+ C\rho \int_0^t \frac{1}{(t-r)^{\frac{1}{2}}} \exp(-(t-r))|y(r)|ds \\
&+ \int_0^t \exp(-(t-r))|h(r)|dr \\
&+ C \int_0^t \exp(-(t-r))|C_1 y(r)|_Z dr.
\end{aligned}$$

This yields, via Young's inequality,

$$\begin{aligned}
&\int_0^\infty |y(r)|^2 dr \\
&\quad \le C\left(|y_0|^2 + \rho \int_0^\infty |y(r)|^2 dr + \int_0^\infty |h(r)|^2 dr + \int_0^\infty |C_1 y(r)|_Z^2 dr\right).
\end{aligned}$$

Hence, for sufficiently small ρ, we get

$$\int_0^\infty |y(r)|^2 dr < +\infty.$$

Hence, $y \in L^2(R^+; H)$ and $y(t) \to 0$ strongly in H as $t \to 0$. Moreover, if we multiply (5.99) (with $w = 0$) by $y(t)$,

$$\frac{1}{2}\frac{d}{dt}|y|^2 + \nu\|y\|^2 \le \alpha_0|y|^2 + |h|\,|y|, \quad t > 0.$$

Hence, integrating on $(0, \infty)$ and using the fact that the $L^2(R^+; H)$-norms of y and h are bounded, we deduce

$$\int_0^\infty \|y(t)\|^2 dt \le C\left(\|y_0\|^2 + \int_0^\infty |h(r)|^2 dr\right).$$

This implies also that $y(t) \to 0$ strongly in V as $t \to \infty$.

This completes the proof of Theorem 5.7.

5.4 The H^∞-Control Problem for Boundary Control Problem

The theory presented above extends *mutatis-mutandis* to boundary control problems. We present below a few results in this direction.

5.4.1 The Abstract Formulation

Let X, U, W be real Hilbert spaces and let A be the infinitesimal generator of C_0-semigroup e^{At} on H, $B_1 \in L(W, X)$, $B_2 \in L(U, (D(A^*))')$, $C_1 \in L(X, Z)$, $D_{12} \in L(U, Z)$. Consider the input–output system defined by

$$x'(t) = Ax(t) + B_1 w(t) + B_2 u(t), \quad t \in R^+ = [0, \infty), \tag{5.101}$$

$$z(t) = C_1 x(t) + D_{12} u(t), \quad t \in R^+. \tag{5.102}$$

Here, $x(t) \in X$ is the state of the system, $u(t) \in U$ is the control input, $w(t) \in W$ is an exogeneous input and $z(t) \in Z$ is the controlled output.

We assume $B_2 \in L(U, D(A^*))')$ and there is $\alpha = (0, \frac{1}{2})$ such that

(i) *The pair* (A, B_2) *is stabilizable and*

$$|B_2^* e^{A^* t} x|_U \le \frac{C}{t^\alpha} |x|_X, \quad \forall t > 0.$$

We have denoted by A^* the adjoint of A and by $B_2^* \in L(D(A^*), U)$ the adjoint of $B_2 \in L(U, (D(A^*))')$, $((D(A^*))'$ is the dual of $D(A^*))$. We denote by $|\cdot|_X$, $|\cdot|_Z$, $|\cdot|_U$, $|\cdot|_W$ the norms of X, Z, U, W and by $(\cdot,\cdot)_X$, $(\cdot,\cdot)_Z$, $(\cdot,\cdot)_U$, $(\cdot,\cdot_W$ the corresponding scalar products. By Assumption (i) it follows that System (5.101), with initial condition $x(0) = x_0 \in X$ and inputs $u \in L^2(0, T; U)$, $w \in L^2(0, T; W)$, has a mild solution $x \in C([0, T]; X)$ given by

$$x(t) = e^{At} x_0 + \int_0^t e^{A(t-s)} (B_1 w(s) + B_2 u(s)) ds, \quad t \in [0, T].$$

The linear H^∞-control theory extends to this case, too. Namely, for a given feedback controller $F \in L(D(A), U)$, denote by $S_F : L^2(R^+; W) \to L^2(R^+; Z)$ the closed-loop operator

$$z = S_F w = (C_1 + D_{12} F) \int_0^t e^{(A+B_2 F)(t-s)} B_1 w(s) ds$$

and the formulation of the problem is completely similar: given $\gamma > 0$, find $F \in L(D(A), X)$, which internally stabilizes System (5.101) and makes $S_F \in L(L^2(R^+; W), L^2(R^+; Z))$ with $\|S_F\| < \gamma$.

Besides (i), we assume also (see (kk) and (kkk)) that the following hypotheses hold.

(ii) The pair (C_1, A) is exponentially detectable, that is, there exists $K \in L(Z, X)$ such that $A + K C_1$ generates an exponentially stable semigroup.

(iii) $D_{12}^* [C_1, D_{12}] = [0, T]$.

We have

Theorem 5.8 *Suppose Hypotheses* (i), (ii) *and* (iii) *hold. Then, there exists an* $F \in L(D(A), X)$ *such that* $A + B_2 F$ *generates an exponentially stable semigroup and* $\|S_F\| < \gamma$ *if and only if there exists* $\widetilde{P} \in L(X, X)$ *with* $B^* \widetilde{P} \in L(D(A), U)$, $P = P^* \ge 0$, *satisfying the Riccati equation*

$$(Ax, \widetilde{P}y) + (\widetilde{P}x, Ay) - (\widetilde{P}(B_2B_2^* - \gamma^{-2}B_1B_1^*)\widetilde{P}x, y) + (C_1^*C_1x, y) = 0,$$
$$\forall x, y \in D(A), \tag{5.103}$$

and such that $A - (B_2B_2^* - \gamma^{-2}B_1B_1^*)\widetilde{P}$ *generates an exponentially stable semigroup. Moreover, in this case, the state feedback* $F = -B_2^*\widetilde{P}$ *is exponentially stabilizing and* $\|S_F\| < \gamma$.

The proof is quite similar to that of Theorem 5.1 and is omitted (see, however, [8] for the proof in the case of an unbounded control system (5.101), (5.102) of hyperbolic type).

5.4.2 The H^∞-Boundary Control Problem for the Linearized Navier–Stokes Equation

Consider the input–output system

$$\begin{aligned}
&\frac{\partial y}{\partial t} - \nu\Delta y + (y\cdot\nabla)y_e + (y\cdot\nabla)y = \nabla p + B_1 w \quad &&\text{in } (0,\infty)\times\mathcal{O},\\
&\nabla\cdot y = 0 &&\text{in } (0,\infty)\times\mathcal{O},\\
&y = u &&\text{on } (0,\infty)\times\partial\mathcal{O},
\end{aligned} \tag{5.104}$$

with the observation

$$z = C_1 y.$$

Equivalently,

$$\begin{aligned}
&\frac{dy}{dt} + \mathcal{A}y = (\mathcal{A} + kI)D_k u + B_1 w, \quad t \ge 0,\\
&y() = y_0,\\
&z = C_1 y.
\end{aligned} \tag{5.105}$$

Here, $D \in L(L^2(\partial\mathcal{O}), H)$ is the Dirichlet map associated with the Stokes–Oseen operator $\mathcal{A} + kI$ (see Sect. 3.3) and $B_1 \in L(W, H)$, $C_1 \in L(H, Z)$. We can write System (5.105) in the form (5.101), (5.102), where

$$X = H, \qquad B_2 = (\mathcal{A} + kI)D.$$

If one assumes that the pair $(\mathcal{A}, C_1)$ is detectable and $(\mathcal{A}, B_2)$ is stabilizable, it follows then, by virtue of Theorem 5.8, that there is a γ-suboptimal solution to the H^∞-control problem associated with (5.105) if and only if the Riccati equation (5.103) has a self-adjoint positive solution $\widetilde{P}$.

A little calculation with boundary-stabilizable feedback controllers designed in Sect. 3.3, for instance that given by Theorem 3.5, reveals, by the same argument as that in the proof of Theorems 5.5 or 5.6, that the H^∞-control problem for (5.105) has a γ-suboptimal solution for a certain $\gamma > 0$. The conclusion is that these feedback controllers are robust in the sense of H^∞-control theory.

As regards the H^∞-control problem for the Navier– Stokes equation

$$\begin{aligned} &\frac{dy}{dt} + \mathscr{A}y + Sy = (\mathscr{A} + kI)Du + B_1 w, \\ &y(0) = y_0, \\ &z = C_1 y, \end{aligned} \tag{5.106}$$

one suspects that one has a similar result as that given in Theorem 5.7, but this remains to be done.

5.5 Comments on Chap. 5

The main result of this chapter, Theorem 5.7 was established in Barbu and Sritharan [25] and the presentation closely follows that work. The state-space approach to the H^∞-control problem became very popular in first years of nineties after the publication of the seminal paper [47] by Doyle, Glover, Khargonekar and Francis. The main advantage of the state-space approach consists in the fact that easily extends to nonlinear control systems (see [8, 77–79] for a few works in this direction). As regards the literature on robust stabilization of Navier–Stokes equations and, in particular, to H^∞-approach to this problem, besides the work [25] we mention also the work of Bewley, Temam and Ziane [34] which treats the robust control theory for Navier–Stokes equation as a differential max-min game of the form (5.61) with finite horizon. The work of Bewley [33] is a very good survey of results and techniques related to this problem at the level of 2001. More is expected to be done on the H^∞-problem for fluid dynamics, especially on computational aspects. For instance, one might expect that under the conditions of Theorem 5.7 a γ-suboptimal solution F_h to the H^∞-problem associated with a finite-dimensional approximation of System (5.41), (5.42), that is,

$$\frac{dy_h}{dt} + \mathscr{A}_h y_h = (B_2)_h u_h + (B_1)_h w_h, \qquad z_h = (C_1)_h y_h + (D_k)_h u_h,$$

is suboptimal too in System (5.52).

References

1. Aamo OM, Fosseen TI (2002) Tutorial on feedback control of flows, part I: Stabilization of fluid flows in channels and pipes. Model. Identif. Control 23:161–226
2. Aamo OM, Krstic M, Bewley TR (2003) Control of mixing by boundary feedback in $2D$-channel. Automatica 39:1597–1606
3. Adams D (1975) Sobolev Spaces. Academic Press, New York
4. Agmon S, Douglis A, Nirenberg L (1959) Estimates near boundary for solutions of elliptic partial differential equations satisfying general boundary conditions. Commun. Pure Appl. Math. 12:623–727
5. Apleby JAD, Mao X, Rodkina A (2008) Stabilization and destabilization of nonlinear differential equations by noise. IEEE Trans. Autom. Control 53:683–691
6. Arnold L, Craul H, Wihstutz V (1983) Stabilization of linear systems by noise. SIAM J. Control. Optim. 21:451–461
7. Balogh A, Liu W-L, Krstic M (2001) Stability enhancement by boundary control in $2D$ channel flow. IEEE Trans. Autom. Control 11:1696–1711
8. Barbu V (1992) H^∞-boundary control with state feedback; the hyperbolic case. Int. Ser. Numer. Math. 107:141–148
9. Barbu V (1994) Mathematical Methods in Optimization of Differential Systems. Kluwer, Dordrecht
10. Barbu V (1995) The H^∞-problem for infinite dimensional semilinear systems. SIAM J. Control Optim. 33:1017–1027
11. Barbu V (1998) Partial Differential Equations and Boundary Value Problems. Kluwer, Dordrecht
12. Barbu V (2003) Feedback stabilization of Navier–Stokes equations. ESAIM COCV 9:197–206
13. Barbu V (2007) Stabilization of a plane channel flow by wall normal controllers. Nonlinear Anal. Theory-Methods Appl. 56:145–168
14. Barbu V (2009) The internal stabilization by noise of the linearized Navier–Stokes equation. ESAIM COCV (online)
15. Barbu V (2010) Nonlinear Differential Equations of Monotone Type in Banach Spaces. Springer, New York
16. Barbu V (2010) Optimal stabilizable feedback controller for Navier–Stokes equations. In: Leizarowitz et al. (eds) Nonlinear Analysis and Optimization: Nonlinear Analysis. Contemporary Math. 513. Am. Math. Soc., Providence
17. Barbu V (2010) Stabilization of a plane periodic channel flow by noise wall normal controllers. Syst. Control Lett. 50(10):608–618
18. Barbu V (2010) Exponential stabilization of the linearized Navier–Stokes equation by pointwise feedback controllers. Automatica. doi:10.1016/j.automatica.2010.08.013

V. Barbu, *Stabilization of Navier–Stokes Flows*,
Communications and Control Engineering,
DOI 10.1007/978-0-85729-043-4,

19. Barbu V, Coca D, Yan Y (2008) Internal optimal controller synthesis for Navier–Stokes equations. Numer. Funct. Anal. 29:225–242
20. Barbu V, Da Prato G (2010) Internal stabilization by noise of the Navier–Stokes equation. SIAM J. Control Optim. (to appear)
21. Barbu V, Lasiecka I, Triggiani R (2006) Abstract setting for tangential boundary stabilization of Navier–Stokes equations by high and low-gain feedback controllers. Nonlinear Anal. 64:2704–2746
22. Barbu V, Lasiecka I, Triggiani R (2006) Tangential boundary stabilization of Navier–Stokes equations. Mem. Am. Math. Soc. 852:1–145
23. Barbu V, Lefter C (2003) Internal stabilizability of the Navier–Stokes equations. Syst. Control Lett. 48:161–167
24. Barbu V, Rodriguez S, Shirikyan A (2010) Internal stabilization for Navier–Stokes equations by means of finite dimensional controllers. SIAM J. Control Optim. (to appear)
25. Barbu V, Sritharan S (1998) H^∞-control theory of fluid dynamics. Proc. R. Soc. Lond. A 454:3009–3033
26. Barbu V, Triggiani R (2004) Internal stabilization of Navier–Stokes equations with finite dimensional controllers. Indiana Univ. Math. J. 53:1443–1494
27. Barbu V, Triggiani R, Lasiecka I (2006) Abstract settings for tangential boundary stabilization of Navier–Stokes equations by high and low-gain feedback controllers. Nonlinear Anal. 64:2704–2746
28. Barbu V, Wang G (2003) Internal stabilization of semilinear parabolic systems. J. Math. Anal. Appl. 285:387–407
29. Barbu V, Wang G (2005) Feedback stabilization of periodic solutions to nonlinear parabolic-like evolution systems. Indiana Univ. Math. J. 54:1521–1546
30. Bedra (2009) Feedback stabilization of the 2-D and 3-D Navier–Stokes equations based on an extended system. ESAIM COCV 15:934–968
31. Bedra M (2009) Lyapunov functions and local feedback boundary stabilization of the Navier–Stokes equations. SIAM J. Control Optim. 48:1797–1830
32. Bensoussan A, Da Prato G, Delfour M (1992) Representation and Control of Infinite Dimensional Systems. Birkhäuser, Boston, Basel, Berlin
33. Bewley TR (2001) Flow control: new challenge for a new Renaissance. Prog. Aerosp. Sci. 37:21–50
34. Bewley T, Temam R, Ziane M (2000) A general framework for robust control in fluid mechanics. Physica D 138:360–392
35. Brezis H (1973) Opérateurs Maximaux Monotones et Semigroupes de Contractions dans un Espace de Hilbert. North-Holland, Amsterdam
36. Brezis H (1983) Analyse Fonctionnelle. Théorie et Applications. Masson, Paris
37. Burns JA, Singler JR (2006) New scenarios, system sensitivity and feedback control. In: God-il-Hak (ed) Transition and Turbulence Control. Lecture Notes Series, NUS 18:1–35. World Scientific, Singapore
38. Caraballo T, Robinson JC (2004) Stabilization of linear PDEs by Stratonovich noise. Syst. Control Lett. 53:41–50
39. Caraballo T, Liu K, Mao X (2001) On stabilization of partial differential equations by noise. Nagoya Math. J. 101:155–170
40. Cerrai S (2000) Stabilization by noise for a class of stochastic reaction-diffusion equations. Probab. Theory Relat. Fields 133:190–214
41. Constantin P, Foias C (1989) Navier–Stokes Equations. University of Chicago Press, Chicago, London
42. Coron JM (2007) Control and Nonlinearity. AMS, Providence RI
43. Crandall MG (1986) Nonlinear semigroups and evolutions generated by accretive operators. In: Browder F (ed) Nonlinear Functional Analysis and Its Applications 305–338. AMS, Providence RI
44. Da Prato G, Zabczyk J (1991) Stochastic Equations in Infinite Dimensions. Encyclopedia of Mathematics and Its Applications. Cambridge University Press, Cambridge UK

45. Da Prato G, Zabczyk J (1996) Ergodicity for Infinite Dimensional Systems. Encyclopedia of Mathematics and Its Applications. Cambridge University Press, Cambridge
46. Deng H, Krstic M, Williams RJ (2001) Stabilization of stochastic nonlinear systems driven by noise of unknown covariance. IEEE Trans. Autom. Control 46:1237–1253
47. Doyle J, Glover K, Khargonekar P, Francis B (1989) State space solutions to standard H^2 and $H^{-\infty}$-control problems. IEEE Trans. Autom. Control AC 34:831–847
48. Duan J, Fursikov AV (2005) Feedback stabilization for Oseen Fluid Equations. A stochastic approach. J. Math. Fluids Mech. 7:574–610
49. Edwards RE (1965) Functional Analysis. Holt, Rinehart and Winston, New York
50. Fursikov AV (2000) Optimal Control of Systems Theory and Applications. AMS, Providence RI
51. Fursikov AV (2002) Real processes of the 3-D Navier–Stokes systems and its feedback stabilization from the boundary. In: Agranovic MS, Shubin MA (eds) AMS Translations. Partial Differential Equations. M. Vishnik Seminar 95–123
52. Fursikov AV (2004) Stabilization for the 3-D Navier–Stokes systems by feedback boundary control. Discrete Contin. Dyn. Syst. 10:289–314
53. Fursikov AV, Imanuvilov OY (1998) Local exact controllability of the Boussinesque equation. SIAM J. Control Optim. 36:391–421
54. Henry D (1981) Geometric Theory of Semilinear Parabolic Equations. Lecture Notes in Mathematics 840. Springer, Berlin, Heidelberg, New York
55. Hormander L (1976) Linear Partial Differential Operators. Springer, Berlin, Heidelberg, New York
56. Imanuvilov OY (1998) On exact controllability for Navier–Stokes equations. ESAIM COCV 3:97–131
57. Joseph DD (1976) Stability of Fluid Motions. Springer, Berlin, Heidelberg, New York
58. Kuksin S, Shirikyan A (2001) Ergodicity for the randomly forced $2D$ Navier–Stokes equations. Math. Phys. Anal. Geom. 4:147–195
59. Kato T (1966) Perturbation Theory of Linear Operators. Springer, Berlin, Heidelberg, New York
60. Lasiecka I, Triggiani R (2000) Control Theory for Partial Differential Equations: Continuous and Approximation Theory. Cambridge University Press, Cambridge
61. Lefter C (2009) Feedback stabilization of 2-D Navier–Stokes equations with Navier slip boundary conditions. Nonlinear Anal. 70:553–562
62. Lions JL (1969) Quelques Méthodes de Resolution des Problèmes aux Limites Nonlinéaires. Dunod-Gauthier Villars, Paris
63. Lipster R, Shiraev AN (1989) Theory of Martingals. Kluwer, Dordrecht
64. Mao XR (2003) Stochastic stabilization and destabilization. Syst. Control Lett. 23:279–290
65. Munteanu I (2010) Normal feedback stabilization of periodic flows in a 2-D channel (to appear)
66. Pazy A (1983) Semigroups of Linear Operators. Springer, Berlin, Heidelberg, New York
67. Pazy A (1985) Semigroups of Linear Operators and Applications to Partial Differential Equations. Springer, Berlin
68. Ravindran SS (2000) Reduced-order adaptive controllers for fluid flows using POD. J. Sci. Comput. 15(4):457–478
69. Smale S (1965) An infinite dimensional version of Sard's theorem. Am. J. Math. 18:158–174
70. Raymond JP (2006) Feedback boundary stabilization of the two dimensional Navier–Stokes equations. SIAM J. Control Optim. 45:790–828
71. Raymond JP (2007) Feedback boundary stabilization of the three dimensional incompressible Navier–Stokes equations. J. Math. Pures Appl. 87:627–669
72. Shirikyan A (2004) Exponential mixing $2D$ Navier–Stokes equations perturbed by an unbounded noise. J. Math. Fluids Mech. 6:169–193
73. Temam R (1979) Navier–Stokes Equations. North-Holland, Amsterdam
74. Temam R (1985) Navier–Stokes Equations and Nonlinear Functional Analysis. SIAM, Philadelphia

75. Triggiani R (2006) Stability enhancement of a 2-D linear Navier–Stokes channel flow by a 2-D wall normal boundary controller. Discrete Contin. Dyn. Syst. SB 8:279–314
76. Uhlenbeck K (1974) Eigenfunctions of Laplace operators. Bull. AMS 78:1073–1076
77. Van Der Schaft AJ (1991) A state space approach to nonlinear H^{∞} control. Syst. Control Lett. 16:1–8
78. Van Der Schaft AJ (1993) L_2 gain analysis of nonlinear systems and nonlinear state feedback H^{∞} control. IEEE Trans. Autom. Control 37:770–784
79. Van Keulen B (1993) H^{∞}-control for Distributed Parameter Systems: A State-Space Approach. Birkhäuser, Boston, Basel, Berlin
80. Vazquez R, Krstic M (2004) A closed form feedback controller for stabilization of linearized Navier–Stokes equations: The $2D$ Poisseuille system. IEEE Trans. Autom. Control 52:2298–2300
81. Vazquez R, Tvelat E, Coron JM (2008) Control for fast and stable Laminar-to-High-Reynolds-Number transfer in a $2D$ channel flow. Discrete Contin. Dyn. Syst. SB 10:925–956
82. Yosida K (1980) Functional Analysis. Springer, Berlin, Heidelberg, New York

Index

V. Barbu, *Stabilization of Navier–Stokes Flows*,
Communications and Control Engineering,
DOI 10.1007/978-0-85729-043-4,